THE MAINTENANCE MANAGEMENT AND TECHNOLOGY HANDBOOK

1st Edition

THE MAINTENANCE MANAGEMENT AND TECHNOLOGY HANDBOOK

1st Edition

Derek Stoneham
CEng MIMechE

ELSEVIER
ADVANCED
TECHNOLOGY

UK	Elsevier Science Ltd, The Boulevard, Langford Lane, Kidlington, Oxford OX5 1GB, UK
USA	Elsevier Science Inc., 665 Avenue of the Americas, New York, NY 10010, USA
JAPAN	Elsevier Science Japan, Tsunashima Building Annex, 3-20-12 Yushima, Bunkyo-ku, Tokyo 113, Japan

First edition 1998

Library of Congress Cataloging-in-Publication Data
Stoneham, Derek.
    The maintenance management and technology handbook/Derek Stoneham. — 1st ed.
      p.   cm.
    Includes bibliographical references.
    ISBN 1-85617-315-1 (hc)
    1. Plant maintenance—Management.   I. Title.
TS192.S7617   1998
658.2—dc21   97-44025
                                                                   CIP

British Library Cataloguing in Publication Data
A catalogue record for this title is available from the British Library

ISBN 1 85617 3151

Published by

Elsevier Advanced Technology
The Boulevard, Langford Lane, Kidlington, Oxford OX5 1GB, UK
Tel: +44(0) 1865 833842
Fax: +44(0) 1865 843971

Printed in the United Kingdom by the University Press, Cambridge

Contents

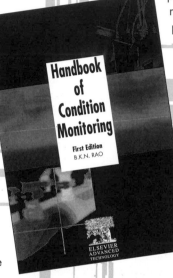

Chapter 1

Purpose

It has been said that the two fundamental forces in the universe are 'Life and Time'. All organisms, all organic creations, all inorganic structures, and all items of equipment and machinery are subject to the degrading effects of use and time. So pernicious is this effect that the role of the maintenance engineer is not so much to prevent degradation as to slow it down; to keep subject items working as long as possible; not merely to replace a broken part or recalibrate a device, but to reduce the effects of a hundred adverse parameters.

In the *Oxford Guide to the English Language* the verb 'to maintain' is defined as 'cause to continue, . . . keep in existence . . .' and this need to extend useful life is the essence of maintenance work. Of all of the accepted technologies, maintenance alone is principally concerned with the preservation of previous creation, whether it is of other species, natural forces, or of the human race.

Maintenance needs arise naturally to oppose the forces of degradation and can be witnessed as interventions following a maintenance plan or an equipment failure. In all cases they are pervasive: wherever there is life or movement, there is a maintenance need. Faced with such a reality, maintenance engineers tend to have sober dispositions, reacting to influences over which they have limited control and preparing in advance for events which operators and other engineers would rather did not occur. For the Maintenance Engineer:

All mature created things are in decline.

This is not to say that maintenance steps are always adopted. In the Western culture, financial justification is primarily concerned with short-term effects, and replacements may be required before maintenance work is needed, or the production process or equipment life cycle may be stopped altogether.

Such thoughts refer to a wider picture than the now outdated traditional view which regards maintenance as inevitable and negative, producing no revenue, often denying an earning opportunity and generating expenses which cannot be avoided. Over the last ten to fifteen years this picture has become much more positive. Maintenance is being seen increasingly in terms of opportunity and of special characteristics; inevitable maybe, but presenting the chance to improve performance and increase earnings.

The maintenance message has begun to change.

An interesting but more pointed maintenance concept is given in BS 4778 (Section 3.1, item 16)

> Maintenance is the process of maintaining an item in an operational state by either preventing a transition to a failed state or by restoring it to an operational state following failure.

Preventive maintenance is carried out to make an item less vulnerable to causal influences by restoring the quality to an acceptable level following operational degradation with time. Restorative action following failure will require some form of corrective maintenance.

In the pages that follow we will refer to the main plant and equipment condition criteria of availability, reliability and maintainability by which maintenance systems are judged and whose targets and requirements result in maintenance work. The emphasis in this manual is on maintainability: the methods and philosophies of maintenance work. Availability features as a major target of maintenance activity, while reliability, an established and separate branch of engineering, contributes to the language used, the means of understanding and many of the analytical outcomes on which maintenance work is founded.

Maintenance and cost reduction

During the last recession many companies, seeing order books collapse, predictably turned to cost reduction as the best means of protecting commercial margins. Maintenance management, for many years the Cinderella of engineering technologies, was quickly identified as a region of high promise, where, ironically previous skill and effort had not been fully matched by positive support. During 1994 over £14 billion was spent on plant maintenance in Britain alone: an important prospect. Based on previous experience, maintenance needs could be reduced, actions more efficiently completed and job cost reductions of 30% or more would not be regarded as unusual.

Maintenance cost reduction is aiding company survival.

Maintenance income

Coupled with the hunt for cost reductions, many companies have changed their organizational structure and words such as 'downsizing' and 'reshaping' have become familiar. Many areas of action and expertize 'traditionally' part of an operator's armory have been transferred to external service companies to accompany a concentration on the core business of the customer company. Such changes present golden opportunities to service companies for major and rapid business expansion. Increasing sales income is one of the first benefits, accompanied by both rising profits and, followed growth and specialization, growing profitability.

Maintenance is a frequent ingredient of these changes, and in addition to the recognition of potential cost reductions it is being seen in some areas as a major source of profit.

The maintenance presence

The present emphasis on minimum facility plant and installation design springs from the high life support costs for remotely quartered personnel coupled with the need to reduce capital and operating costs of prospective new installations. By this approach, equipment and facilities strictly necessary to operational purpose are included, while standby capacity and load sharing duplication are frequently left out. The effect on maintenance is to make the job more difficult, especially where critical equipment fails and the need for a rapid response is more important than ever.

Such forces are part of a vicious circle: present commercial realities require that

A semi-submersible drilling rig testing in the North Sea.

operational costs are reduced before long-term investments can be contemplated, and all operational cost contributors come under close scrutiny. Maintenance cost reduction is both target and opportunity, fuelling the development of new philosophies, systems and techniques, without which the commercial presence could not be sustained.

On many sites, apart from operations itself, maintenance is one of the few activities constantly in the working environment, and maintenance personnel are often the only regular visitors to remote and quietly working equipment. Present changes make this even more likely, coupled with the detailed technical knowledge of the machines themselves Maintenance technicians have a unique source of information and understanding, making them potentially valuable contributors to construction or project work of many kinds. For instance, many construction projects require that plant trains or machines be isolated and that lifting appliances, sources of heat, pressure, power and inherent danger be identified and made safe. The maintenance group is ideally qualified to provide this type of contribution, and taking both preceding aspects into account:

As well as preserving, maintenance contributes to change.

Aim of the manual

As well as reflecting the present maintenance environment, this manual is intended to highlight the meaning of different features or procedures and act as a signpost for a maintenance manager facing an existing installation with a 'what do we do now?' question. For those needing more detail, these pages will reinforce some requirements while questioning others.

For these reasons, the order in which subjects are presented and their emphasis may seem strange. Chapter 2 concentrates on the immediate situation. Work scheduling, prioritizing and frequency are all included because of their relevance to short-term decisions, and it is not until Section 5.3 that the future is formally included.

Chapter 3 discusses the different philosophies which are in use at present. They are sometimes applied together in one operation, helping to underline reasons for their use and to assist managers contemplating change. Chapter 4 follows with a glance at where such methods are in place.

Chapter 5, entitled 'The maintenance toolbox', pinpoints the principal facilities that the maintenance group employs to get the job done. The list, of course, cannot include all that are used, and some (such as in Section 5.10) are applicable to all manufacturing and processing industries.

The main tool for all major maintenance systems is of course the computer. Although not always specified here, many maintenance elements and procedures are handled by the computer and the features in Sections 5.1–5.4 are based on some 'in-house' developed systems with reference to some newer proprietary offerings.

Chapter 6 is intended to refer to the main users of the different techniques discussed with the prospect of cooperation, or perhaps shared resources in non-competing areas of enterprise, where the business may be different but the problems are the same. Chapter 7, 'Maintenance and Finance', is definitely not a treatise on financial affairs; it merely considers some of the financial questions facing maintenance managers together with a few of the tools that may be useful. Chapters 8 and 9 consider some of the key influences on maintenance, underlining why some decisions are made and what should not be overlooked.

Maintenance quality and maintenance safety, covered in Chapters 10 and 11, refer to major separate areas of business activity, and, like Chapter 7, they are considered from a maintenance point of view and not as an attempt to explain their full and detailed ramifications in any depth. Other sections reinforce the present and practical nature of the manual, with the exception of Chapter 12, which refers to the dominance of 'safety-critical assessment' in present maintenance development thinking; this merits an immediate explanation.

The golden triangle

One representation of integrated maintenance management is shown in the simple diagram of Figure 1.1 and reflects present debates on safety-critical assessment. Work

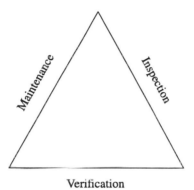

Verification

Figure 1.1
The golden triangle.

in this area is of course being paralleled by several North Sea oil companies, reflecting the current preoccupation with changes in the regulatory stance following the Piper Alpha tragedy and the recommendations of the Cullen Report[1]. The triangle

---

1. The following is a quotation from 'A guide to the Offshore Installations (Safety Case) Regulations 1992 – Introduction'.

On 6 July 1988 an explosion on Piper Alpha offshore installation led to the loss of 167 lives. The recommendations in Lord Cullen's report were influenced by HSC/E's experience of regulating major hazards onshore under the Control of Industrial Major Accident Hazards Regulations 1984 (CIMAH) (SI 1984/1902). The CIMAH Regulations and the European Community Directives they implement were a response to certain major accidents that took place during the 1970s, notably the Flixborough accident in the UK in 1974 and the disaster at Seveso, Italy in 1976. They require the demonstration of safe operation and certain installations are also required to submit a safety report to the HSE. In the first instance the safety report is a means by which manufacturers demonstrate to themselves the safety of their activities but it also serves as a basis for the regulation of major hazard activities.

(See Chapter 11.)

clarifies the main elements of effective maintenance work, and while these are currently confined to offshore North Sea operations, this thinking and the resulting methods will probably permeate to all activities with a maintenance ingredient (see Chapters 11 and 12).

The triangle encapsulates the three key ingredients of an effective maintenance system and illustrates both the separateness of each component and their dependence on each other. It also helps clarify questions of 'what has to be done' and 'who does what', as effective systems depend on different methods and different teams whose interests often overlap.

Maintenance

Effective maintenance is the key purpose of the three 'triangular' ingredients taken together. It is the work-centred or executive side of the triangle, carried out either by the operating company or by a separate company specifically contracted to them for this purpose. Such companies may be legally responsible for the maintenance of the whole plant or they may specialize in specific equipment or part of the process.

As we will see in the following pages, maintenance work is present in different forms.

(1) *First line maintenance*

Often conducted by non-maintenance personnel, this includes early maintenance work such as cleaning, greasing, adding lubricant, zero setting and recording key measurements. Operators will report observed equipment deterioration that cannot be corrected by first line maintenance to the maintenance group.

A maintenance technician works on a major repair. Picture by courtesy of Chevron (UK) Ltd.

The content and formality of first line maintenance varies according to the equipment and the plant. It must not be regarded as comprehensive, as so many items requiring maintenance are not in the orbit of operations personnel. In fact, one of the less obvious purposes of pre-planned maintenance (PPM) is to bring an experienced technician into 'contact' with the machine.

(2) *Planned maintenance (see Sections 2.5 and 3.3)*
As the name suggests, both the timing and the content of planned maintenance work is known in advance.

(3) *Shutdown maintenance (see Section 5.8)*
A version of planned maintenance; work of this kind can only be done when the production facility is shut down.

(4) *Breakdown maintenance (see Section 3.1)*
When an item of equipment fails to meet its primary function it has broken down, and maintenance work, which may include repairs, replacements or adjustments, is necessary.

(5) *Emergency maintenance*
Once identified by a breakdown or an inspection, emergency maintenance requires immediate work.

Inspection

All plant equipment, machines and structures are subject to a regular inspection programme, timed to detect performance or safety problems and by routine ensure that all items receive required maintenance.

A control room operator accesses a platform computer system. Picture by courtesy of Chevron (UK) Ltd.

Inspection is usually conducted by the operating company in general, but there are some notable exceptions which statute[2] requires to be treated by recognized third party inspectors.

(1) Lifting equipment: appointed lifting equipment companies.
(2) Helicopter refuelling equipment: Civil Aviation Authority (CAA).
(3) Platform lifeboats: Department of Transport (DOT).
(4) Process pressure vessels: Certifying Authority, such as Lloyd's Register or Det Norske Veritas, appointed by the Health and Safety Executive (HSE).

Verification

There is no escaping the fact that, however sophisticated the maintenance system is it has no value whatever if it produces no action. Maintenance systems are result-dependent, and *effectiveness is more important than efficiency*. This is not to say that efficiency is not useful; most systems compare estimated plus standard hours 'recovered' with total team 'attendance' hours to calculate the hours efficiency. This practice is not always welcomed by maintenance crews, as it smacks of monitoring people, whereas the usual intention is to use an efficiency measurement *to see if system changes lead to improvement*.

This leads to the question of verification (sometimes confused with inspection), verification is a process, not a measurement, and its two primary purposes are:

(1) To check that the maintenance work is being done.
(2) To confirm that maintenance standards are in existence and being met.

With the best of intentions it is very difficult for in-house verification to be completely impartial, and traditionally internal appraisals are supplemented by specific third party inspections. The term 'inspection' is used loosely here, and refers to the judgements about the system, the results achieved, and the overall disposition of the installation and selected safety critical systems (see Chapter 12).

Computer-based systems take various collected numbers and turn them into usable statistics. Although this process is sometimes overdone, useful measurements are frequently referred to in the verification process.

(1) Number of maintenance team manhours available for work in a specified tour.
(2) Number of hours recovered resulting from maintenance work done in a specified period.
(3) The maintenance work backlog.
(4) Plant or equipment availability.

There are other figures used in the verification process, but number 4 above is one of the most powerful. Availability, as the name implies, is a measure of the plant's or equipment's ability to carry out useful work, and refers to the reliability, maintainability and maintenance support.

---

2. As Chapter 12 indicates, the roles and responsibilities of the bodies mentioned and others concerned with statutory requirements are changing and the outline given here is gradually but currently being replaced.

One of the less attractive features of regular scheduled maintenance is the opportunity to close work orders with work not done, which may be necessary because of following high-frequency jobs or genuine maintenance decisions not to proceed. This practice is difficult to avoid, because by preventing the maintenance backlog figure from rising it disguises the true value of how much maintenance work is being completed; yet if relevant work orders were allowed to remain open, duplicated jobs would effectively show as outstanding.

The manual's foundation

It is true that many examples and descriptions included here are based on upstream oil production operations, particularly offshore. This is partly the author's particular background, but it also recognizes extensive maintenance experience in the industry and a generally uniform approach to maintenance management. Several influences have combined to produce the current environment and explain the continuing interest in maintenance processes of all kinds:

(1) The current modest price of oil and its effects on commercial finances and organizational structures.

(2) The changing nature of recent extraction operations to smaller oil fields requiring different production technology.

(3) The influence of the economic recession on sales volumes and the resulting need to constrain capital investment and operational costs.

(4) The present maturity of some installations, with both oil fields and production equipment approaching the end of 'normal' operational life.

(5) The occurrence of major incidents, such as Flixborough and Piper Alpha, the requirements of the Health and Safety Executive (HSE) and the ongoing changes in the statutory regime.

(6) The growing availability of low-cost computing power, allowing the processing and analysis of high volumes of electronic data plus the introduction of effective and commercially attractive standard computer packages for maintenance management.

(7) The growth of technically reliable means of predicting equipment failure, reducing production downtime and the incidence of repairs.

(8) The development of management philosophies which challenge the blanket approach to maintenance work and target activity to best effect and lower cost.

These influences have resulted in a climate of excited attention concentrating on maintenance efficiency, operational safety and financial viability.

It will also be clear that many of the examples described in these pages and the constant preoccupation with production downtime refer to a continuous production operation where the most obvious problem is the immediate effect on company revenue, and hence opportunity cost, when critical items have to be shut down. Conversely, fewer processes are at work and fewer products are involved, resulting in both operational and maintenance requirements being comprehensively understood. The maintenance work itself is more immediate because the 24-hour operation provides no reason for delay, and where shift-working is in use the means are often at hand to get the job done.

In practice, many operators of continuous processes run only a skeleton crew at night, with a callout option if daytime personnel are accommodated on site. Under these circumstances, night-time maintenance for either continuous or batch production is often more expensive when higher labour rates apply in such cases.

Non-continuous production has the advantage of allowing the maintenance of important production equipment when it is not in use, when different resources or production lines are working or when the processes have been switched off at the end of the working day. The different processes tend to be more numerous, with maintenance work various but less demanding, particularly as major unit exchanges or visiting specialist teams from the Original Equipment Manufacturer (OEM) can be more readily arranged.

Increasing equipment complexity and specialization, coupled with organizational changes, have resulted in changes in the way that maintenance work is conducted. In the following pages there are references to the use of experts or specialists and the reasons for choosing 'in-house' teams, 'on site' accommodated teams or 'manned up' technicians flown in for the completion of a single job. The makeup of the teams is also changing, featuring groups made up of several specialists who major on certain machines and provide general support to the team when their own particular speciality is not the subject of current work. In the Third World countries, labour and hence maintenance work is less expensive and the need for training and work experience a frequent necessity, yet the content of the plant itself depends on the production process wherever it is located. Under such circumstances the work team will include more support staff, watchkeepers and roustabouts than equivalent teams in Western applications.

We should also reflect on the thrust of the manual and its preoccupations and limitations. Although it refers strongly to areas of parallel business activity it is prepared from a practical maintenance perspective and presents signposts to areas of interest and concern, not expert treatises. It is intended to describe maintenance management areas of attention, problems, action and decision; to point to those factors which face maintenance managers in the day-to-day business of an operating plant; and hopefully help them avoid some of the pitfalls in their work. They have a vitally important job to do.

The maintenance manager

We could also observe that there are two major causes of maintenance work which reflect the major twin causes of degradation:

(1)   Deterioration of equipment caused by the environment to which it is exposed.
(2)   Deterioration of equipment arising from its actual use.

Both of these influences are considered in the following text. Various maintenance systems are discussed as are the different causes and effects of failure in use. Together with pre-use maintenance (Section 2.5) these cover the main causes of degradation.

We must also consider the manager himself, because if his temperament, experience and knowledge are not right, maintenance will be dominated by the short-term demands of other operating departments and the longer term well-being of the whole operation will be severely dislocated. As a combination of manager and engineer he occupies a position which can only be met by persistence and decision. His propositions will be unpopular, apparently ill timed and open to disagreement and contradictory experience. There are two bitter understandings to master:

(1) Customer departments for maintenance do not wish to buy and will offer sound reasons why downtime tasks should be delayed.
(2) Even when favourable conditions present themselves, do not expect immediate requests for maintenance work to be carried out.

If the manager can accept these limitations he will find a fascinating and rewarding field of engineering which uniquely touches all other branches and in contradiction to many earlier assertions unmistakably shows where things have gone wrong and can highlight future dangers and opportunities.

Chapter 2

Maintenance work management

It is widely accepted that the way in which work is managed has a profound effect on its costs and its success. Like so many things in business life it cannot be left to manage itself, and yet this is exactly the method so many are tempted to use when dealing with maintenance work. The parents of this approach are the thoughts that maintenance represents a small proportion of work overall and that constant fire-fighting leaves no chance to improve.

These arguments are both seductive and wrong. The reverse could readily be argued; that because of the devastating effect of poor maintenance on production, it is a most important activity to get right. The production manager, aiming at near-impossible targets, is subject to enough uncertainty already, without adding low output quality and excessive downtime resulting from poor maintenance.

This is not to say that the maintenance teams are at fault or that they are regularly failing; ironically, it may be that through unending action they are keeping things running, and such special efforts obscure the fact that there are problems. It may further be the case that the maintenance teams themselves, focusing on today's immediate problems and raised in the climate of disappointment, are simply disinclined to promote the case for more effective working. In addition, maintenance work has peculiarities of its own, and while methods of effective management are real, they need to cater for special aspects.

2.1 The conditions of work

Almost all maintenance work concerns second-hand machines. Replacement parts may have to be manufactured for older equipment, and components become corroded, mis-shapen, broken, distorted, subject to creep, temperature cycles, water ingress, chemical attack and misuse, all of which degrade the machine on which the work is to be done. Often one of the first actions required is cleaning, applied to both the machine and its working environment, the latter intended to prevent contamination of exposed parts and allow examination, measurement and subsequent reassembly. Such steps are prior to and in addition to the repair of the machine itself, and one of the early features of maintenance is that comparison with other work can be very misleading.

Unlike factory manufacturing, maintenance work is necessary in many different environments, which will change what are nominally the same pieces of work. For instance, draining cold oil from a reciprocating engine at an Arctic -25 °C is very different from the 'same' task in the western deserts of North Africa, where hot metal and hot oil present a handling problem.

Similarly, the same task conducted at night in the open or offshore would usually take longer than the same task carried out in a well-lit, warm factory or workshop on land, with special tools, power, parts and information all to hand.

The offshore/onshore difference in location is an interesting example. Almost all materials for offshore work are transported by boat and lifted aboard by crane. Both transit and lifting are affected by weather conditions, while material storage and working facilities are confined at the work destination

In some other parts of the world, maintenance, production and construction are all conducted without support from the locality. The requisite materials, tools, technical information and work specialists are ferried from external sources, often by plane. For the maintenance team this is especially difficult, as the extent of the work and the need for a specific component may be unknown until a strip-down has been completed. The result is delay, extended job durations, multiple journeys for specialist personnel, increased job costs and increased downtime for affected production systems.

When maintenance work requires the inspection of gas or fire detector heads it may apply to units mounted into a ceiling forty feet above the floor. Such a disposition requires the erection of temporary scaffolding or the use of special man-lifting or climbing equipment; this is often done by specialists and changes the content of the job and the nature of the work. Similarly, the repair of an electrical junction box, after locating the correct item, may require the removal of steel or concrete floor panels, the use of special lighting and the attendance of a Safety Officer.

As a further example, the replacement of a jammed float-operated level switch inside a waste fluid storage caisson would require special access, safety lines and breathing apparatus to achieve entry and handle the effects of toxic fumes.

There are many such situations suffering the effects of high temperatures, radiation, contamination, difficulties of access or movement and it is important to remember that, although the maintenance work requirement is considered during the installation design,

the place where the maintenance work is done is not designed for maintenance work.

Other machinery

On many plants, construction sites or buildings, work is affected by the structure, other assemblies and machines nearby. The effects on the working environment of personal access, noise, heat and vibration may often be sufficient to prevent an apparently minor task being undertaken until the disturbing source is shut down or the job retained to coincide with other work on the offending item.

There are also tasks which cannot be done while the subject machine is part of a running system. For instance, the replacement of a leaking control valve in the fuel line of a compressor motor requires the motor to be stopped unless a bypass line is available. In some cases such dependency is less obvious and more dangerous; for instance, the replacement of a larger valve in the water supply of the helideck fire monitor (normally switched off) could pass undetected until an emergency arose and the monitor was turned on. We should note immediately that fire monitors are usually supplied via a ring main, and although formal arrangements prevail such items may often be individually isolated and removed without interrupting supplies to other items.

Energy in place

There is always the likelihood of maintenance work being attempted on equipment which contains, or has a route to, aggressive energy. The possibilities are electrical energy, high pressure (fluid or pneumatic)[1], high temperature, displaced weight, elastic force (such as stretched steel cables), inherent kinetic energy and all cases of potential energy. The presence of high energy is often invisible and highly dangerous: the effects of breaking into a high pressure steam line, for example, or rupturing a steel cable loaded in tension by several tons, can be deadly and dramatic. However, even personnel experienced in associated technical disciplines can be present in normal operating conditions without realizing that latent and lethal dangers are present. The avoidance of such risks requires experience, extra work, extra care, extra cost and *attentive supervision*.

> *Energy is the hidden destroyer. Often present and usually invisible it can kill in an instant and without warning.*

Effects on production

Although careful status reporting and modern condition monitoring techniques are increasingly improving diagnosis, until the covers are finally off no one knows for certain whether additional work will prove necessary. Assignment of time and effort to the task may be inadequate and the materials marshalled for the original work may prove incorrect or insufficient. Also, the act of disassembly for the initial examination may itself be the cause of unexpected damage, introducing the aptly titled *maintenance-induced failure*. Whatever the cause, maintenance work is more likely to suffer the damaging effects of the unexpected, reducing production teams to a mixture of wrath and despair.

Maintenance tasks can be roughly grouped under production equipment or supporting services, and because of the revenue-earning quality of production operations the most important maintenance tasks inevitably affect production. For this reason, key plant and production systems are often supported by built-in standby

---

1. Even among engineers it is not widely realized that for equal hydraulic and pneumatic pressures in a pipe or pressure vessel the latter is very much more dangerous. This is referred to in an HSE paper 'Safety in Pressure Testing' Guidance Note 4 (General Series), which specifically recommends that pressure tests using air, gas or steam are to be avoided, and, where unavoidable, stringent precautions are to be used. If we consider the higher energy input required to raise a gas to working pressure, then the possibility of explosive decompression during maintenance is a major safety consideration.

machines intended for operation and hence continued production during maintenance or repair of the parent plant. Unfortunately for the maintenance team, this only applies to special cases, and the present emphasis on minimum facility installation designs while containing initial construction costs is also reducing the advent of installed standby equipment.

There is a particular problem when production can be readily switched to the standby for the parent machine to be shut down and repaired, but this requires that the standby will run up. Bearing in mind that standby running hours are deliberately kept low to prevent age-related problems appearing in both machines together, the risk of non-starting is real. *When a machine is used infrequently the chance of invisible failure is much greater.* Perversely, the very existence of the standby machine encourages a sense of operational security, increasing the dangers that the standby was meant to avoid. *For a standby to do its job, it must work first time!*

There is a halfway house employed by the plant operator which provides for unmounted new or reconditioned equipment, available at short notice for replacement of an offending item. The latter, having been removed, is shipped to the repairer and then returned to stock.

This method, or 'rotational regime', can be usefully applied to a group of similar machines requiring regular inspection and reconditioning, such as oil platform Christmas Trees (well head pressure containment and valve assemblies), where sixty trees on a platform would not be regarded as unusual. By regularly replacing a modest number each year the entire group can be reconditioned over a few years, and statutory and maintenance requirements met with minimum effects on production.

There is a snag: as the age of the equipment group will mainly date back to the initial installation, all members will deteriorate at broadly the same rate, leading to the need for reconditioning of several units at the same time. For simultaneous failures to be avoided, some change-outs will take place before an individual machine's condition would dictate.

Immediate replacements are used to reduce production downtime or increase safety, but their use adds complications to the maintenance task. In particular, the standby or replacement machine must be ready for immediate use and there is a real risk that these non-working machines are overlooked by the maintenance programme. Also, small differences not mentioned by the supplier, such as mounting bolt sizes, overall dimensions and coupling shaft diameters, can cause mayhem if they are only discovered after the parent equipment has been removed.

In addition, the changeover itself must be managed. At some point, for example, the old machine has to pass the new one. In a confined work space this may be impossible, and the replacement machine must remain on the deck of the supply boat until the installed item has been removed. Alternatively, the act of removal may reveal unexpected coupling damage, causing a sudden race to find a new part.

There is a further drawback, in that replacement assemblies held in stock or nearby are much more difficult to test than installed or standby units, increasing the possibility of invisible failure; ironically, the very situation that the use of the replacement was meant to avoid. Routine participation in the work programme, within the load constraints of the machine, will often improve the performance of the installed unit and the use of the replacement method requires careful thought.

When such arrangements can be used, the assigned maintenance task achieves a significant reduction in expensive downtime at the expense of a minor increase in work scope. Where they cannot be used, the prospect of lost production is a potent reason for delay, which sadly means that the work often takes place when it can no longer be avoided, at a time when machine damage and resultant downtime are greater.

Postponing the evil day often makes maintenance matters worse.

2.2 Work frequency

It is an ironic feature of maintenance work that *major jobs happen less frequently*. It may be obvious that the routine tasks of oil checking, visual inspections, filter changes and others happen regularly and often, while major strip-downs or repairs are expected to be years apart. These differences follow the nature of the work, firstly checking the requirements and adjustments of sound running and secondly making repairs, replacements and corrections arising from use and time. Major tasks are more likely to require dedicated personnel, tools and materials for completion, and in every organization people of the calibre required are most in demand: they are likely to be promoted or transferred and sometimes seek employment elsewhere. Whatever the reason, the result is scarcity, which, combined with the effect of infrequency, means that major jobs are sometimes done by personnel who have rarely attempted that piece of work before. It is a silent testimony to the resourcefulness of workers that jobs of this type are completed regularly without incident, *but that is not always the case: the possibility of disaster is real*. For this reason, careful job preparation includes a training element which increases confidence and reduces working time and the risks of error or delay.

Instead of their own employees, many companies use technicians supplied by the equipment manufacturer or from a specialist contractor, paying higher prices for the labour hours used but making job duration and downtime savings, arising from skills born of higher job frequencies and special training.

Jobs of this kind would include changing the slew ring on a pedestal crane, completing a hot gas path inspection on a turbine or carrying out a wall thickness inspection on a process pressure vessel. Each would require the presence of a small team, perhaps four or five people depending on the job, and would present the operator with safety, life support and accommodation matters to attend to. Because of the latter, jobs of this kind receive significant interest and attention, but it is important to note that this is not always justified by cost. The natural assumption is that because major jobs are expensive, they are the *most expensive*, but this is not necessarily true. When comparing the costs of a routine daily task with those of a major five-yearly examination we should remember that the former will nominally be actioned over 1800 times more often during the same period, magnifying the total job costs by the same factor.

There is also the matter of when a piece of work should be carried out. The manufacturer's recommendations, designed for maximum protection of both the equipment and the manufacturing company, are more frequent than Operations can permit or 'in-house' maintenance would recommend, while production targets often assume that the item is never taken out of service and the frequency is correspondingly very low. In addition, maintenance technology itself refers to the mean operating time[2] between failures (MTBF) as a useful basis for decisions of this kind, without underlining that this is a statistical value depending on the equipment population and the number of previous failure experiences. Where users or fleet operators have large numbers of very similar units, such as aero engines in use by a major airline or lorry tyres required by a road haulage company, the calculated MTBF is a very useful guide, and this is often expanded by gathering industry wide experience to inflate the number of items composing the equipment population (see Glossary). Unfortunately, for the majority of users this is not the case, for the following reasons.

(1) The number of relevant machines is small, such as two or perhaps three turbines, or one or two pedestal cranes.

(2) The MTBF is needed for major failures, which are known to be the most infrequent, so when the figure is most reliable and based on the highest frequency it is valuable but needed least.

(3) When the failure frequency is low the time between failures is extensive; it is likely to be several years between relevant major events.

(4) The entire maintenance system is aimed at preventing the occurrence of failures. This is acknowledged as a dilemma. Without the failure we do not have the knowledge, but even with the resultant uncertainty, ask any production manager which he would rather do without.

(5) Throughout many different sites, operators and applications, the demands on equipment and corresponding failures vary hugely.

This is not intended to disregard the MTBF; in fact, it is an almost natural outcome of examining the equipment maintenance history. However, it should be used with discretion, since maintenance timing based on a low-population MTBF can lead to serious errors of timing.

In cases where the MTBF is considered look at both the statistical spread as well as the number of samples available. Where a tight grouping of numerous readings prevails statistical significance can be confirmed, and by using the standard deviation the error attaching to different selected values can be estimated.

When making judgements about the timing of maintenance work consider the following:

(1) Consider whether a failure can be tolerated. If it can, the equipment itself

---

2. Notice that the definition refers to 'operating' time for the MTBF figure. However calendar time is often used by maintenance engineers when considering MTBF especially if the operational duty of the machine has not changed. Also if the item is in continuous use operational time and calendar time are the same.

will provide the information at failure. If, however, the consequences are severe, err towards the conservative value.

(2) Look at the maintenance history. Where it contains many readings of similar situations the MTBF is a useful guide.[3]

(3) Consider the manufacturer's recommendations and reflect that these will be conservative.

(4) Remember that failure may not have occurred. The timing of maintenance is chosen so that failure can be avoided.

(5) Include operational and maintenance inputs. First-hand experience is very pertinent to each installation.

(6) In important cases, sample industry-wide experience and contact other users with similar installations.

(7) Consider also the maintenance interval derived from optimum analysis (see Section 7.4)

In addition, it should be remembered that the frequency should not be regarded as fixed. Changes in condition, use and performance can effect the maintenance frequency and the influences listed above should be referred to at times of change during the equipment's life.

The two main criteria recognized in the work box (Section 2.3) are timing *and content*, and even where regular PPMs are in use, both should be recognized as variable. We have noted in Section 5.3 (The maintenance plan) that the selection of the frequency helps to define the content (for example, instrument recalibration neatly fits the three-monthly PPM), but when equipment performance requires, specific tasks may be transferred to a PPM of different frequency, i.e. the task frequency is changed.

Each PPM is therefore affected, and together with variations required according to equipment age and duty, the content steadily changes. Failure to recognize both timing and content as variable will result in less PPM adjustment with time, less use of valuable work experience and less relevant work routines.

2.3 Maintenance work box

At first sight it would appear that all work management eases when job preparation expands. The marshalling of parts, materials and tools, coupled with team building, information gathering and target setting, improves work efficiency, reducing errors and time taken. This is particularly true of major maintenance tasks, often separated by several years, when the organization gathers little experience and generates no skills. However under PPM (pre-planned maintenance), as tasks become more frequent they also get smaller, while organizational knowledge and skill both improve. When the work required is frequent it is the sheer volume of work that becomes the problem and the concentration of attention on one high-profile task can invoke calamity by overlooking the less interesting remainder. Ideally the work management systems

---

3. Where an item is normally changed out at failure, the MTTF (Mean Time to Failure) is used, as there will be no failure date and time in the maintenance history for the currently installed item.

chosen cater for the different needs of different types of task, but the reality is that different systems are often used together, each addressing those work groups where their speciality is most effective.

At this point three main messages need to be underlined:

(1) The marshalling of jobs is particularly important when they are numerous.
(2) Effective work preparation delivers work cost reduction.
(3) Job savings made increase as work scope and work intervals increase.

The nature of every piece of maintenance work is heavily influenced by the two main variables of timing and content. It may also be obvious that these two features are related, especially when a machine is approaching failure, when each passing minute increases the damage and the content of subsequent repair. The first essence of maintenance work timing is simply to know *when the work will be done*. We are not concerned here with work time, estimates, targets, durations or the effects of manning on such figures (those are the more detailed questions of job management); we simply wish to know – when.

This is not to dismiss the question of how long the job takes to do, the assumption being that most jobs are modest and arranged to be so. However, this is a serious matter; in some cases the interruption of production increases the opportunity cost and decreases the chances of releasing the item for maintenance or repair in the first place. Use of MTTR (mean time to repair; see Glossary) offers a statistical value based on previous experience, but it is a mean, with the usual questions of value and spread; also, for first repairs previous information will not exist, and neither will the MTTR.

In some areas of productive work, such as manufacturing or construction, 'when', is not seen as a fundamental variable, the timing decision being included in a sequence or a schedule. In a more casual world, 'when to maintain' would be answered by memory or impulse, but there are limits to human memory and many sites see 20 000 maintenance interventions per year, where such an informal approach is not possible.

Operating companies have spent huge sums of money attempting to answer this and other key questions. Such methods as introducing computer-based calendar PPM systems, calculation of machine MTBFs (mean time between failures) from historical records, the more recent employment of condition-based monitoring techniques and the continued routines of much first-level and statutory inspection, all either respond to or attempt to answer the timing question.

The underlying reason for this concern is simple: provided capital outlays are comparable, *costs are lowest when machines never stop working and last for ever*. We engage in inspections, condition monitoring and maintenance to safely extend the period of efficient operations, and so are vitally interested in work timing.

The other main consideration is of course '*what*'. If we know this, we can prepare, by gathering materials, tools, information and expertise. Some of the methods outlined below cater for this and simplify the resulting work management process; others will specify the time of an intervention, leaving the content to be detailed as the work proceeds. As mentioned above, timing and content are related, particularly in practical circumstances when we wait too long and the machine suffers extra damage as a result.

The work content is usually the combination of six main factors:

(1) The work required to return to immediate production.
(2) The original equipment manufacturer's recommendations.
(3) The formal experience of the maintenance team.
(4) The operator's specified requirements.
(5) Requirements of safe working practice.
(6) The demands of relevant statutory instruments.

Item 3 will be a collated maintenance history which will include records of all work carried out, details of failures, use of replacement materials, dates and times, details of disciplines required, estimated labour hours needed, details of special tools and skills employed, evidence of defects and faulty components, and sources of performance behaviour. Usually held on a computer database, such a history is composed of retained work order records of previously completed work sorted and presented in reports or screen displays. There are of course other considerations, such as time and resource availability, access to the workplace and parallel project work, but these will usually modify or postpone the work required rather than change the job's objective, though long delays will increase the work scope as extra tasks become necessary.

When choosing which maintenance philosophy to follow it often helps to consider the work required in the form of the 'maintenance box': you can see where other similar work is located and assign the job by the most appropriate philosophy.

	A	B
1	Timing known Content known	Timing unknown Content known
2	Timing known Content unknown	Timing unknown Content unknown

A1 *Timing known, Content known*
Preplanned Maintenance (PPM)
Planned shutdowns
Routine inspections
Scheduled changeouts

A2 *Timing known, Content unknown*
Statutory surveys
Third party inspections

Condition-based maintenance (*Note*: Careful interpretation of condition-monitored results coupled with complete maintenance histories is leading to improved predictions of both the fault timing and its nature, giving greater job content knowledge. When this applies, CBM belongs in box A1.)

B1   *Timing unknown, Content known*
     Anticipated maintenance work
     Contingency work awaiting shutdown
     Run to destruction

B2   *Timing unknown, Content unknown*
     Breakdown maintenance
     Immediate repairs arising from inspection
     Run to failure

From a work management point of view, work characterized by Box A1 is the most welcome, while work in Box B2 is the least welcome, and it is clearly advantageous to transfer work to a more manageable box or at least to mimic desirable properties where possible. The most obvious members of B2 are of course equipment failures. By using an effective maintenance regime the intention is to keep this box as empty as practicable. Notice that run to failure (B2) and run to destruction (B1) are different, as the former is expected to be repairable while the latter requires a replacement action, and thus the work content is known. They are also separated by the MTTR, which refers to replacement times only in the latter case.

Careful evaluation of maintenance histories and shutdown programmes for equipment in B1 can be used to narrow the timing uncertainty. Such work, followed by full job preparation, can provide for a very rapid response when the situation finally requires it.

The condition-based maintenance referred to in A2 follows itinerant or portable condition sampling, and sophisticated interpretation of sampled results can place the group in Box A1.

Requirements attaching to other work in A2 can often be anticipated with care once operations are routinely established and comprehensively recorded.

2.4 Customers for maintenance

Many of these notes refer to methods adopted by major operators, particularly those in North Sea upstream oil operations. This is not to imply that different types of application are less interesting or representative; it is merely that the spread of technical, organizational and statutory problems is sufficiently wide as to give many examples of pitfalls and useful solutions. There is also the question of commercial size, and there is an irony here. Maintenance is a measure of crucial importance to major operators, who are generally well aware of this, and is a measure of future survival to many smaller companies, who are not. The following table attempts to highlight the effect on a small user of a 100 hour repair required to a critical piece of main production plant.

For a typical 100 hour repair to a key item of plant:

Equipment status	Plant downtime	Likely installation
Built in standby	1	Major plant
Site replacement	5	Minimum facilities
Single unit for on-site repair	100	Small production unit

This simple example underlines the need for maintenance management in commercially small equipment users, in situations where it is most likely to be missing. The table shows plant downtime; if, however, the comparison is made against proportional company income, the result is even more graphic.

2.5 Pre-use maintenance

It may seem incongruous but some items of equipment have to be maintained before they are ever used, underlining the fact that maintenance becomes a need once an item exists rather than from the time it starts work. Although certain groups of equipment deteriorate faster than others, it is not easy to say categorically which items will be affected, as the need for pre-use maintenance depends as much on the situation as it does on the equipment itself.

There are six particular characteristics of pre-use maintenance which do not apply to 'normal' maintenance work.

(1) If pre-use maintenance is applied before commissioning, then the equipment concerned *will never have been run by the operator before.*

(2) A final performance test may have been witnessed at the factory, but the operator has no direct experience of its use unless he has identical equipment already.

(3) The principal aim of pre-use maintenance *is to prevent deterioration.* When the equipment is finally commissioned, ideally, it will be as new. In the majority of cases conventional maintenance work is mainly concerned with restoration, i.e. to return to higher equipment condition achieved previously.

(4) The work usually applied to non-running equipment is generally confined to checking of fluids, the 'barring over' of rotational assemblies, ensuring the movement of moving parts, rotation of bearings, the greasing of vulnerable metal surfaces and the checking of protective films, packaging and surface finishes.

(5) In addition, conditions will be constantly reviewed, such as a change of season, lower ambient temperatures, the onset of wet conditions, changes in daylight hours, disposition of tides, and windblown sand. Also, the state of the site will change: other equipment will be delivered or moved and lifting and transport machines may be in local use, together with contractors' earthmoving, concreting and piling activities.

(6) Recognizing that maintenance work is often conducted in unsuitable working conditions, *the location for pre-use work is usually worse,* both for the technician and the equipment. At a storage location, lighting, power and

normal working services are usually absent, and in the open, environmental contamination is eventually certain.

Somewhat surprisingly, there are seven main situations where the need for pre-use maintenance can arise.

(1) Local storage by the original manufacturer prior to shipment.
(2) During factory to warehouse or warehouse to site transportation.
(3) During customer warehouse storage.
(4) During storage on site prior to installation.
(5) In the time gap between equipment installation and the start of operations.
(6) In the second time gap between operational start-up and the start of regular maintenance.
(7) When equipment is classified as on-site non-built-in standbys omitted from the maintenance programme.

All installed equipment in normal operational use will of course deteriorate and the seven situations listed above are those areas where

equipment deterioration can occur beyond the scope and detection of the regular maintenance system

Item 1 above refers to those occasions when equipment is stored before shipment is called for. In most cases, the customer's goods-inwards inspection will detect problems of poor condition upon arrival (Item 2), but where equipment is made to order and stored at the factory at the purchaser's request, pre-delivery maintenance may be necessary.

Many manufacturers will regard storage at the factory as unusual and their own systems may not include the regular checking of undelivered customer-owned equipment.

Some customers will despatch their own inspectors to the factory, especially when goods are required for a major construction project. Whoever conducts the inspection,

the needs and actions for pre-delivery maintenance must be checked.

Manufacturers are well acquainted with shipment problems and employ crating and packing arrangements to avoid them. Damage can still occur, however, and long periods in transit or many different handling occasions merit extra attention (see Sections 10.1 and 10.2). Even if the customer does not face the responsibility, without pre-use checks problems will be discovered at the time of shipment or on arrival, and any delay for rework can ruin a project or maintenance programme.

Pre-use maintenance will replace losses and broken parts, restore damaged surface finishes, and replace lost lubricant and coolant, while Item 3 is aimed primarily at preventing deterioration rather than correcting damage and

often results in a pre-use maintenance programme.

Item 4 refers to a more serious situation. It is known for equipment to be delivered to site well in advance of its installation date, and sometimes before the site is ready to receive it. Heavy machines manufactured to close limits, constructed of closely machined parts and accurately moving surfaces, can be offloaded on to soft ground and left to face the destructive qualities of the environment prior to uplifting and repositioning in the correct location.

(1) The static deflections of heavy metal parts will distort machinery casings, mountings and bed plates, especially when loaded out of true on a soft surface.
(2) Both weight and temperature effects will disturb alignments, inter-machine couplings and the subsequent transfer of torsional power.
(3) Rotational shafts will deflect between supporting bearings, affecting rotor balance and forcing lubricant away from the points of contact.
(4) Coolants and lubricants will leak or evaporate, or can separate into dense and less dense constituents.
(5) The effects of solar radiation and temperature cycling will embrittle plastic parts, changing the plasticity of 'O' rings and interfacial seals.
(6) Environmental temperature effects will only be partly offset by equipment self-heating (unless the item is run regularly), and the thermal properties of different materials will create stresses on vulnerable parts.
(7) External finishes, electrical contact and switching surfaces, accurate faces of meshing and sliding parts can all be degraded by corrosion and other chemical effects.
(8) The presence of water will rot wooden and paper packing materials, attack both site and equipment mounting surfaces, corrode exposed machined faces and introduce galvanic action between adjacent dissimilar metals.

It has been known for major parts to be delivered three years before they were needed, and others made redundant by plant design were resold to another operator as 'unused'.

Item 5 (previous list) refers to the time between installation and the start of operations. Because of the extent and complexity of some plant building projects, items can be installed and unused for several months. Further, some equipment may be installed at the outset and not required for early operational use or until significant output expansion has been achieved. This can lead to problems. Any claim by the customer under warranty for subsequent problems of condition or performance will not be welcomed by the manufacturer, and repairs or corrections required will have to be paid for; in the worst cases of non-use, equipment may be reduced to ruin.

The commissioning stage of the construction project prior to the handover to operations will run the equipment after installation to confirm its effectiveness and condition.

After commissioning engineers have demonstrated effective running and the equipment is accepted at handover, operations take responsibility and *the maintenance system starts its formal work*.

This section refers to the questions of maintenance before use, but three situations need to be considered, i.e. before use, normal use *and after use*.

The second group is dealt with throughout the manual, and the third group is frequently overlooked. After-use maintenance does not refer to equipment temporarily out of service, but to those items which will not be used again. Quite apart from the difficulty of deciding whether an item is truly and finally out of service, there are questions to be considered which are often regarded as outside the maintenance orbit.

(1) Does the unused equipment represent a source of danger, (e.g. residual power, pressure or electrical energy)?
(2) Will special cocooning or preservation be necessary?
(3) Will extra work be required for item resale to another customer?
(4) Is equipment after-care needed in the maintenance programme?

Many of the subjects concerned are the same as those referred to under the pre-use heading, and it must not be overlooked that:

whether before or after, lack of operational use does not mean a lack of maintenance responsibility.

2.6 Work scheduling

The work schedule discussed below has its foundations in the work plan, and it is important to consider the schedule as a necessary tool of work management, whatever system of job selection is in use, for the identification, coordination and control of jobs by on-site maintenance supervision. Having established the inclusion of pre-planned maintenance jobs into a work plan, most computer-based systems harvest the plan for the next predetermined work period and include the tasks triggered in a work schedule. We could also usefully consider the work schedule as a computer program which selects and presents the maintenance jobs to be done from a prepared database plus a selection from existing work orders for jobs that remain outstanding. The computer-driven work schedule program would work to a careful set of rules and allow insertions and deletions by the maintenance planning engineer.

The scheduling programme has the following properties:

(1) Defines relevant work site and week numbers.
(2) Identifies the maintenance work crew applicable.
(3) Extracts triggered work from the work plan database.
(4) Adds activated work orders to the schedule when selected by maintenance planning.
(5) Accepts rescheduled work from previous work tours which has not been reassigned and not started.
(6) Includes details of jobs already in progress.
(7) Adds a work priority number pertinent at the time of the schedule for each job entry and allows revision by maintenance planning or supervision (see Section 2.7).
(8) Provides details of each job scheduled, including disciplines required, estimated manhours per discipline, subject equipment identification (such as tag numbers) and resource totals required.

(9) Will run once per work tour without special intervention prior to the onset of the work period and will not run retrospectively.

(10) Refers to the rules of work association before including planned, non-planned or rescheduled work (see Section 8.3).

The most common periods are equal to the work tour of three weeks or a fortnight, varying according to the operating company's work organization.

Figure 2.1 shows the 1996 Red and Blue crew rota for the Ninian Central Oil Platform. It is based on a three week rota. The maintenance planner, depending on the system being used, will prepare the work schedule for issue to the next crew as they come on board, and computer run timing is shown in Figure 2.2 as R1 to R5. The diagram shows a section of the work plan for a two-week rota from weeks 22 to 35, with two crews on site for three two-week periods each. So Red crew is working on site for week numbers 24–25, 28–29 and 32–33, while Blue crew works in weeks 22–23, 30–31 and 34–35, and similarly throughout the rest of the year. Notice that, for the reasons described in the work plan (see Section 5.3), weeks 26 and 27 have no planned work assigned to them.

The following maintenance jobs will be included in the work schedule.

(1) Regular maintenance tasks which fall due in the chosen work period according to the work plan and triggered by the computer schedule run.

(2) Regular maintenance tasks scheduled for previous work periods which, not having been completed or cancelled, have been rescheduled to this subsequent work period.

(3) Activated equipment breakdown or defect jobs (detected by PPM routines, operational reports or equipment failures).

(4) Special additional maintenance jobs assigned to the period by statutory authority or safety department instructions.

Item 3 above sometimes refers to equipment defect jobs which could be corrected by a later, lower frequency PPM job not currently due. The maintenance planning engineer may decide to replan this PPM to make it a present event while cancelling the Defect Work Order as a simple means of clearing the work needed.

We should remember that the work plan handles predicted maintenance jobs of regular inspection and routine work, and these are transferred to the work schedule when they apply to the work period in question, together with activated defect or repair work orders outstanding and opportunity work orders awaiting the requisite situation. Some systems treat the planned maintenance schedule and the defect (unplanned) outstanding list separately, which is usually a mistake if the same work crew is used in such cases. There are of course, repair tasks which require the attention of external contract specialists, but all work required is coordinated at one place and will be contained in the schedule.

Notice also that Defect or Repair Work orders included in the schedule are those 'activated' by maintenance management. Although the intention is to complete work of this sort as quickly as possible, it is not always possible to do this, and delays

NINIAN CENTRAL PLATFORM 1996 PPM CALENDAR

Day	DE	January					February				March				April				May					June		
Wed	27	3	10	17	24	31	7	14	21	28	6	13	20	27	3	10	17	24	1	8	15	22	29	5	12	19
Thu	28	4	11	18	25	1	8	15	22	29	7	14	21	28	4	11	18	25	2	9	16	23	30	6	13	20
Fri	29	5	12	19	26	2	9	16	23	1	8	15	22	29	5	12	19	26	3	10	17	24	31	7	14	21
Sat	30	6	13	20	27	3	10	17	24	2	9	16	23	30	6	13	20	27	4	11	18	25	1	8	15	22
Sun	31	7	14	21	28	4	11	18	25	3	10	17	24	31	7	14	21	28	5	12	19	26	2	9	16	23
Mon	1	8	15	22	29	5	12	19	26	4	11	18	25	1	8	15	22	29	6	13	20	27	3	10	17	24
Tue	2	9	16	23	30	6	13	20	27	5	12	19	26	2	9	16	23	30	7	14	21	28	4	11	18	25
PPM Week	1	2	3	4	5	6	7	8	9	10	11	12	13	14	15	16	17	18	19	20	21	22	23	24	25	26
Crew	R	R	B	B	R	R	B	B	R	R	B	B	R	R	B	B	R	R	B	B	R	R	B	B	R	R

Day		July					August				September				October					November				December			
Wed	26	3	10	17	24	31	7	14	21	28	4	11	18	25	2	9	16	23	30	6	13	20	27	4	11	18	25
Thu	27	4	11	18	25	1	8	15	22	29	5	12	19	26	3	10	17	24	31	7	14	21	28	5	12	19	26
Fri	28	5	12	19	26	2	9	16	23	30	6	13	20	27	4	11	18	25	1	8	15	22	29	6	13	20	27
Sat	29	6	13	20	27	3	10	17	24	31	7	14	21	28	5	12	19	26	2	9	16	23	30	7	14	21	28
Sun	30	7	14	21	28	4	11	18	25	1	8	15	22	29	6	13	20	27	3	10	17	24	1	8	15	22	29
Mon	1	8	15	22	29	5	12	19	26	2	9	16	23	30	7	14	21	28	4	11	18	25	2	9	16	23	30
Tue	2	9	16	23	30	6	13	20	27	3	10	17	24	1	8	15	22	29	5	12	19	26	3	10	17	24	31
PPM Week	27	28	29	30	31	32	33	34	35	36	37	38	39	40	41	42	43	44	45	46	47	48	49	50	51	52	53
Crew	B	B	R	R	B	B	R	R	B	B	R	R	B	B	R	R	B	B	R	R	B	B	R	R	B	B	R

Figure 2.1
Red and Blue crew rotas for the Ninian Central Oil Platform (reproduced by courtesy of Chevron Oil (UK) Ltd).

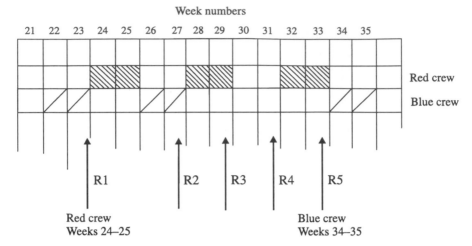

Figure 2.2
Part of the work plan for Red and Blue crews (reproduced by courtesy of Chevron Oil (UK) Ltd).

often occur while maintenance await the arrival of replacement parts, special skills, tools or a working opportunity.

In addition, rescheduled work is not as obvious as it seems. Some routine jobs 'triggered' by the work plan and included in the schedule, for a variety of reasons, are not done, and the planning engineer will decide whether a job of this type will be closed out as 'not done' or rescheduled to a forthcoming work tour. Such decisions are not easy, and it is usual practice that when a reschedule is required for a non-critical job it is rescheduled to a forthcoming work tour of the same crew, to avoid crews accusing each other of work dumping. Critical jobs are another matter, and if one crew cannot clear them it falls to the next on board for immediate attention.

When a maintenance planning engineer faces a decision about rescheduling a job not done, he will consider the following:

(1) Is it a high-frequency job due to be repeated in the next work tour anyway?
(2) Can the work be rescheduled without destroying the arrangement of the work plan, or will replanning be required?
(3) Has the job been delayed or closed out 'not done' recently? (Check performance indicator no. 9, p. 315.)
(4) Are there alternatives, like a forthcoming shutdown or a closely related defect work order?
(5) Are there statutory work elements in the job or specific authority or safety instructions related to it?
(6) Is part of the maintenance programme in jeopardy whereby a response by the authorities is to be expected?

(7) Can the job be included given the constraints of work association?

Although mentioned above, we should underline the main difference between the maintenance work plan and the work schedule:

the maintenance work plan aims at the ideal future, detailing which tasks are to be done and when.

As noted in the work box (Section 2.3), for a job to be included in the work plan both the 'what' and the 'when' of the job are required, which means that defects and repairs are not included. Some computer systems produce separate lists of work scheduled and defects outstanding, and as computer-generated printouts look similar the opportunities for confusion are clear. By including all *work scheduled, rescheduled and defects outstanding* in the work schedule, a single printout allows all jobs to be conveniently managed.

The work schedule is all-embracing, including both planned and unplanned work. The latter includes current defects or repairs, activated opportunity work orders and other jobs outstanding.

2.7 Work prioritizing

Usually represented as a single digit between 1 and 9, priority numbers are listed in the work schedule being assigned to planned and outstanding maintenance or repairs in addition to work already in progress. Such priorities are intended to indicate the operating organization's work preference, whereby in the absence of different supervisory instructions, priority 1 jobs will be treated first, followed by priority 2s and so on, in the sequence indicated by the priority numbers. In the work schedule for Blue crew, in weeks 5 and 6 there could be a total of 40 jobs, of which three are priority 1s, six priority 2s, and so on. For some systems all jobs are assigned a priority until all 40 are classified. In addition, the sudden failure of a critical piece of plant will introduce an additional priority numbered job, effectively delaying all lower priorities previously identified and raising the number of jobs to 41.

We should note at once that some maintenance operations assign priorities to a few key jobs only to ensure that they are done first, and rely on sufficient team resources to complete the remainder of the schedule during the work period[4,5].

The importance of priority numbers lies in the clear and repeated focusing of attention on the most urgent jobs. Unfortunately, attributing a high priority does not mean that the job will be done. There are many constraints, of which shortage of materials is the most obvious. Several jobs may share the same priority, each urgently required and perhaps being worked on by different disciplines.

Although assessed by maintenance during the planning stage, priorities are

---

4. One of the advantages of the method described here is that when additional jobs are added the priority of previously assigned jobs in the schedule does not change.

5. Because of the action of the priority modifier, those jobs assigned a priority number in one schedule and transferred to a subsequent schedule without being started will have their priority numbers modified.

important to all operational groups and usually assigned by a previously agreed method, reflecting different departmental purposes. Equipment function and status are immediate properties used during the assessment, with other factors considered during selection by the 'on-site' maintenance supervisor.

One simple method of assessing priorities is described here, and other methods are in use.

There is sometimes confusion between 'priority' and 'criticality', and as the following assessment table shows, the latter features at the top of the priority table. Criticality, as described here, refers to the severity of an effect and its probability (see Chapter 12 and the Glossary), and for some items their function is loosely described as 'critical'. Priority is a transitory qualification which refers to a work schedule and expires when the work is finished.

The existence of a priority number follows a tacit acceptance that resources are limited and that not all jobs can be done at once. Also, if resources were unlimited access and opportunity would limit the work that could be done (see Section 11.2).

There is a further danger: priorities can shoulder aside the requirements of the maintenance schedule and the benefits of efficient work management will be lost. Priority work is exceedingly compelling and it has advantages when existing team motivation is poor. If normal work routines are being dominated by priority work, either the maintenance team is too small or troublesome equipment needs to be replaced (see Section 8.9).

Selecting priority numbers

From the listed descriptions, select the target priority which applies and then correct the figure according to listed modifiers which apply.

Job priorities can be established in the manner described below, but should be suitably modified for the specific application.

(1) From Table 2.1, assign the noted priority to each job in the work schedule according to the nearest description. Should the job concerned meet two different entries, choose the higher priority of the two.

(2) Note that in this model priority numbers ascend as the urgency diminishes, so the highest priority number shown here is 1.

(3) Notice also that the graduations in Table 2.1 are selectively broad, partly to ease the assignment process and partly because of the purpose itself. It is not intended to make refined judgements or detailed descriptions, but to focus the work schedules on urgent tasks first. Priorities are not used to replace the need to match resources available with those required. Although some jobs on the schedule may be rescheduled when they remain undone, it is expected that resources will generally be in balance.

(4) Although planned work will represent the main portion of jobs in the work schedule, breakdowns and repairs will also be included, together with work already in progress. Supervisors should not be expected to consult several reports or computer screens to establish the current picture. Not all systems

TABLE 2.1 PRIORITY ASSESSMENT

Plant/equipment condition requiring job	Priority number
1. Immediate threat to human life, safety critical equipment or safety system failure Such systems are required to be in a state of constant readiness and include fire and gas detection, fire-fighting equipment, escape systems, emergency shutdown systems, site communication and public address systems, potable water and food processing or storage which are all applicable to the safety of operators, maintenance or site personnel, support staff, contractors, site visitors, residents and workers, and the public near to site boundaries or geographically wider. Often the result of on-site safety officer or safety authority immediate instruction or condition monitoring report of impending failure (see Section 3.6 and Chapter 11).	1
2. Statutory body instruction Direct instructions from a statutory body such as the HSE (Health and Safety Executive), DOT (Department of Transport) or the CAA (Civil Aviation Authority) to an operator defining specific action required, usually reinforced by authority correspondence.	2
3. Production-critical equipment failure Where operational downtime and consequent loss of revenue results from production equipment failure and production restart will require equipment repair. May result from a third party equipment examination (such as of lifting equipment) reporting immediate requirements.	3
4. Production-critical equipment scheduled work Where no equipment failure exists but maintenance downtime or safety hazard is specified in the current maintenance schedule.	4
5. Production or services non-critical equipment failure Work needed on failed equipment which will improve production operations without affecting downtime.	5
6. Scheduled maintenance for production non-critical equipment The main content of the maintenance work schedule for each current period and published immediately prior to the onset of each work tour, includes computers, non-critical electrics, standby batteries.	6
7. Scheduled maintenance of service equipment Major content of the current maintenance work schedule as above. Refers to site cranes (non-lifting element work), materials handling equipment, power generation, normal electrical distribution, telecommunications, workshops.	7
8. Scheduled life support services Includes all equipment/systems maintenance not covered above and cooking, cleaning, laundry and entertainment equipment.	8

present in this way, and merging of the contributory information may be required to avoid confusion.

(5) Use of priorities for work selection is 'demand' or 'user requirement' driven and is not the most efficient when considering overall work management. More work can usually be done in a given time when jobs of content or procedural similarity are grouped together. For this reason, priority numbers may be omitted by the supervisor for less urgent jobs and more usual methods of job selection used.

In some priority lists, numbers 5, 6 and 7 in Table 2.1 are arranged differently. There is a strong case for this when the platform is crowded and failures of life support systems or equipment are followed by immediate demanning. Selection of priorities is never easy, and can easily become the focus of argument and disagreement. It is important to remember that the list is not a statement of importance; it is an action sequence. All groups are important or they would not be listed at all, and one purpose of the list is not to eliminate argument, but by its creation ensure that disagreements are settled before the site responds to a crisis.

There are some influences or 'modifiers' (Table 2.2) which will change the priority number as it is assigned, according to the local situation. Priority numbers from Table 2.1 should be changed as indicated wherever the 'modifiers' apply. Many modifiers *increase the priority number chosen* as the higher number indicates less urgency.

TABLE 2.2 PRIORITY MODIFIERS

Relevant condition	Change to priority number
1. Existence of built-in standby Available for immediate use.	Increase by 1
2. Machine one of a group of identical machines Where machine members are in current fault-free operation.	Increase by 1
3. Rescheduled work from previous work tour Repeated on each occasion the job is rescheduled and progressively lowering the priority number, i.e. increasing the priority.	Decrease by 1
4. Opportunity work released by higher priority High-priority work on different but related equipment introducing the opportunity to conduct opportunity work.	Decrease to match parent machine
5. History of equipment reliability From previous completed maintenance or repair work orders: High reliability Low reliability	 Increase by 1 Decrease by 1
6. Impending shutdown Where medium-priority non-scheduled work can be shortly replaced or merged with shutdown work.	Increase by 2

The system described should be refined to suit the existing maintenance system without removing its basic simplicity. Priorities will change quickly and the revised response must be known immediately.

Six priority modifiers are shown in Table 2.2, and others are relevant in different circumstances. Number 3 is particularly noteworthy, as its repeated and disciplined use will ensure that apparently low priority work will not be overlooked.

Chapter 3

Main methods and philosophies

This chapter describes the most important maintenance philosophies and their relevance to different situations. Some, like 'breakdown maintenance', have been in use, often by default, for centuries when considering buildings or structures, and from the mid-seventeenth century with reference to machines. The other systems described have mainly been developed after the Second World War, partly in response to perceived shortcomings in methods already in use and partly in response to technical opportunities. Although there has been a progression of developments, all philosophies remain in use, each more effective in different areas and often employed together as complementary techniques. When contemplating the extension or introduction of a maintenance management system it is important to establish which philosophy it follows and that of the method to be replaced. It is tempting and natural to see the advantages of the new technique in relation to the disadvantages of the old one, rather than the other way around. Such choices can be much more difficult than they seem, and there are several factors to remember:

(1) No one philosophy holds all the advantages. If this were the case there would be only one method in use.
(2) All philosophies have disadvantages, not always mentioned, which need to be considered.
(3) The value of every philosophy varies according to the user organization, where and how the philosophy is interpreted and used, and the nature of the team engaged in the work itself.
(4) Human behaviour is the most potent influence on the effectiveness of the different philosophies employed.
(5) There is no such thing as 'no philosophy in use' in an operational situation. It may be ignored or informal, in which case breakdown maintenance applies.
(6) When considering new philosophies, remember the work itself, how it is to be managed, and its effects on the organization and on personnel.

Pneumatic Handbook

Eighth Edition

Recognised as the standard reference work on modern pneumatic and compressed air engineering, this new edition has been completely revised, extended and updated to provide essential up-to-date reference material for engineers, designers, consultants and users of fluid systems.

CONTENTS

1. Basic Principles
2. The Compressor
3. Energy and Efficiency
4. Compressed Air Transmission and Treatment
5. Applications for Compressed Air
6. Sensors and Controllers
7. Control Valves
8. Actuators
9. Seals
10. Applied Pneumatics
11. Vacuum and Low Pressure
12. Engineering Data

ISBN 1 85617 249X   Price: £135*/US$217/NLG351

☐   Please enter my order for the
Pneumatic Handbook [PIH41+B1]
ISBN 1 85617 249X   Price: £135*/US$217/NLG351

Name: —————————————————
Job Title: —————————————————
Address: —————————————————
—————————————————————————
—————————————————————————
State: ——————— Zip/Postcode: ———————
Country: —————————————————
Tel: —————————————————
Fax: —————————————————
E mail: —————————————————
Nature of Business: —————————————————

Return to:
Alistair Davies
Priority Orders Department
Elsevier Trends Division
PO Box 150, Kidlington
Oxford OX5 1AS UK
Tel: +44 (0)1865 843181
Fax: +44 (0)1865 843971
E-mail:
a.davies@elsevier.co.uk

PHB97

**ELSEVIER
TRENDS
DIVISION**

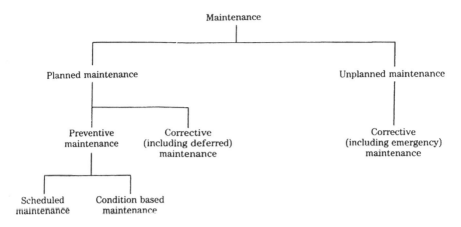

Figure 3.1
Forms of maintenance (from BS 4778).

(7) All situations are different. Apparently identical requirements, such as two factory units producing the same products on different sites, may have access to different maintenance resources, different labour rates, response times and material shipment capability.

Most philosophies assist or formalize the decision-making process, which helps to select and time a maintenance action but has less regard for the work itself or its organization. Some methods, such as PPM or condition-based maintenance, for example, can cause disbelief in the workplace, leading to argument and inaction, while breakdown maintenance does not suffer from this problem[1]. For this reason the two former methods do not always work well with TPM (total productive maintenance) because of the latter's emphasis on team empowerment and personal motivation. We could assume from this that breakdown maintenance via a technical *fait accompli* is free of dissent – not true. The argument, however, changes from 'is this piece of work really required' to 'why can't I have it yesterday?'. This of course is the familiar argument between maintenance and production, the long-term and the short-term question.

Recently the phrase 'core business' has become very topical, often referring to major companies around the world, the extent and nature of their subsidiary holdings and the concentration of their efforts and resources in carefully selected areas. Such thinking has also been applied within operating organizations, and, under various names, has resulted in the transfer of previously clear 'in-house' functions to external specialist companies.

---

1. There is an interesting diagram showing the forms of maintenance in Section 3.1, subsection 6, BS 4778 1991, reproduced here as Figure 3.1. Notice that some of the titles used are different as the philosophies discussed above centre on the various methods of maintenance management. (Refer also to Section 2.6 and the sections of this chapter.)

Maintenance has often featured in such changes, and companies specializing in maintenance work have found themselves responsible for huge increases in their work 'portfolio', sometimes in areas of uncertain knowledge and experience. Facing these new responsibilities with a mixture of excitement and alarm, maintenance companies are encountering positive long-term commercial prospects with short-term operational problems.

The customer internal expertise engaged in the past to provide technical effort, plus the work management functions are disappearing fast. Maintenance companies accustomed to being told when, where and how to act are increasingly required to take the management initiative on their own. The engagement of specialist personnel previously employed by their customers and transferred with little choice to a new employer has helped such companies, and, aware of these problems, major company customers are using periods of transition and steady handover.

Unfortunately, questions of philosophy return with some force. Maintenance companies may be responsible for more than one site and more than one customer, each having different maintenance philosophies and different operational problems. The coordination of the various requirements is proving to be difficult and the clear enunciation of the different philosophies, their replacement or change is essential and urgent.

3.1 Breakdown maintenance (maintenance work box B2)

Description

As the name implies, no maintenance work is conducted until the subject equipment exhibits a functional failure, whereupon repair and maintenance work are immediately instigated.

In the work box diagram (p. 20) breakdown maintenance would nominally appear in box B2 (Timing unknown, Content unknown). However, this is deceptive, as the subject equipment, the maintenance history and the experience of the site team, all known in varying detail, will have an influence on this method when in use.

For example, some machines are known to fail first in a simple and non-destructive way, like the chain failing on a bicycle. When this failure is corrected the machine receives further attention at the same time and more serious failures are prevented.

Alternatively, some breakdown failures are expected, the content being known without knowing when, and these would fall in Box B1. A typical example would be the exhaust system on the family car.

Breakdown maintenance covers a wide range of user approaches; it may be carefully applied to selected items or it may be the result of no philosophy at all. It must not be assumed that breakdown maintenance implies disaster, but the possibility is there.

Advantages

(1) Breakdown maintenance is extremely simple to introduce. It can be applied quickly with limited resources and information.
(2) Breakdown maintenance can be used as a useful foundation, allowing the addition of other techniques.
(3) In the hands of an experienced maintenance team, used on a carefully selected basis, breakdown maintenance can be very effective. This is especially true when applied to non-critical items known to fail in a safe manner when the failure is anticipated and prepared for.

Disadvantages

(4) Applied as a general approach to all equipment and circumstances, breakdown maintenance can lead to a commercial nightmare. Of all the philosophies, it is known to be the most expensive. The damage to machinery and the interruption of production can incur huge consequential and contractual costs. At any time large and unexpected expenses can destroy the budget and cause a business extinction.
(5) In some cases, the consequences of the breakdown are not confined to the subject machinery. Apart from production losses, parallel equipment and processes can be affected, and in a few well-known cases like the Flixborough chemical plant explosion (caused by similar but different reasons), whole plants can be destroyed.
(6) Breakdown maintenance is not safe. This philosophy exaggerates uncertainty, increasing danger and the risk to human life. When applied to an operator's staff, ignoring a known danger is morally wrong and contractually indefensible. When it results in the destruction of people in the street, innocent and unaware, as in Bhopal in India, it becomes a financially ruinous national and moral scandal.
(7) Work arising from an unexpected breakdown is much more difficult to manage and will often be corrected at a high price. The use of special skills, employment of less expensive working periods, careful job preparation and the acquisition of materials at competitive prices are all jeopardized when immediate action is required.
(8) During the period of early plant and installation life, equipment tends to be more reliable, and the breakdown method leads to equipment being undermaintained. Later in the life cycle, the incidence of failures increase and equipment is shut down more often. The absence of a carefully planned maintenance regime can allow the dangerous gathering of failures and the real risk of total plant shutdown for high and unmanageable numbers of outstanding urgent repairs.

Comment

Breakdown maintenance should never be followed by default. The downside of its use, although unlikely, is far too severe. However, there are occasions when it can be usefully employed in carefully selected and monitored applications.

Note also the word 'exhibits' in the opening description. As all maintenance engineers will testify, unseen or, to the operator, *'invisible' failures* repeatedly occur. These become visible when equipment classed as available, but switched off, fails to operate when switched on. Breakdown maintenance increases the incidence of such failures. Although statistically trivial, the effects are disastrously magnified when they apply to safety and standby equipment. When the plant is on fire, the emergency shutdown system must work *then, not later*. Breakdown maintenance will not do.

3.2 Run to destruction (maintenance work box B1)

Description

Under this philosophy, selected equipment is used normally until it fails, when it is discarded and replaced. In the work box diagram (p. 20), run to destruction would appear in Box B1 (Content known, Timing unknown). The work required to switch off machinery, perhaps isolate, remove and replace the old unit, and connect and switch on the new one, is known and can be prepared for.

Advantages

(1) Preparation for such work can be detailed, merely awaiting the requirement to proceed.
(2) Work management of such tasks is straightforward, as materials, technical content, duration, targets, disciplines and tools required are all known.
(3) With the careful selection of subject equipment and circumstances, such a philosophy can prove less expensive.
(4) Stockholding can be reduced, the replacement item, plus fixings, couplings, and connections only being held. Stockholding, inspection and handling costs are reduced.
(5) When relevant performance knowledge is based on a usable statistical population with equipment life and time of likely failure known, the run to destruction philosophy may be modified to act using a replacement calendar, effectively moving the revised approach into work box A1 (Timing known, Content known): the most beneficial work management profile.

Disadvantages

(6) When applied to the wrong equipment, run to destruction is expensive and dangerous. It suffers from all of the disadvantages listed under breakdown maintenance.
(7) The full effects of equipment destruction cannot always be predicted in advance.
(8) Use of this method, even more than breakdown maintenance, encourages site personnel to disregard warning signs of impending failure and increasing danger.
(9) There is a strong temptation to apply this technique in regions where it is just not suitable. Limitations of available resources or supervisory attention can lead to the use of this philosophy by default.

Comment

Usually regarded as a version of breakdown maintenance, this philosophy does have features of its own, particularly the usual absence of any form of evaluation, which means that failures are unlikely to be diminished and there is little possibility of moving this type of work into an A1 work box.

It is important to stress that, like breakdown maintenance, run to destruction must never be introduced by default. Its use must be confined to carefully selected equipment and the consequences of failure known and accepted in advance. Also, the occurrence of ultimate failure can introduce danger into the workplace, and run to destruction can easily lead to other more serious problems.

3.3 Pre-planned maintenance (maintenance work box A1)

Of all the work management philosophies, pre-planned maintenance, or PPM, is the most famous and the most widely used. Formally developed following experience gathered during the Second World War, it has recently suffered some adverse and ill-informed comparison with newer philosophies, such as reliability-centred or condition-based maintenance. The truth is that all of these approaches have something to offer. The use of an effective system in the wrong place will reduce the user's inclination to employ it in those areas where it is most effective, damaging its reputation unjustly and encouraging the use of a different system in a different wrong place[2].

Pre-planned maintenance divides the working calendar into discrete separate elements and assigns PPM jobs to them. Each job is identified by its frequency, as well as often being named according to the task itself. So a five-yearly pressure vessel internal inspection describes what the job will do and how often it occurs. Similarly, a six-monthly lifting equipment examination details these two aspects of the job in question.

It is usual for the least frequent period (or *periodicity*) to be five years (often quoted as 60 months) and the highest periodicity, or the most frequent, to be daily, with a wide range of selectable frequencies in between. The most common are:

60 months	(five-yearly)
24 months	(two-yearly)
12 monthly	(annual)
6 monthly	
3 monthly	
monthly	
biweekly	
weekly	
daily	

Because maintenance work is actioned according to the dictate of the calendar, *irrespective of machine condition*, it is often argued that this method requires work to be done that is not necessary. For example, due to production constraints a water

---

2. Planned maintenance is detailed in BS 4778 SI 1991, Section 3.2 Subsection 16 as: The maintenance organized and carried out with forethought, control and the use of records to a predetermined plan.

injection pump is switched off immediately after its annual PPM and remains unused for the next twelve months, whereupon the next 12 monthly PPM falls due after a period of no work. This is clearly a nonsense, and a good PPM system will ensure that such incidents do not occur. Three methods stand out:

(1) *Application of condition monitoring to critical equipment.*

As we will see when discussing condition-based maintenance, the use of condition monitoring can be more expensive in some cases and relies on special technical skills. Although its use is growing, in a nominally PPM environment it is assigned in selected critical cases only. Warnings of deteriorating condition or impending failure are used to trigger the inclusion of the appropriate PPM in the work programme[3].

(2) *Monitoring equipment 'hours run' where there is a clear hours of use/condition correlation.*

In some cases there is a clear correlation between a machine or component's use, and its need for maintenance work, corresponding to the mileage on a car, time in use of a platform crane or running hours of a bilge pump. Collected readings are used to estimate the rate of use and predict when the next PPM will fall due.

(3) *Use of daily equipment status reports to influence the work demands of the PPM system.*

Reliable daily status reports are vital to the success of any effective system. When equipment is reported as turned off or assigned to standby, the PPM programme is reduced to protective actions only. When the item is returned to duty the PPM programme is restarted at the appropriate point in the PPM programme. Remember that the equipment may be returning to duty following a major repair or refurbishment, and the programme clock is effectively set back to zero.

The PPM method has a *unique characteristic in that the work required is partly defined by the frequency itself*, the total work required being split down into more manageable pieces. However, the first ingredient of the content refers to actions that suit the frequency, for example examination of the maintenance history shows that the brushes on certain electric motors wear out every 58 weeks, so each annual PPM will include their replacement. If pressure gauges drift during use and require recalibrating every 6 months, then this work is included in the six-monthly PPM.

This method of subdivision is highly relevant to the maintenance of a facility consisting of many identical elements, such as d.c. battery backup or the low-power electrical distribution system. Suppose that in the former case the whole backup facility consists of 24 identical d.c. batteries requiring to be

(1) Inspected annually.
(2) Discharged and recharged annually.

---

3. This use of condition monitoring does not have universal approval, as by tying CBM to the PPM the principal advantage of CBM, that of conducting maintenance when condition requires, is undermined.

Two PPM tasks could be issued each calendar month, the first to discharge and recharge two batteries, and the second to inspect two different batteries, the only change each month being the identity of the batteries for the month. The sum of 24 separate PPMs over the year neatly completes the maintenance requirements for the whole facility.

Advantages

(1) Because of the regularity of PPM work, both *work timing* and *work content* are defined, simplifying work management and allowing preparation in advance (see item 8 below).

(2) PPM is ideal for high-volume modest-value important tasks which cannot justify the assignment of expensive measurement equipment.

(3) PPM allows the use of all work management techniques, such as work planning, resource scheduling, discipline selection or job grouping.

(4) The repetitive nature of many PPM tasks leads to the generation of work skills, increased job knowledge and the creation of in-house expertise.

(5) The use of several smaller more frequent maintenance tasks as opposed to less frequent major ones increases the chance of detecting emerging problems.

(6) The use of PPM encourages the use of regular hours and less expensive out of hours working.

(7) The use of PPM routines and regular hours underlines supervisory actions, such as the delegation of tasks, inter-task coordination and workplace control.

(8) The knowledge of work to come allows the full range of preparation to be completed before the work starts, including:
   - Communicating and underlining the requirements of safety measures and specific safe working practices.
   - Selection from competing work priorities.
   - Establishment of supervisory steps, task delegation and progress and reporting requirements.
   - Setting work priorities and task coordination.
   - Defining work objectives, methods and targets, identifying special needs of equipment shipment, handling and local storage.
   - Marshalling of special tools and materials on site.
   - Identifying and gathering relevant disciplines and expertise.
   - Establishing lines and means of communication.
   - Allocating extra resources and supporting technical services.
   - Creating work teams and team identity.
   - Gathering technical information, maintenance history and relevant work experience.
   - Defining isolation and shutdown requirements.
   - Completing team training and task evaluations.
   - Allocating logistical support, life support, transport and accommodation.
   - Restating budgetary and financial limitations.
   - Confirming usable working facilities, workplaces, means of use access and special local practices.

3.4 Reliability-centered maintenance

RCM has had a profound effect on the way maintenance requirements are understood and the language that is used to describe them. Although we refer to it as a philosophy, it is more accurately a collection of methods of formal assessment. It refers to equipment function, reliability and the statistics of failure, resulting in recommended methods of maintenance and a closely defined understanding of the role and importance of each physical asset.

Although RCM analysis, step by step, takes regard of the whole plant, it structures its evaluation by examining each single main asset in the equipment database, implementing each of the following stages in a review.

Asset selection

RCM first considers what assets are concerned, how they are defined and whether they should be included in the review. Some assets, for example, may have been shut down or are assigned to a different maintenance responsibility, and the *intended function* effectively changed to 'remain safe' or 'remain secure'.

There are fundamental asset dispositions to consider, as some of the machinery on site will be out of service for several reasons[4].

(1) Machines specifically assigned to discontinued products or processes.
(2) Machines temporarily not required because of low output requirement.
(3) Machines awaiting replacement.
(4) Machines awaiting an installation change.
(5) Machines awaiting removal.
(6) Machines on standby duty.
(7) Machines which have failed and await on site repair.

It has repeatedly been the case that non-operational machines have not been properly shut down or decommissioned, retaining stored energy, access to power supplies, and environmental features, such as trapped fumes, combustible materials, and stored chemicals.

At this stage RCM asks what needs to be done so that the selected asset will *deliver its intended function*.

Statement of functions

Using formal questioning and recording, RCM first interprets and then defines the *primary and secondary functions* of the selected asset. Primary functions usually follow the operation they serve, are readily definable and refer back to the original design

---

4. Items 3 and 4 refer to 'replacement' or 'installation change' and can require different treatments in RCM. 'Replacement' describes the process of removing or de-installing an asset to allow the installation of a new or reconditioned item specifically the same. Apart from equipment age and position on the maintenance cycle there is no change to the maintenance requirements. An 'Installation change', however, can be many different things, such as the installation of a new version, the installation of two units in place of the one existing, or reorientation of the installed unit, changing piping and access. Similarly, we could also include item 5, but it has been separately listed.) Changes to the maintenance arrangements inevitably follow.-

specification, such as a process pressure vessel's primary function of reducing flow rate to introduce liquid separation, or the pipeline flow valve allowing the progress of fluid along one of two pipes, A or B.

However, time and circumstances change: fluid mixes vary, new constituents or recipes apply, and machines may be derated because of age or overload demands, and the primary function may not be so obvious after all.

Secondary functions are both more obscure and more numerous, such as the effective storage of 85 cubic metres (3000 cubic feet) of crude oil, or the containment of fluid at 10×10^6 newtons per square metre (1450 PSI), both secondary functions of the pressure vessel above. Similarly, a secondary function of the pipeline flow valve referred to may be the denial of any fluid flow in the unselected pipe A or B. Although secondary functions are sometimes disregarded, they should never be forgotten; such a course can lead to danger and calamity.

During the Second World War, the primary function of the German railway was the transmission of the fighting machine, and a secondary function was the transmission of maintenance crews and materials for its own repair. Persistent bombing by Allied air forces approaching the invasion was effective, particularly against the railway's secondary function, making maintenance and repair almost impossible and ultimately causing failure of the primary function.

The facts of functional failure

RCM reviews the possible failure of both primary and secondary functions, itemizing the ways such a failure can occur. A function failure may be a total breakdown or the inability of a piece of equipment to meet the functional standard required. For example, when the drive end bearing of a water injection pump seizes, the drive rotor stops spinning and may catch fire, and all pumping action ceases because of total functional failure. Alternatively, accumulated deposits on the impeller reduce the fluid delivery from the required 10 cubic metres per minute to a degraded 8. The latter is less dramatic and creates less work, but the pump is failing to meet its primary function, and functional failure has occurred in both cases.

Identification of failure modes

Once functional failures are identified the different *modes and causes* of each are identified.

> The failure mode establishes **how** an asset must fail to create a functional failure.

This is a serious step, and some selection is usually applied to discard some of the least likely modes of failure. Remember that, as an offshore oil platform is likely to introduce about 15 000 failure modes altogether, some means of reduction is clearly necessary. Failure modes can only be disregarded when they are highly unlikely (being struck by a comet), beyond redress (being hit by a nuclear missile) or because the failure mode is already more effectively addressed elsewhere in the analysis.

RCM emphasizes failure modes because preventive maintenance tasks are developed to address each one of them and reduce or prevent the resultant failure (see FMEAs below). So, if the stem of a fuel line ball valve shears during closure the failure mode could be shown as 'sheared valve stem' or more accurately 'stem shear due to low-temperature embrittlement' or perhaps 'stem shear due to lack of lubrication', depending on what is known. The maintenance task may be to grease the valve stem assembly, or the construction/design task to extend the steam tracing or pipeline cladding for warmer operation. The important thing to remember is that *maintenance addresses the mode of failure*.

Additionally, the assessment of failure modes sometimes has to change, partly because of external changes, such as geology or weather, partly because of changes in understanding and partly because of changes in applicable human knowledge.

An interesting example of the last category occurred in the mid- to late 1970s during the design and deployment of North Sea oil platforms standing on the seabed in 60–90 metres (200–300 feet) of water with module undersides well clear of the surface.

Weather and shipping evidence indicated that the incidence of a 30 m (100 ft) wave was, at most, five or six times in a platform's lifetime. When, after the onset of operations in the Northern Sector, three such waves had occurred in six months, changes in cellar deck (lowest level, nearest the splash zone) operations and ongoing platform designs hastily followed. An apparently unlikely and perhaps disregarded failure mode for smashed or drowned equipment on the cellar deck required different operations because of changed information.

Effects of failure

In technical respects the *effects* of a *failure mode* are often easier to identify than the failure mode itself. For example, the loss of lubricant in a generator gearbox (failure mode) will often lead to seized bearings, stripped drive trains, high temperatures and sheared shafts (all failure effects). These two steps in RCM analysis are bracketed firmly together under the single title of *failure mode effect analyses* (the documents are called FMEAs), and help provide the foundation for determining the appropriate PM task[5,6,7].

Such failure effects have several important properties:

(1) 'Effects' define what happens when no action is taken (assumed for the purpose of analysis).
(2) All 'effects' occur should the failure occur.
(3) 'Effects' may be separated in time but all are dependent on the failure.

One of the first benefits of knowing the failure effects is to help remove any doubt that the failure has occurred, bearing in mind that a recognized failure is the most

---

5. Fault mode and effects analysis is defined in BS 4778, Section 3.2, 1991 as: 'A qualitative method of reliability analysis which involves the study of the *fault modes* which can exist in every sub item of the item and the determination of the effects of each fault mode on other sub items of the item and on the required function of the item'. Notice the use of the word 'fault' instead of 'failure'. A further note in the standard refers to the latter as 'deprecated'.

6. Failure mode and effects analysis is also developed in BS 5760, Part 5, 1991.

7. Refer to Annex A: Maintenance Planning Analysis, BS 6548, Part 4, 1993.

potent motive for a maintenance job and opposes the natural human reluctance to carry out unnecessary work. Any certainty will help to eliminate argument, confirm priorities and complete the schedule.

Failure effects often come to the rescue, ensuring that the supervisor's authority and the technician's efforts are not wasted. One of the most obvious failure effects is sound. A gradual loss of steering at over a hundred miles an hour in a high-speed road car was a heart-stopping mystery until transmission fluid was heard boiling in the steering box, after previous strip down examinations had revealed no cause. Most technicians are familiar with the sickening low-frequency rattle of broken parts or the rising shriek of a runaway motor that has shed its load or is driving a broken shaft.

There are many other effects, such as surface discoloration, local temperature increases, escaping gas or steam, transmitted vibration, lubricant contamination, and performance degradation, which provide warnings and confirmation.

Response

Although not a formal part of the RCM analysis, the response of the operating organization is treated as vital to the effects and consequences of the failure and of course to the subsequent maintenance or repair work required. Failure is often considered as the last step in a sequence that the maintenance systems are intended to prevent and the analytical work intended to improve. It is, however, also sensible to consider 'failure' as the first event of a series of effects and consequences, each logically following one after another unless they are interrupted. It is the *response* of the organization which provides this interruption, preventing failure effects from becoming increasingly serious. Should the failure introduce wider consequences beyond the operational site then external public or perhaps emergency services will respond for similar reasons.

Consequences of failure

The 'effects' of failure, such as a sheared gear tooth or a fractured drive shaft, though often serious, are directly linked to the failed asset or the process system of which it is a part. 'Failure consequences', however, are far wider in their impact, can be mild or devastating and are far more difficult to predict. The majority of consequences concern the operation or site where the failure occurs, so, for example, the failure of a delivery pump in the potable water system will gradually reduce the drinking water available, and, as consequences, cause demanning of the site and suspension of other work until the system has been corrected.

Unfortunately, we cannot assume that all consequences are so restricted. A cloud of escaping toxic gas may, as a consequence, cause the suspension of unrelated work nearby, but if the same gas crosses the site boundary, entering a crowded high street, the consequences:

(1) Become even more serious.
(2) Are far more varied.
(3) Are much more difficult to predict.
(4) Start the 'consequence' clock timetable.

In a recent example, a valve in a cross-country oil pipeline started to leak, allowing light oil to escape onto roughly horizontal ground. Because the site of the leak was remote, usually unmanned and the leak was small, it remained undetected, continuing to leak for several months. Across a steadily forming pool of oil ran two parallel railway lines, which allowed occasional trains to pass each other as they travelled in opposite directions. Action of the wind and sun generated a light vapour across the surface of the oil and at the deepest point the railway lines lay just beneath the surface. On one fateful day, two trains travelling towards each other with their wheels splashing in the oil struck each other as they passed and the contact of steel on steel caused a spark. The spark ignited the oil, the whole pool caught fire, both trains were engulfed and all passengers and crew perished as a *consequence of a valve failure*. The leak was only discovered after the two trains failed to arrive at their respective destinations.

Failure consequences are affected by circumstances, by organizational response, by type of failure and other things besides.

(1) *Consequences are affected by operational circumstances*
A major airport working at night during the period of non-operational quiet was resurfacing the runway and its approaches before normal airport operations during the day. The resurfacing machine, weighing several tons, failed and refused to restart. Failure consequences were initially limited to the work loss (the failed machine was largely undamaged and the prospective repair was minor), but the clock was ticking. Unfortunately, when the time came for normal airport operations to restart the failed machine had not moved, and at this point the consequences became more serious. Flights were diverted or cancelled, visits were postponed, appointments missed, and many airport staff left idle. Between nighttime and daytime the operational circumstances changed and new more serious consequences applied.

In many situations, consequences will be mostly operational, as other machines in the same system and other systems on the same site are affected. Processes of continuous or near-continuous production are particularly vulnerable: machines and systems are dependent on each other, capital investment is high, and any loss of output is extremely expensive.

(2) *Consequences depend on failure*
It may seem obvious, but with no failure there are no consequences of that failure. When, as in many cases, the consequences are horrendous, the use of failure detection and prevention systems is the more painless and economic option available.

Two of the purposes of RCM analyses are:
(a) To provide proof that the systems are beneficial and economically worthwhile.
(b) To ensure that the systems define and follow the optimum (most effective) way of maintenance organization and action.

(3) *Consequences vary with different types of failure*
     If the electrical power supply to the fire monitor, for example, is interrupted
     by dirty connections, consequences vary, and depend on the type of failure:
     (a) When the failure is known[8], effects and restoration work are minor and
          there are no consequences.
     (b) When the failure is invisible the failure effects are still minor, but it is
          possible that the failure will remain undetected until the pump fails to
          deliver firewater during an emergency. The consequences in such
          circumstances are severe[8].

(4) *Major consequences can arise with minor failure effects*
     As the preceding example underlines, very serious consequences can result
     from technically trivial failures, especially from invisible failures or failure
     of non-failsafe devices.

(5) *Consequences always exist*
     Once a consequence of a failure effect has been identified it remains as one
     of a range of possible outcomes unless a structural change takes place, even
     when consequence prevention methods are in place. For example, one method
     of ensuring that a pressure vessel does not explode under high pressure (i.e.
     to avoid a range of serious consequences) is to fit pressure relief valves which
     vent vessel fluid when internal pressure becomes too high.
         Such a technique is ideal provided that the pressure relief valve is set cor-
     rectly and functionally operational. *If the device or protection mechanism fails[9]
     then the consequence returns.*

(6) *Consequences can be delayed*
     When considering timing, the only certainty is that consequences postdate
     the failure. They may be dormant or commence at the time of failure, with
     extended durations.
         Survivors of many life-threatening situations report psychological problems
     extending over many years or triggered much later by apparently innocuous
     events.

(7) *Consequences are not confined to the location of the failure*
     Explosive pressure surges may be transmitted along a pipeline; electrical
     disturbances through distribution systems, power transmission lines and metal
     structures; and corrosive vapours, toxic gases and radiation particles borne
     on the wind can cause damage in a different country or a different continent.
     Failures on moving vehicles, trains, ships and aircraft can result in accidents
     and consequences in distant places. There are many examples.

---

8. If the failure is discovered by maintenance inspection before an emergency occurs, the invisible failure
becomes a known failure.
9. When the protection device fails, operation and maintenance teams still subconsciously assume that
the vessel is protected. Coupled with invisible failures, a double danger applies and inaction is sadly common.

Categories of consequences

Although consequences are highly varied they can be conveniently considered under the headings of 'location' or 'characteristic'.

Consequences by location:

(1) *Single asset*
    For example, an asset may be so severely damaged following a failure that it has to be replaced.

(2) *System-wide*
    Effects on an asset may have consequences on other assets of the same system. For example, the contamination of a filtration unit in the potable water system requires that all assets in that system are shut down.

(3) *Plant-wide*
    The discovery of harmful bacteria in the feedstock for a bottling plant requires that the plant is shut down for cleaning.

(4) *Overall site*
    Many sites are host to more than one plant which share the same location and some common services. There are many situations which can have consequences for the whole site.

Consequences by 'characteristic':

(1) *Operational consequences*
    Within the physical outlines described above, consequences frequently influence operational activity.

(2) *Safety consequences*
    Previous accidents have shown that although safety consequences are centred on operational personnel they can readily migrate to adjacent plants, sites and public places.

(3) *Environmental consequences*
    Pollution and contamination of the air we breathe, rivers, water systems, sea, rain reservoirs etc. are immediate examples of consequences influencing wide reaches of the Earth, frequently remote from the problem source.

3.5 Total productive maintenance (TPM)

In the Western world since the late eighteenth century many production systems and procedures have been based on the operational principle of the

division and specialization of labour

whereby total output was multiplied when the work required to produce an item was broken down into small definite steps, which yielded increased output as different skills were developed. In a similar manner, organizational growth compartmentalized such skills, introducing separate steps in the production process and enhancing specializations, while reinforcing each discipline in the sequence.

This approach has been spectacularly successful, converting the methods used by craftsmen to that of production workers and increasing output by several hundred times. During the last twenty years or so, things have begun to change. It is recognized that tasks or processes which pass through several departments encounter barriers to communication, coordination and authority, which have to be reconciled each time a transfer takes place. When the products concerned are comparatively simple, few such transfers are required and the gains arising from departmental skills more than outweigh the difficulties of repeated transfer. However, products and processes have become increasingly complex, increasing the number of separate specialist influences, while workers themselves are motivated by a widening range of interests, needing a wealthier variety than that provided by the earlier models.

Effective organizations vary according to the technology they use, and as production processes become more intricate, fresh departments are needed, increasing structural width. When coupled with the removal of some tiers of management in the pursuit of reduced costs, the result is a wider and shallower organization pyramid. Although such arrangements reduce the corporate overhead and improve top to bottom communication by shortening the command chain, they actually make interdepartmental transfers more numerous. The two key tools used to manage these difficulties and release the latent talent of company personnel are:

(1) teamwork
(2) empowerment

and both are endemic in the TPM armoury.

> The TPM method encourages quality, safety, operations, maintenance and other disciplines to work together in a team environment to improve 'overall equipment effectiveness' (OEE) and to use the increased authority of the team members to remove barriers, simplify processes and achieve substantial productivity improvement.

The initials TPM stand for total productive maintenance, and like many good titles it does not mean what it says. Although maintenance is a key component, there is no magical maintenance process, and some writers have commented that strictly speaking it is not a maintenance technique at all. All of the maintenance steps triggered by the use of TPM, like 'critical assessment', 'maintenance prevention', 'condition monitoring' and many others also feature in other maintenance systems. The difference lies in the way in which such steps are used, and at this point two qualities of TPM stand out:

(1) *TPM is a production-centred system*
     The targets are higher quality fault-free processing, greater availability and increased rates of production, all aiming at improved output by raising the overall equipment effectiveness.

(2) *TPM negates maintenance isolation*
     By expressly involving the use of team-working, the method ensures that all

departments affecting operational performance have direct influence on decisions affecting maintenance work. Acceptance and understanding of actions required is improved and many arguments averted.

A question of name

The question mark over the TPM title arises because of the different objectives used when compared with other maintenance philosophies, most of the latter having a bias towards the status quo, aiming to correct deterioration and restore failing equipment to full performance. RCM, for example, looks at machine functions to see if they are being met, and condition monitoring watches for deterioration, extrapolating findings into the future; both of these require judgements within a framework of existing duty.

TPM does not face these limits. By looking at overall equipment effectiveness, it refers to *all sources of performance constraint*, whether they be poor production rates, high levels of equipment failure, process downtime, high rates of re-work, low product quality, poor safety performance, or high set-up and restart times, some of which would escape interrogation under the conventional approach.

By including concepts of organizational quality, TPM fosters *team methods of creating continuous improvement*, using the following action framework.

(1) Introduce the TPM message to mixed discipline teams and underline teamwork.
(2) Emphasize the measure of *overall equipment effectiveness* and record achieved results.
(3) Employ the techniques of *continuous improvement*, looking forward to new opportunities and backwards to demonstrate progress achieved.
(4) Establish the operators' process of first line maintenance.
(5) Include existing PPM upgrades in the continuous improvement process.
(6) Embody key techniques of condition monitoring, predictive maintenance and RCM to upgrade the quality of maintenance prevention.

What may not be immediately clear is that the TPM/quality improvements approach uses the technique itself to generate needs and opportunities. TPM does not make formal judgements against previously defined standards; it refers to its own measures of OEE. In other words, it does not ask the question, 'what should it be?'; it looks further and asks, 'how can we improve?'.

Data recording

TPM-initiated changes refer to the measures of overall equipment effectiveness, and when improvements are made the OEE shows a corresponding rise. Although results will be clearly visible, subjective assessment will take longer to see an improvement, and a simple but formal method of measurement is needed. The TPM team themselves can best define how OEE should be measured, and fastidious recording of resulting values is essential. We should remember that not all changes made will yield improvement, and any such steps need to be quickly detected and reversed. Also, the opening steps of continuous improvement may centre on small changes taking time to produce positive results. In addition, several changes may be introduced together, and the correct identification of improvement sources will enable them to be repeated or expanded.

TPM requires data collection and interpretation with measurement of OEE following recorded changes.

The quality climate

Human beings respond to a climate of effective organization: we are all positively influenced by evidence of operational order and confidence. Actions are more assured, operations more efficient and the incidence of error diminished. Some commentators have referred to this approach, applied to plant and equipment, as 'asset care', emphasizing properties of

Operational cleanliness
Neatness in the workplace
Formal activity arrangement
Respect and self-discipline
An atmosphere of order

Such an environment can be achieved by the progressive effects of TPM in a work programme applied in these four main regions of activity:

(1) *Autonomous maintenance*
    This will include
    (a) Regular equipment cleaning.
    (b) Machine workspace tidy, whether in use or on standby.
    (c) Development and use of formal lubrication regime.
    (d) Operation of routine maintenance inspection.
    (c) Containment action for emergent machine problems.
    (f) Accurate and complete fault reporting.
    (g) Full autonomous maintenance.

(2) *Equipment improvement*
    (a) Discovery and treatment of abnormalities.
    (b) Defining and setting of optimal equipment conditions.
    (c) Maintaining optimum conditions.
    (d) Apply continuous improvement to increase availability.
    (e) Steadily increase performance targets.

(3) *Quality maintenance*
    (a) Apply continuous improvement to reduce and then eliminate equipment defects and failures.
    (b) Monitor and improve the quality of maintenance work done.
    (c) Eliminate progressive machine deterioration.
    (d) Ensure that maintenance work is developed through the TPM team.

(4) *Prevention and reduction of future maintenance*
    (a) By use of planned maintenance.
    (b) Enhance equipment maintainability.
    (c) Attend to work management and work procedures.
    (d) Aim for maintenance-free future equipment by feedback to the design process.

Values and effects

By the use of teamwork and empowerment different disciplines find that by working and winning together they discover talents and abilities that they did not know they had. The working team becomes an increasingly effective force, achieving targets and understanding in a highly effective atmosphere. The nature of the team changes from one of collected disciplines to that of a cohesive unit. Group members become identified as 'our people', effectively 'owning', the equipment under its umbrella and constantly aiming to increase availability and reliability. With increasing achievement, rivalry and suspicion disappear and the group cooperates to win.

Steered by the use of TPM methods, teams will identify six causes of performance loss (the six big losses):

(1) Equipment breakdowns leading to unplanned downtime.
(2) Losses through changeovers, setting up and adjusting.
(3) Output loss through idling and minor stoppages.
(4) Slow running of equipment.
(5) Start-up losses.
(6) Quality rejects, scrap and rework.

These losses lead to a poor working environment, an increase in late deliveries, more interrupted jobs, poor general performance, deteriorating company image and a decrease in the overall equipment effectiveness. One relief from this grim scenario results from a rise in surplus capacity attendant on falling demand, often the unwelcome result of poor performance such as this.

The positive work

To prevent the emergence of these problems, TPM teams include these steps in their decisions and preparations:

(1) Restore equipment before attempting to improve the OEE.
(2) Constantly aim for IDEAL conditions.
(3) Eliminate the six losses listed above.
(4) Always address cleanliness.
(5) Repeatedly raise the following five questions:
    Why doesn't it work as intended?
    Why can't we improve reliability?
    Why don't we know the consequences of failure?
    Why can't we set optimal conditions for the process?
    Why can't we maintain optimal conditions?

The three cycles of TPM operation

By these queries and techniques, teams develop and introduce three cycles of TPM operation:

(1) Condition.
(2) Measurement.
(3) Improvement.

Condition appraisal

Assessment of equipment criticality.
Implementation of refurbishment programmes.
Preparation of future asset care and maintenance schedules.

Measurement

Assessment and definition of what and how to measure.
Use of 'overall equipment effectiveness'.
Reference to equipment maintenance and duty history.

Improvement

Constantly attend to removal of six losses.
Follow best practice.
Pursue the problem-solving routines.

The continuing results

The TPM method makes a major improvement in the persistent maintenance problems of

- Poor relations with operational departments.
- Repeated reluctance to release equipment for maintenance.
- The isolation of most maintenance work.
- Improper understanding of maintenance needs.
- The strong links between maintenance, safety, quality and operational performance.

As such problems are diminished and the positive results described above materialize, performances following successful application can be truly astonishing. In the author's view, TPM is best considered as a powerful corporate companion to other maintenance systems and to refer to it as a maintenance system alone is to unnecessarily limit its potential for wider applications.

3.6 Condition-based maintenance

Of the different maintenance philosophies discussed in this handbook, condition-based maintenance is almost certainly the most dynamic. It has erupted on to the maintenance scene, growing from an interesting scientific application around 15 years ago to a proven and established technology today.

Condition based maintenance relies on the detection and monitoring of selected equipment parameters, the interpretation of readings, the reporting of deterioration and the vital warnings of impending failure.

Warnings and interpretation of equipment defects lead to the choice of actions and preparation, the whole package being described as condition-based maintenance. As the name suggests, it is concerned with the monitoring of equipment condition, and we should quickly distinguish between this and 'process monitoring', leading to process control, plus 'status monitoring' (see Section 5.7), used to describe the regular checks of equipment states[10].

Monitoring of some parameters, such as rotational velocity, vibration at key positions and temperatures of selected components, is also conducted using transducers installed by the equipment manufacturers for the purposes of *machinery protection*. Strictly speaking this is not condition monitoring: transducer output signals do not subscribe to recorded history but are compared with preset alarm or trip levels, which, if exceeded, cause the shutdown of the machine or trigger alarms in the control room. Use of such events is carefully considered during plant design, as the monitored machines are usually part of a process system whose sudden shutdown may cause problems elsewhere.

There is a main division of condition monitoring methods between installed and portable techniques, both of which have their adherents and different advantages for their use.

The installed condition monitoring system is characterized by high sampling frequencies (down to milliseconds), impressive computing power and high capital spending. A cost of a million pounds for a major single-site system, including installation, although at the expensive end of the price range, would not be regarded as unusual. Systems are based on installed computers, hard-wired transducers and modular electronics. Most recent designs employ an open system architecture which can run on personal computer platforms and use windows-type software. Multi-user, multitasking and multi-location features are employed, and operators' time is used efficiently by engaging computational power, remote access and configuration or treatment changes at the touch of a button, through software. Recently introduced analytical tools are impressive and they can be used to accomplish rapid system reconfigurations and sophisticated analyses from simple data (Fig. 3.2).

The alternative is the non-installed system, which employs a portable hand-held data collector. Readings are taken at preset points on a data collection tour of the monitored plant at selected times, usually a week or a month apart. The condition monitoring engineer carries a small range of transducers (refer to the description of accelerometers below and Figure 3.4) with him, the same device often being used each time and in several places as he passes around the sampling route.

Modern collectors are extremely powerful and carry extensive reading history and

10. Condition monitoring is neatly defined in BS 4778, Section 3.1, 1991 as: 'The continuous or periodic measurement and interpretation of data to indicate the degraded condition of an item and the need for maintenance'.

Figure 3.2
*'Kompass': an automatic sampling and interpretation system (reproduced by
courtesy of Brüel & Kjaer (UK) Ltd, Stanmore).*

interpretive software available for interrogation at the time and place of the sample.
Once the data collection has been completed, readings are downloaded to the
system's main computer, which is used to record readings, calculate trends, prepare
condition summaries and predict future performance or impending failures (Figure
3.3).

Although arguments about the merits and suitability of these two systems are
well known, at the time of choice there is usually little disagreement as they are both
more useful in different situations. Major installed systems are often part of an original
plant design and part of its commissioning and installation programme, they centre
on 'on line' monitoring and frequent analysis of constantly flowing data, their application to critical production machinery is particularly applicable if the plant is
highly productive with high downtime costs.

Portable systems have the advantage of being usable almost immediately, particularly
with equipment which is already in regular use. This is significant, as major systems
can take months to install, especially when there is so much frantic activity on a
new site. In addition, capital costs are negligible, and when downtime or repair savings are achieved, early cash flow improvement is possible. The use of fewer transducers limits reading variation, and data collection by experienced personnel is an asset.
The portable method is particularly useful for small and older installations, and ironically has a role in the gathering of 'early use' data on major sites before the installed
system becomes available.

The most frequently raised drawback is the sampling frequency; unlike the installed
systems, the detection of all faults is not possible before the event, although 'hit

Figure 3.3
A Brüel & Kjaer data collector used in portable applications as
described in the text and claimed to make and store measurements
in seconds. It is an impressive example of a large competitive
range of portable data collectors. Notice the spectrum display – an
attractive feature of this particular model (reproduced by courtesy
of Brüel & Kjaer (UK) Ltd, Stanmore).

rates' of 85–90% are likely with well-run systems. Unfortunately, the one way to increase the detection rate is to raise the sampling frequency, which, for portable systems, is directly coupled to the cost. Raising the frequency from monthly to fortnightly for example increases the basis of site visit costs from once to twice a month.

Main techniques

Condition monitoring engages many different measurement and detection methods, chosen according to the variable being measured and the relevant working environment. The following are among the most widely used.

Vibration detection

The measurement and analysis of vibration is a technique frequently applied to rotating or reciprocating equipment, such as pumps, compressors, motors and turbines. The detection of acceleration, velocity or displacement of moving parts, usually transmitted through housings or mounting frames and its subsequent analysis, can locate the source of out-of-balance forces, assess their severity, indicate damage done and forecast the progression to failure. This may seem a little odd, as the effects on any mechanical component could be more readily calculated if we could measure the forces acting on it directly. Unfortunately, this is both difficult and expensive (or even impossible) and we are left with the indirect monitoring of the effects, one of the most prevalent being vibration. Bearing in mind that machines are designed to run smoothly and in balance, the out-of-balance forces which cause vibration most frequently result from faulty machinery.

The two types of transducer used to monitor vibration are the *accelerometer* and the *velocimeter*, which produce electrical output signals according to the value of monitored variable experienced (see Figure 3.4).

The unit of acceleration is metres/second/second (often expressed as metres/second^2 (or in imperial units feet/second^2), and some accelerometers give their output measurements directly in metres or millimetres/sec^2. Others quote mV/g (millivolts per g) or pC/g (picocoulombs per g)[11]; it depends on the type of accelerometer used. Some units employ internal amplifying electronics to give an output in millivolts, while others produce a change in electrical charge in the monitoring circuit and are usually connected to a charge amplifier to increase the signal. Resultant acceleration outputs may be converted to velocity and then to displacement by successive integration, and some modern accelerometers are capable of producing the different types of signal.

Most of these devices employ the physical properties of a piezoelectric crystal, whereby the component's electrical charge changes when it is compressed by an external force. Transducers using this principle include a tiny seismic mass which applies varying forces to the crystal according to the acceleration acting on it. Values of acceleration and velocity will vary according to the excitation frequency, and a flat or unchanged output value over the operating frequency range is a feature of good accelerometer sensitivity (see Figure 3.5). So if there is a slight out-of-balance at one point on a main pump shaft spinning at 3000 rpm we would expect to see a

---

11. The use of *g* as a measure of acceleration is widespread. It is the acceleration experienced by a body in free-fall towards the centre of the Earth under the effect of gravity (9.81 metres/sec^2). However *g* varies across the surface of the Earth, and SI units (metres/sec^2) are slowly replacing it as a measure of acceleration. Also, picocoulombs are a measure of electrical charge equal to 10^{-12} Coulombs. These units are shown here in Figure 3.5 on the *y*-axis of the sensitivity curve.

Figure 3.4
Typical piezoelectric accelerometers used to provide an
electrical signal equivalent to the acceleration experienced
by the device. Transducers of this general group can be
manufactured to provide velocity or acceleration outputs
as required by the customer, while some of the more
elaborate units (not shown) can do both simultaneously.
The items are shown full size. The upper black component
is a triaxial version simultaneously monitoring three
mutually perpendicular axes. (Reproduced by courtesy of
Environmental Equipments Ltd, Newbury.)

sharp rise in acceleration at this point on the frequency response curve (50 Hz =
3000 rpm), with corresponding increases at rotational multiples of this number.

This type of diagram (or spectrum) of velocity or acceleration (whose value is
shown as amplitude variations in the *y*-axis) plotted against frequency (shown as
divisions of the *x*-axis) and their study is a valuable source of information in detecting faults and emergent change (Figure 3.6). It is the interpretation of such spectra,

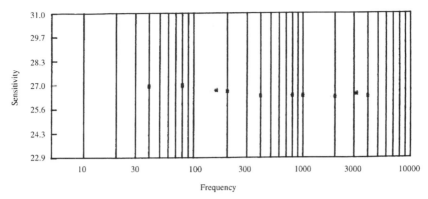

Figure 3.5
The curve shows the sensitivity of an accelerometer, the vertical axis plotted in
picocoulombs/g (pC/g) and the horizontal axis showing the frequency range in Hz
plotted logarithmically. See also footnote 11. (Reproduced by courtesy of
Environmental Equipments Ltd, Newbury.)

coupled with knowledge of the source, which is used to detect errant vibrations and
their causes. When spectra are taken at the same point sequentially, changes of
measured value at the same frequency can indicate a trend of change which is
extrapolated to make predictions about future failure.

Lubricant analysis

It has often been observed that the lubricating oil is the only component of a machine
which is in contact with all the moving parts. The analysis of the lubricant contrasts
with transducer-gathered information in that, at the outset, it refers to the machine
in general. The process used is very straightforward: oil samples are taken at prescribed
intervals from *exactly the same position* in the lube oil circuit and analysed, usually
in a laboratory. The oil is checked for both its own properties, such as viscosity and
density, and contamination by foreign material, such as carbon or metal particles.
Here is the first dilemma: degradation of the oil can with careful analysis identify
the source of contamination and point to that part of the machine which is faulty[12],
but extra use of the oil will lead to more certain analysis, whereas the machine
deterioration in the short term could be quickly reduced by replacing the oil.

Lubricant and hydraulic fluid analysis is also practised by the RAF, who since
1968 have operated a Spectrometric Oil Analysis Programme (SOAP) which can now
deal with 16 types of aircraft, analysing around 16 000 samples annually. The system
has been localized on a network of PCs employing the latest software. Using the
newly named COAP, the team monitors wear metal trends against warning levels

---

12. Some operations are working to compare identified contaminants with a database which cor-
relates details of the machine's materials with the exact components that include them. The DRA (Development Research Agency) is working on methods to produce digital data from wear particle shape image
analysis (particle morphology).

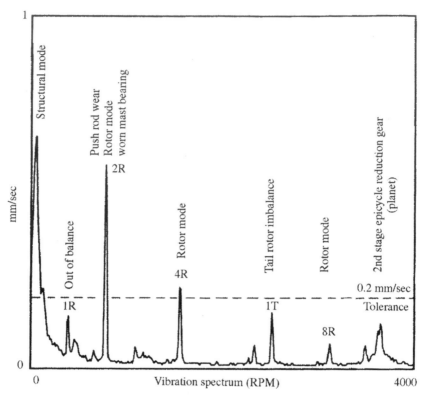

Figure 3.6
Velocity spectrum of a typical helicopter vibration. The vertical axis shows velocity
amplitude in mm/sec, while the horizontal axis shows frequency plotted in rpm.
(Reproduced by courtesy of Environmental Equipments Ltd, Newbury.)

set by the design authority or based on historical data. The RAF also considers the
pattern of wear metal production, the rate at which wear is increasing, the dilution
of levels arising from oil additions, the duty of the component and the relevance of
any equipment changes.

Most condition monitoring processes keep a detailed watch on the trend of the
readings taken from successive samples. One of the problems of off-site fluid analysis
experienced by industrial users is the maintenance of a strict oiling and sampling
regime. The unscheduled topping up of an engine oil, particularly from a different
source, can play havoc with the resulting trend.

There is also current interest in particle detectors for online detection of debris
in fluid circuits. Of the three main methods of detection (magnetic, electrostatic and
inductive), testing currently focuses on the latter. Ferrous particles down to 70 μm
and non-ferrous particles (e.g. phosphor bronze) down to 150 μm can be detected.
These readings are applicable to a pipe 12 mm in diameter, and it is expected that
corresponding readings for smaller pipes, benefitting from higher flux densities, will

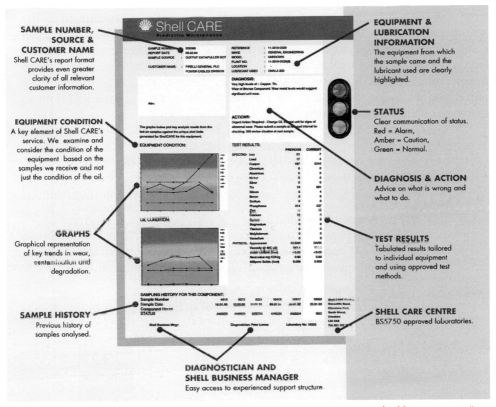

Lubricant analysis using a proprietary 'traffic light' system. Figure supplied by courtesy of Shell UK Ltd. (Downstream Oil)

produce higher responses. The impact of these particle sizes varies according to where the lubricant is at work. In general, for engines it is necessary to detect particles 100 μm or smaller in size, while for gearboxes the figure 1000 μm is less stringent.

Thermal imaging

The operation of machines, equipment and processing produces heat, which in some cases can result in hot spots and infrared radiation, detectable by thermal imagers. The causes of such radiation can be corrosion, electrical energy loss, friction from moving mechanical parts, escaping hot vapours, or fluids and process effects, like electrical switching or hot exhaust fumes.

Portable thermal imaging cameras are used for thermographic inspection to locate the sources of such radiation and can detect heat in local or large areas, often without any disturbance of the subject process. The imagers used convert any heat received by the camera, both sources of heat and reflection of heat, into pictures showing differential heat emissions. A semiconductor is used in the detector to convert thermal energy into an electrical signal, which is immediately processed to produce a television picture. This picture can be examined at the time using a picture monitor or recorded as a video image for later study, computer enhancement and attachment to the maintenance history. The thermal image, initially shown in black and white, can distinguish between temperature differences as low as 2 °C. Pictures show white for high temperatures, black for cold and grey for the range in between. More recent systems are presenting thermal images in colour, increasing the value and impact of the picture and easing the interpretation. Figure 3.7 shows the thermal image produced from an electrical connection malfunction.

Like other condition monitoring disciplines, the interpretation is important. An experienced engineer can gain much from careful image interpretation, especially when computer enhancement has been applied to the pictures. (Refer to the Condition Monitor No. 87, p. 4, McKinnon & Clarke.)

Thermal imaging is a widely used technique which has several positive features:

(1) *The process is mainly non-invasive*
    In most situations equipment and processes must be operating for the thermal emissions to be created. Apart from the opening of a cage or cabinet protective doors to ensure a clear vision of the subject, little preparation is needed.

(2) *The method is applied in general and in particular*
    Thermal imagers can be used for surveys of main sections of plant, such as

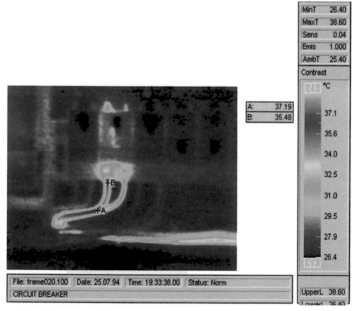

Figure 3.7
The thermal image produced from an electrical connection malfunction, as recorded and presented on the Matrix Resource Maintenance Computer System (reproduced by courtesy of Matrix Resource Management Ltd).

electrical distribution systems or steam tracing on pipework, and also be used for local examination, such as a single machine or an electrical switch.

(3) *Thermal imaging requires no installation*
The portability of the method is a key advantage. Thermal imaging equipment can be used on all systems and, if needed, on different plants. Capital costs are more readily justified, but proper record maintenance is essential.

(4) *Thermal imaging can highlight dangers or reduce costs*
The detection of dangerous hot spots, such as electrical arcing and sources of ignition, is the most dramatic use of this technique, but the routine identification of energy loss is also a valuable feature that leads to regular operational savings.

Acoustic emission

Acoustic emission (AE) is a purely passive technique involving the detection of high-frequency stress waves within the range 50 kHz to 1 MHz arising from physical processes, such as friction, impacts and metal removal, which occur when machinery is in distress. In the absence of stress-related processes, high-frequency background signals are low, giving acoustic emission an excellent signal to noise (S/N) ratio and allowing a clear distinction between good and bad machines.

The better known vibration techniques avoid low signal to noise problems by conducting analyses in the frequency domain; in other words, the signal amplitude varies with different frequencies and a graph of this effect would show frequency along the x-axis. This technique works well and is well established, but it is more complex and requires more sophisticated treatment.

AE, however, achieves high signal to noise ratios directly in the time domain, giving a simpler approach which is especially useful for portable monitoring. As a result, it is not usually necessary to use complex signal processing in order to extract the AE signal from inconsequential background noise, and the approach allows the following:

(1) Instantly categorizes machines as 'all right' or 'suspect'.
(2) Ranks suspect machines in terms of problem severity.
(3) Monitors the rate of degradation of 'suspect' machines.
(4) Sets levels at which action needs to be taken.
(5) When coupled with datalogging and appropriate analysis, can identify exceptions, create trends and support failure predictions.

AE will be particularly attractive to those users who require rapid answers and are disinclined to make major capital investments in online systems, yet who employ critical production equipment in their processes. When used for portable monitoring, AE is different from other portable methods as a machine's condition can be assessed on the spot, and further analysis may prove unnecessary. For judgements of this kind, machine-specific historic data is not needed; it is possible to acquire the immediate condition of the machine, and details like shaft diameter, speed or bearing type are not required.

Although machine 'distress' is closely correlated to its 'condition', a further measurement of dB is necessary at each reading to create trends for suspect machines and make forecasts of future problems. For this reason, multipoint dataloggers have been introduced (one is shown in Figure 3.8) which record regular readings, taking advantage of the features described above to allow trending, failure forecasting and the successful monitoring of usually awkward slow-speed machinery.

Present use

The current constraints of the economic climate have underlined the natural aversion to overmaintenance which can arise from calendar-based methods, and these have steadily been modified to include predictive maintenance techniques. The case for greater emphasis on maintenance management is reinforced by constant fresh evidence of organizational changes, ferocious commercial competition and increased demands for quality products and improved performance. The role of condition monitoring is *to watch over and hence protect equipment*, and the following few highlights illustrate why this particular method has proved to be so attractive. However, one quality stands out:

Figure 3.8
A 1000 point route mode datalogger, which, from ease of use, clarity of recorded signals and PC-based analysis software, permits trend plots, exception reporting and the archiving of acquired signals to be included. (By courtesy of Holroyd Instruments Ltd, Matlock, Derbyshire.)

Condition-based maintenance offers quick and predictable improvement in user business finances.

Now the following highlights will help explain why this statement has been made.

(1) CBM methods and their results can be introduced quickly.
(2) CBM can be used to complement existing methods.
(3) CBM can be introduced gradually and checked for effectiveness before other methods are discontinued.
(4) CBM is readily applicable to widely different industries.
(5) CBM is particularly relevant when applied to production-critical equipment.
(6) CBM finds the balance between over- and under-maintenance.
(7) Condition monitoring addresses the surveillance problem arising from reduction in site manning.
(8) CBM amplifies the fault detection power of experienced human beings and introduces powerful additional methods of detection and interpretation.
(9) Condition monitoring is essentially required for normally unmanned facilities.
(10) By using modern computer-based interpretation, operational situations may be interpreted automatically and experienced resources assigned where they are needed.

The latter point deserves some expansion. CBM systems can be machine-specific or system-specific. For a chosen bearing, acceleration may be continuously monitored on a known axis of vibration and constantly watched for change at selected frequencies to give early warning of impending failure. Alternatively, health checking may be applied to a whole machine community to indicate which need immediate attention, which are gradually deteriorating and which are running without problem.

Various methods may be similarly applied to:

(1) Conduct wide-ranging surveys of electrical distribution systems by thermal imaging.
(2) Check the overloading or displacement of structures by use of strain gauges.
(3) The detection of leaking fluids, valves passing and pipework erosion in fluid storage and transfer systems by acoustic monitoring.
(4) The detection of metal to metal abrasion, rotational bearing failure, oil degradation or contamination by using lubricant analysis.
(5) The detection of out-of-balance forces, rising acceleration and likely forthcoming failures in rotary equipment.
(6) The rapid grading of machinery into 'sound' or 'suspect' condition by use of acoustic emission.
(7) The detection of rotary misalignment and sources of inaccuracies, energy loss and poor machine performance by use of optical alignment checking.

These are just seven among thousands of the applications under the condition-based maintenance heading, which makes it difficult to select a single attribute more dominant than the rest. It could be labelled as a means of improving labour efficiency, which it can do, or as a technique for raising machine reliability, which it can also do, or perhaps be billed as a means of extending equipment life, which naturally results. These are some of the things that follow condition monitoring and condition-based maintenance, allowing a wide variety of choices to be made.

A basic question

Asked to explain why condition monitoring occupies a role of growing importance in many operating installations, most of us would cite technical merit, growing reliability and the need to make confident forecasts of the future. Most of the responses would be equipment-centred rather than refer to any business reasoning, and in this respect they would be wrong.

Why do users invest money and time in the purchase and support of condition monitoring systems and services?

If this question were asked of the user he would probably agree with the technical attributes and then add that there are 'other things' of greater importance. It is important to know what these 'other things' are.

The concept of total cost

Operational management has increasingly underlined the difference between total price and total cost. Earlier purchasing methods based on formal systems of lowest price have been steadily modified to reflect the working life of costs of:

(1) Purchasing.
(2) Storage and handling.
(3) Operations.
(4) Maintenance.
(5) Repair or replacement.

and apply them separately to major items of production equipment. An effective condition monitoring system can help reduce the figures for items (3), (4) and (5), and the resultant total cost. Even more importantly:

(1) By extending equipment life, condition monitoring can allow the amortizing of lower total costs over an increased production volume.

Although such an effect is not solely applicable to condition monitoring, it is extremely compelling. Lower product unit costs can change sale price, competitive position and likely profit, and argue most powerfully in favour of this approach.

Simpler work selection

The selection of priorities from among the host of work competing for resources and attention is simplified by using the condition monitoring system.

(2) CM underlines those tasks that will not wait.

There is both danger and benefit here. Condition monitoring systems are not applied to the whole equipment population. First, attention tends to focus on faulty members of the monitored group rather than less demanding commitments for ordinary equipment. However, simple faults can rapidly progress to emergency failures; repeatedly omitted routine tasks can spell disaster.

(3) The condition monitoring system at work with vital equipment allows human eyes more time to watch the apparently mundane.

Effective work management

In most work situations there is an uneasy balance between planned and random work. The latter, if unchecked, leads to constant fire-fighting and chaos. The maintenance team's ability to handle the unexpected is still needed, but the aim is to confine it to a small proportion which can be handled without disruption.

Immediate work demanded for a task hazily remembered inevitably leads to inefficiency and error, especially as many maintenance tasks occur infrequently and do not include the skills acquired through constant repetition. For major work, even when part of a planned programme, it can be several years since the job was last done, if indeed it has been done before. During this time, knowledgeable staff may have been promoted, transferred or even changed employers, leaving a team new to the job required.

In this situation:

(4) Condition monitoring helps to make the random more predictable, improving the quality of work done.
(5) Condition monitoring helps change the balance of planned to random work in the work programme.

Both of these items facilitate good work management through:

- More effective evaluation of alternatives.
- Closer affinity of the work plan and the work situation.
- Higher planned content of an expanded work programme.
- More effective preparation and collection of parts, materials, tools and skills required.
- Clearer definition of work objectives.
- More comprehensive job research and training.
- Reduction in conflict of competing needs.

The work itself

Via the agencies described above, condition monitoring assists the achievement of the following improvements to the work itself:

- Work more rapidly completed.
- More efficient use of material and labour, leading to lower prime costs.
- Lower use of other supporting resources, such as tools, equipment and platform skills.
- Less interference between different jobs requiring the same workspace or other resources.

(6) Condition monitoring will help prepare for efficient completion of the work itself.

Management and supervision

There is evidence to support the contention that human teams work better when planned actions are clear and they are competent at the tasks required. Condition monitoring, via the mechanisms described above, will help to bring this about, releasing talent to achieve results which are often a surprise.

Teams will also enjoy greater confidence in both themselves and their own supervisors, the latter imbued with greater effectiveness, created from a strengthening of initiative and knowledge. In addition, relations between supervision and management will benefit from the increase in certainty, diminishing the need for hard choices and the scope for disagreement.

(7) Effective condition monitoring through the effects of good management and supervision will contribute to:
   (a) clearer work objectives
   (b) improved management and supervisory initiatives
   (c) improved team, supervisory and (in turn) corporate confidence
   (d) improved staff morale and motivation
(8) A good condition monitoring system will instigate attention and action.

Inspection frequency

The regular inspection of large process or mechanical items often requires some disassembly before judgements about their internal condition can be made, and it makes economic sense to carry out some preventive work at that time. Most work schedules allocate jobs to specific times, such as 'day number' or 'week number', and the work required is broken down to suit the scheduled time.

This method is very effective for handling large numbers of modest tasks, but is less useful for managing high-cost infrequent jobs, such as opening up major process pressure vessels for wall thickness checks. Tasks like this can be decoupled from the fixed time slot, provided an effective condition monitoring method is available.

As most effective maintenance work is done before failure, conventional maintenance schedules, working to a blind calendar, will be more conservative than corresponding decisions based on condition monitoring.

(9) Condition monitoring can achieve savings by requiring major inspection less often.

Also, programme costs fall as the ratio of planned to unplanned work changes; expense is less dependent on the calendar, and a greater portion of cost is assigned to work shown to be necessary.

(10) Condition monitoring improves the effectiveness of money spent on inspection and maintenance.

Let us play with numbers for a moment and assume that 12-monthly inspections can be deferred by this method for three months and other major frequencies *pro rata*. Additionally, assume that this applies to only 25% of the work required. Then for a conventional budget of 100, the new total becomes

$$(25 \times \frac{12}{15}) + 75 = 95$$

a conservative reduction of 5%. An estimate like this justifies analysis using actual figures.

Maintenance and production figures

Any reduction in the maintenance budget will, of course, be welcome. but will have less impact on company finances than the corresponding changes in production. Reduction in inspection frequency and improvements in work efficiency combine to decrease equipment downtime, necessarily resulting in increased production.

(11) Condition monitoring reduces equipment downtime and increases production.

If the user revenue is 20 times greater than the maintenance budget, then the 5% reduction in downtime produces extra revenue equal to the entire maintenance budget.

(12) Condition monitoring can offset most or all of the maintenance budget when it helps create small increases in production.

Only condition monitoring allows such inspection savings without the increasing risk of failure.

Different techniques

Preventive maintenance based on time-based work schedules has been employed for many years, with condition monitoring often used in a supplementary role to confirm the effectiveness of the schedules and provide 'equipment failure insurance'. While the schedules remain, however, most good equipment does not approach failure, and condition monitoring warnings mainly refer to equipment whose scheduled inspection frequencies are too low.

(13) When used only to support a time-based schedule, condition monitoring does not meet its full potential.

Management information

Widely used formal management information systems suffer from problems of expense, accuracy and a near total preoccupation with activities in the past. Condition monitoring is one of the few methods of forecasting which can be represented in both current information and future work.

(14) Condition monitoring can contribute currency and prediction to information systems.

Task priority

As well as identifying those tasks that will not wait, condition monitoring will also help prioritize those defined as necessary but not urgent. Legitimate but differing perceptions of priority are prevalent in the workplace, and jobs which remain undone will rise up the list of priorities, becoming more urgent as time progresses.

Condition monitoring can enhance a formal assessment process, ensuring that important work is not shouldered aside by convenience, fashion or political argument.

(15) Condition monitoring systems contribute to work priority assessment, simplifying choices and underlining unpopular requirements.

User maintenance reputation

Many partners and investors in major plant operations sadly see all maintenance activity as a costly loss of production. Any equipment failure which reduces output is greeted with dismayed demands for emergency action and a return to full output.

Such classic short-term thinking leads to inefficient working, quick fixes, overtime hours, and pressure on employees and interpersonal relationships. Managers suffer loss of credibility and difficulties in communication, while supervisors impose doubtful decisions on a bewildered workforce.

Condition monitoring can aid the anticipation of failure and turn unpleasant news into forecasts requiring action. The initiative passes to the operator's management, who can treat equipment failures as confirmation of effective operational control.

(16) Use of condition monitoring restores the operator's initiative, allowing him to forecast situations and report actions taken instead of making excuses.

Improved operations

Unexpected equipment failures will occur in any work situation. The use of condition monitoring means fewer unforeseen failures and improved monitored equipment reliability.

(17) Condition monitoring leads to enhanced equipment reliability and improved operational performance.

Operational quality

One of the classic advantages of condition monitoring is the resulting improvement in product quality, but according to some operators improvement in quality is not a bonus: it is the reason. Quality targets are becoming increasingly stringent, and many operations are part of TQM (total quality management) organizations. Most service organizations, for instance, depend heavily on the smooth operation of their machinery for reliable performance (*Condition Monitor* magazine, No. 99 – Royal Mail).

(18) Condition-based maintenance can be a key ingredient of total quality management.

Temporary situations

Changes in monitored equipment can occur quickly and dramatically (e.g. rotational equipment embodying high revolutions and power transmission through gears, belts, cranks etc. can be exceedingly destructive when things go wrong). Equipment which can catch fire or rupture under excessive pressure or the effects of sour gas; and materials that shear or burst under excessive load, all lead to danger and damage, and all can benefit from CM warnings, however short.

There are periods when such occurrences are more likely, and it is these that justify a temporary installation. Rapid response to changes in the equipment environment is vital. CM systems are there to warn, and warn more stridently in times of higher risk.

The risk of failure

The implied assumption is that the risk of failure over time is constant, which for most types of equipment is clearly not true.

Recalling the 'bathtub curve' of failure against time as an example (refer to 'ageing failure' page 350 glossary), during the period immediately following installation and at the end of equipment life, the risk of failure is higher. Occasions of high work ing loads, introduction of standby units, local project work, severe environmental circumstances etc. embody other times of greater risk.

There are many situations like these, and the use of temporary condition monitor-ing can be used to 'cluster' the samples taken around the chosen times.

Sampled data

Some circumstances change so quickly that operators and maintenance personnel can only respond by sharply increasing the sampling frequency of CM systems already in use. In some plants CM systems apply to the main equipment only, and a switch to standby units means less, not more, sampling information, unless portable data collectors are quickly introduced.

Whatever is done here, data based on patient sampling over months can be completely disrupted by the influx of short-term readings, and trend statistics can be distorted long after the temporary situation has ended. The CM purpose for these short periods is to greatly increase the sampling frequency applied to endangered equipment. Many operators believe that the immediate advantages offered by port-able systems are steadily offset by the difficulties of arranging high sampling frequen-cies when non-site personnel are to be used. The answer is often additional portable data sampling in the 'short term' and a temporary hard-wired installation where periods of months or more apply.

It is often regarded as convenient if the existing routines are undisturbed and ad-ditional temporary designations and sampling applied to subject equipment, with both sets of readings interpreted during this danger period. In fact, most systems include extra unused circuits available for expansion or circumstances like these.

(19) Condition monitoring can be applied quickly to temporary situations.

The resulting conclusion

The world is rapidly changing, and many of the old economic certainties no longer apply. Customers are reducing the costs of maintenance, together with increasing plant availability, product quality and total output. What they demand are systems and equipment on which they can rely, that are accurate, have long service life, that require less attention, have lower running costs and can be adapted quickly to meet changing needs. Also, a dependency on remote monitoring is growing; this is partly because there is no real alternative and partly because many processes occur in hostile places. Cost also has an influence, particularly on transducers. It is of course necessary to be able to rely on remote performance, but long life and reliability are key characteristics of any device which requires a plant shutdown if it fails.

The attributes of condition monitoring described above show individual benefits which, when taken together, become collectively dramatic. Many businesses are now operating with no financial slack and can only generate fresh investment funds by reducing costs. Condition monitoring can help achieve this and the restoration of management initiative without penalizing existing levels of production. We should remember:

    (a)   Major production plants are dedicated to high-volume output of a small range of products.

    (b)   Reconfiguration of plant is expensive and time-consuming, with devastating losses in production.

(20) Use of condition monitoring can dramatically improve figures from existing production, encouraging users to take firmer control of their own businesses.

Disadvantages

Many users are guilty of employing condition monitoring in isolation – as a 'watching brief' – merely to confirm the current operating status of a machine, without using the data to forecast performance. It is believed that a main inhibitor to a more in-depth use of condition monitoring techniques is the high level of technical competence that is required. This is exacerbated by the lack of in-house resources and the high costs associated with utilizing external consultants.

Condition monitoring is also a technique that does not work well with simple comparisons, and although effective and specific data is always valuable, the major commercial benefits of its use – reduced downtime and lower maintenance costs – will not show unless the forecasting step is taken.

Effective use of condition monitoring data is hampered further by organizational delays in its translation from the measurement step to the initiation of timely, corrective actions. Such problems often arise when the condition monitoring function is carried out by technical specialists, whose recommendations are carried out by a separate team. This seemingly disjointed approach undermines the advantages of doing the monitoring in the first place.

Trend analysis provides users with a powerful and widely accepted technique that allows both desirable and undesirable changes to be detected, without the need to postulate absolute values. If repeated readings show unexpected variations of a monitored parameter, such as an increase in motor field winding temperatures or an increase in acceleration at a certain frequency, then either the installation environment has changed or the machine condition is deteriorating. However, over a large number of measurement points, this technique generates vast amounts of data which must be interpreted.

Use of computers and the reduction of the large number of readings into trends will remove this apparent disadvantage, and where sophisticated systems are in use software tools for automatic interpretation are available.

3.7 Campaign maintenance

When considering the way that it is done, campaign maintenance is a current title for a well-established technique; the real emphasis lies with the reasons for its adoption in some sectors, plus the accompanying changes in management philosophy. As its name suggests, campaign maintenance is aimed at well-defined equipment groups and short-term maintenance targets, with work conducted with an almost military approach, often by task forces brought in specifically for the purpose. The contents of work packages will vary according to the installation, the operational and statutory needs, and the resource availability, but all packages consist of:

well-defined maintenance tasks simultaneously completed in a short time.

So a typical group of tasks could be the replacement of all main header valves on an oil platform, or perhaps the on-site reconditioning of all pressure relief valves on a given production train. The maintenance team drafted in for the job, armed with all equipment needed for disassembly, testing, measurement, reconditioning and reinstallation, would be experienced in and selected for the particular task. Using portable workshops, special-purpose handling equipment and a small quantity of replacement items, a rolling pattern of disassembly, replacement and on-site reconditioning would be established to complete the work package in the shortest possible time.

Once the work package and the maintenance 'campaign' was complete, the task force would leave the plant until recalled for a new task or would be replaced by a different group with different tasks to do.

Although this approach is very reminiscent of shutdown maintenance, there are two main differences:

(1) Campaign maintenance is aimed at short-term necessity.
(2) Campaign maintenance replaces rather than supplements regular maintenance programmes.

So, there we have it: campaign maintenance is aimed at economic rather than technical targets, which is initially difficult to understand because, with the exception of breakdown maintenance, other philosophies are financially more efficient. The reasons lie in the situation where campaign maintenance is used, and in particular the approaching end of overall plant life, as a sort of twilight scenario. The extraction industry, producing products like oil, coal, stone, slate, gravel and natural gas, is already familiar with the concept of finally taking a plant out of service and decommissioning it, of shutting the plant down and throwing away the key. There are many stories of fields being played out, reserves which are impossible to recover, of coal or gold seams running out, geological faults and other reasons why a plant will become idle. However, the longevity of many plants, particularly those in manufacturing, tempts us to regard them as permanent when this is clearly not the case, and some of the reasons why a plant will shut down are discussed below.

(1) *Exhausted field*

In all extraction operations resources are finite, and once exhausted the local plant will shut down (unless a separate third party use is found). One of the detailed evaluations based on geological surveys and computer modelling, prior to major investment in extraction plant, is the careful calculation of extractable reserves. The subsequent design aims for the life of the plant to exceed the life of the field[13].

(2) *Market forces*

The emergence of a less expensive alternative, a reduction in market demand for the end product or its over-supply by other producers can all result in a plant being permanently shut down. The manufacture of town gas from coal, a product of major importance, was largely abandoned when cheaper and more plentiful supplies of natural gas became available from the North Sea. The production of shale (oil-bearing rock) is already constrained by high production prices when compared with more conventional supplies of oil and the existence of 'clean' energy from wind turbine, solar and hydroelectric facilities. Similarly, older and smaller oil refineries have shut down when faced with surplus refining capacity and competition from larger and more efficient processors.

(3) *High and rising costs*

When we consider field exhaustion there are two different quantities at work. The first is the geological or technical limit (that point at which further output becomes impossible), and the second is the economic limit (that point at which further production is simply not financially justifiable).

(4) *Political or statutory changes*

Sometimes the legal or social environment in which a plant has operated undergoes change by which continued operations are not permitted. Examples are the use of scarce resources or materials in wartime or a change in pollution laws referring to waste products or processes which become environmentally unacceptable.

---

13. Plants are often changed during their working lifetimes to take advantage of improved methods of extraction which were not available at the time of the initial design.

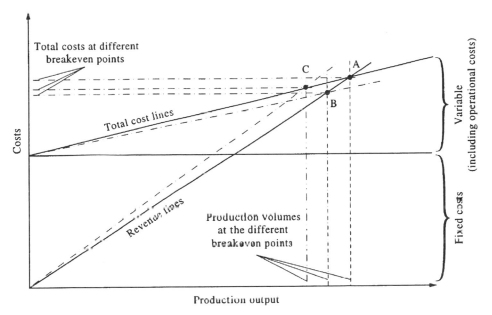

Figure 3.9
Breakeven point presentation.

It is not clear to the outside observer that even the 'non-extraction' or more 'permanent' plants are involved in change as new product lines replace old ones or new machines and techniques become available. So the impact of a plant closure when it does come makes it even more dramatic, especially in the locality of the closure. The UK North Sea oil sector is well advanced in preparation for the shutdown and decommissioning of some older platforms now approaching exhaustion after starting work in the mid-1970s. We will witness the introduction of brand new platforms like Chevron Alba and the Chevron/Conoco platforms for the Britannia gas field as old facilities are decommissioned and others undergo major platform refurbishment to extend their working lives.

Before turning to the reasons for closure we should consider the *economic balance* applicable to a particular plant. Financial personnel often resort to breakeven analysis to describe the influence of major economic forces on an organization[14]. A typical one is reproduced in Figure 3.9, showing main costs and the location of the breakeven point. Most managers will be familiar with this form of presentation, and it is not intended to develop an essentially financial idea in a maintenance handbook. There are, however, three points to emphasize:

14. Breakeven analysis uses a simplified graphical representation to show the relationship between plant total costs and total revenue. The breakeven point defines the amount of production needed when total costs and revenue are equal.

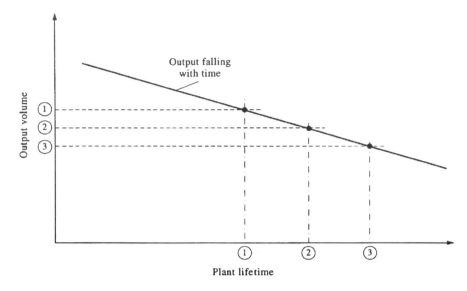

Figure 3.10
Small reductions in production volume required for breakeven can extend plant lifetime
even when output is falling.

(1) From the initial position at A, the breakeven point moves back when costs
     are reduced, i.e. less production output is needed (see point B).
(2) The breakeven also moves back when the revenue is raised, again less produc-
     tion output being needed (see point C).
(3) When revenue falls the breakeven moves forward, i.e. *more* production is needed
     to achieve breakeven, and a similar result follows a cost increase[15].

All of these pertain to the campaign maintenance approach. Remember that when
short time-scales only are considered, modest variations in either cost or revenue
graph lines can have considerable impact on breakeven timing (Figure 3.10).

We should also revisit the question of plant age, where in spite of the best atten-
tions of the maintenance teams over the years, older plant deteriorates and
maintenance costs will rise.

The case for campaign maintenance

Advantages

Of the four reasons (page 75) for plant shutdown described above, it is number three
which leads to the use of campaign maintenance, and the following describes the
reasons for its adoption.

---

15. Although savings are often possible, for large cost reductions, cost and revenue are linked, reductions in costs often introducing falling revenue and also reducing or eliminating improvements in the
breakeven position.

(1) *Surplus production resources*
    The levels of production output and hence the operational equipment needed to support it reach a peak in extraction operations. Although this is often offset by compensating changes, production overcapacity leads to some operational equipment being progressively underused as production falls. This means that older more fragile equipment may be deliberately de-rated, extending maintenance periods and equipment life, or duties may be switched between duplicated machines. The latter can lower the cost of running the out-of-use machine or they can be cannibalized into one reliable unit[16].

(2) *Lower downtime costs*
    As production output falls the cost of a downtime hour is less, ie, the absolute opportunity cost is lower. Maintenance actions which extend equipment functional life but cost much less in downtime pounds become much more attractive.

(3) *Lower operational costs*
    There are several contributory reasons for cost reduction, and this is the principal attraction for the operation of aged plant.
    (a) As production output falls, both operation and maintenance costs are reduced as team sizes decline.
    (b) Although the plant is getting old, many items of equipment will be relatively new and likely to be usable well past the final shutdown date.
    (c) As the maintenance programme is reduced, many items relegated to permanent standby or disuse will be removed from the maintenance programme.
    (d) The equipment replacement budget, plus replacement purchases for more efficient alternatives, are removed from capital budgets.
    (e) Many items of plant will have an achievable residual value once the final shutdown date has arrived.

(4) *Reduced risk*
    Although revenue figures are lower, the capital investment figure is commensurately lower and takes account of the shortened production period remaining and corresponding lower total revenue. For these reasons, both the period of exposure plus the capital amount at risk are lower and the overall risk is very much reduced. Operational and market forecasts pertaining to a closer horizon will be more accurate, with financial recovery and forecast profits more likely to be realized.

---

16. As the extraction of oil from a reservoir falls, the injection and processing of water increases. As a result, the total fluids processed by an oil platform may steadily rise even when oil output is falling.

(5) *Assigned resources only*

Comprehensive maintenance programmes require work conducted in parallel over a wide range of different machines, services and locations. Many interventions will be of minor work content, and their importance lies in their frequency and regularity. This returns to the maintenance man's concept of keeping the plant in sound, safe and reliable operational condition. Such a philosophy is only achievable with operational and maintenance resources in attendance or on call to constantly ensure sound operational status.

Campaign maintenance, although often invoking parallel work programmes aiming at short-term targets only, does not follow this concept, and by avoiding the consequences of medium-term neglect can ignore some actions usually regarded as essential. The result is the funding of task-dedicated resources and actions only, instead of the wider support normally required.

Disadvantages

Many maintenance managers react to the philosophy of campaign maintenance with undisguised dismay, and it has been dismissed by some as little more than fire-fighting, largely irrelevant to the balanced choice between serious philosophies.

Although such reactions seem harsh when we consider that the basis of all maintenance work is to preserve, then it is sometimes argued that:

campaign maintenance is closer to fire-fighting than routine maintenance

and the maintenance managers' reactions, which conjure up images of plunder and asset stripping, can at least be understood. This first and possibly greatest disadvantage of campaign maintenance lies in a problem of communication between maintenance professionals and customers for this approach. The latter can, with some justice, maintain that because campaign maintenance is applied in circumstances approaching final shutdown they provide a valuable service in declining circumstances, allowing subject plants to continue for longer than would otherwise be the case. Furthermore, this method requires professional maintenance people both to take part in and to face the harsh reality of final plant shutdown, where the use of campaign maintenance is the recognition that this is to occur.

(1) The philosophy emphasizes communication problems between users of this approach and maintenance professionals.

The next problem is how to gather committed personnel needed for the work from the ranks of professional maintenance personnel.

(2) Maintenance personnel assigned to these projects often regard the work as temporary or bounded by the limits of specific job contracts.

This naturally leads to questions of motivation, already blighted by the prospects of closure. The short-term culture applies at all levels of the user's hierarchy and engages contract or short-term survival targets rather than longer term maintenance purposes.

(3) The corrosion of quality, safety and working standards.

Both quality and safety performance depend heavily on the positive participation of human beings and we have already referred to difficulties arising when working teams do not see the direct benefits of their efforts but take comfort from continued performance within the user's wider organization. This last comfort is denied to the practitioners of campaign maintenance in the circumstances described here, and the reductions in the standards of all work, the quality of work management and supporting services plus the rigours of safety procedures and the striving to reach longer term targets are all impossible to maintain.

(4) The blurring of maintenance and decommissioning work.

A moment's reflection will emphasize the fundamental differences of two separate working cultures, the first dedicated to keeping an operation running with plant in sound condition, while the second is dedicated to taking it apart. Because maintenance personnel are on site, familiar with the equipment and technically capable, they are natural additions to decommissioning teams, who will usually be simultaneously active while campaign maintenance is being conducted. In addition, key sectors of preceding maintenance programmes are removed from maintenance schedules, liberating experienced maintenance personnel for attachment to these decommissioning duties. There is also a technical need in some cases, as selected items of equipment and machinery, young by the standards of the overall plant and in good condition, are removed for sale and shipment elsewhere.

(5) Permeation of the short-term culture.

As with all organizations, principal objectives become part of the working culture, affecting the manner in which work is conducted and the relations between the user or employing organization and its employees or directly contracted personnel. Having embraced the short-term approach in major organizational purposes it is clearly impossible to expect working employees to do anything else. 'Shop floor' or working relations are almost certain to suffer.

(6) Higher costs of work done.

Although budgets and total costs can be shown to be less, the problems of work management and motivation described above lead to working turmoil and less efficient work practice. Once the 'endgame' scenario has been accepted, only those jobs which must be done receive the funding they need. There is no opportunity to defer a piece of work to another time or to avoid emergent problems by judicious rescheduling. *There is no other time, the work must be done or we fall.*

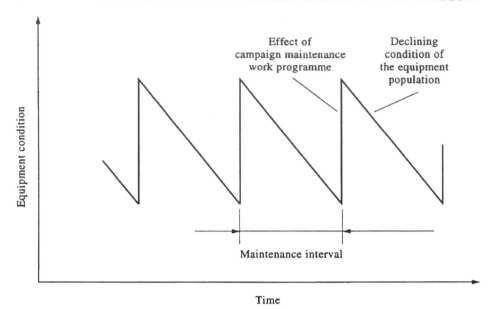

Figure 3.11
The condition of equipment under campaign maintenance.

The importance of positive industrial relations can now be highlighted, and conversely opportunities for disruption and indiscipline will be openly inviting in already difficult circumstances.

  (7) Reduced equipment condition.

      Quite apart from the equipment degradation which can easily follow from the problems described above, there is a technical basis for regarding campaign maintenance as responsible for delivering poorer conditions of operating plant. It must be said that these comments can also apply to shutdown maintenance, but unlike the former this approach usually complements an overall maintenance programme rather than being active in isolation.

      If we regard newly installed equipment after early adjustment and debugging as the peak of technical condition (not operational; this comes with regular use and steady improvement) and regular maintenance as constantly working to return to such conditions, then the time span between them or the maintenance periods witnesses a gradual deterioration in equipment condition. We could regard this like a sawtooth graph, where equipment condition sharply improves with maintenance and then gradually declines until the next maintenance intervention (Figure 3.11).

      Here is a way of regarding the maintenance periodicity. Remember that, forgetting maintenance-induced problems, the deterioration in an item of equipment results from many factors, but principally from its use.

Campaign maintenance routines are most effective when they are applied to an equipment population and, unlike formal maintenance programmes, treat the whole group at the same time. Scheduled maintenance, however, deliberately treats members of the equipment population at different times so that they do not all deteriorate together, and while the nominally 'worst' is dealt with the condition of the other members varies *so that they cannot all fail together* and in the intervals work is balanced among different teams. When the campaign maintenance philosophy is applied to such an equipment group in isolation the work commences *where the whole equipment population is at its lowest condition.*

Chapter 4

Regions of application

The technical processes of maintenance and repair of similar machines or equipment are themselves nominally similar, but changes in use, application and circumstance inevitably affect the point at which it is needed and the manner in which the work is conducted. As the RCM discipline underlines, it is generally the failure of a machine to meet its function which most often introduces a maintenance need, rather than evidence of wear or shortfalls in performance. What may be tolerable in one circumstance may be completely unacceptable in another, such as a small leak in an oil refinery hydraulic line, or a similar leak in a cream pump in the food industry, where the rigid avoidance of product contamination changes the timing and need for maintenance action in respect of an apparently trivial effect.

Similarly, the welding repair of a cracked structural steel member is a conventional task when required in a land-based power station, but requires divers, undersea torches, diving suits and special safety arrangements when applicable underwater to an offshore oil platform steel jacket.

So, when we consider differences of maintenance practice we also need to refer to the industrial uses of the plant as well as the region in which it is located.

It should also be apparent that there are a whole range of work variables which change according to the region of application. In some countries the whole plant could be considered as 'outside' of the work and social fabric that surrounds it. Maintenance tasks are bigger, as they include many of the work elements usually embraced by operators or other technicians and installation management increases the job work content to justify the high expense of bringing in teams from distant and expensive cultures.

In addition, such plants are often isolated, remote from the usual matrix of support and representing the export of a key but singular element of an industrial hierarchy or infrastructure which has grown in a foreign place. The result is few local services or associated businesses, extreme difficulties in obtaining replacement parts, tools or representative materials, and many of the processes of fabrication, inspection, measurement and repair are only obtainable from both external and imported sources.

If the plant and the process themselves are strange, then maintenance is even stranger. A lot of what we would regard as natural actions are themselves subject to a different nature. Responses which in some places would be common in others would be unusual. The interpretation of symptoms may lead to different requirements, while other situations, being less important, would produce no reaction at all. The incidence of failure types would change according to the environment, such as temperature failures in warm climates, and run to destruction would be more prevalent when replacement items were simply not available.

In Western countries many industrial developments are located close to population centres with local residents often employed by the plant (a policy encouraged by operators to foster local support and contain costs). These people have more political and legal authority and a greater ability to communicate, with access to expert opinion and information, than is available to their Third World counterparts. Coupled with the financial means to take disagreements to court, disputes and damages can be exceedingly expensive to the operator in terms of settled claims, production losses and additional costs. It is worth reflecting that had the Chernobyl explosion occurred in a densely populated Western country the reaction of national authorities, the employees and the local population would have been very different and very much more expensive.

In the text that follows we also refer to the environment and such obvious influences as high or low temperature, rigging work at great height, and work in high winds, at night in cramped conditions and high noise levels, which (similar to the examples above) affect both maintenance decisions and methods adopted. Although it is intellectually well understood, this theme will occur again and again: *maintenance work is like no other*; even though the machines and the processes are the same, maintenance is affected by geography, by local environment, by the processes in use, by the work regime, and even by the culture of the application.

When the word 'environment' is used, as in the previous sentence, engineers think of the physical influences referred to, largely because such influences result in technical changes and maintenance actions. However, the dictionary describes it as 'Something that surrounds. . .The aggregate of circumstances. . .', and with this in mind other aspects require attention whenever the maintenance workings of an application are considered. Some of the obvious ones are:

- The climatic and local weather environment.
- The geographical disposition.
- The national, political and cultural background.
- The commercial, business and legal situation.
- The regional infrastructure, transport and communications.
- The safety, statutory and employment structure.
- The regime of accepted work practices.
- The range of technical ability and support.
- The availability of experienced staff.

International companies would immediately refer to the applicable tax regime, operational costs, financial expectations, sources of profit and many other things besides. For the maintenance operation all of these influences will have some effect on the work and the way it is done, unlike the needs of the original construction, which are affected by these same influences but are shorter and more isolated. Maintenance has to support the plant in its ordinary day-to-day detail, and is heavily implicated in all aspects of its existence and for all the years of subsequent operation.

Comparisons with other applications can be extremely illuminating or totally misleading. Figures for working times, the amount of stock required, the expected working life, maintenance frequency and a thousand other considerations will vary according to the specific circumstance. It is not unusual for the labour hours required for a known maintenance task conducted on an offshore oil platform to be three times greater than the equivalent figure for the 'same' work on the beach. Use of any such information must be treated with great care: both its source and its treatment need to be evaluated and modified before it can be used. When the mean value of labour hours for a particular piece of work is published, the source statistics are relevant and the number and spread of the component figures and the work conditions from which they arise need to be known so that estimates for the subject installation can be realistically made.

There are some general aspects of this thinking to be borne in mind when work management methods are being chosen for an application. It is important to remember that selecting the arrangement at the outset is far easier than modifying the work pattern later.

(1) Similarities reside in technical detail, differences reside in the nature and region of application.
(2) Operators and their contractors manage maintenance of specific installations.
(3) Work management methods, team sizes and the use of specialists vary with regions and types of application.
(4) Manufacturers and equipment specialists are experts in specific equipment and machinery.
(5) The final arbiter in each situation is the maintenance manager on the spot.

Assistance may come from many sources: operators in the same country, other companies in the same industry, the manufacturers of the machinery in question, other users of the same equipment, manufacturers of the same product, other managers in the user company and, of course, other members of both maintenance and operational teams. Finally, however, the methods used arise from the maintenance management decisions of the company itself. There are many factors to consider, some relevant, some not. The decisions made will refer specifically to the requirements of the single subject operation.

CHOOSING AN ASSET MAINTENANCE SYSTEM ISN'T A MATTER OF LIFE AND DEATH.

OR IS IT?

You know that safety is one of the most critical factors for the well-being of your employees, and your entire operation. That's why you owe it to yourself to learn more about MAXIMO's powerful, integrated safety features. MAXIMO® is the first asset maintenance system to offer a breadth of features for distributing automated safety information directly to users, and enforcing proper procedures through technology. With MAXIMO, you can document hazardous procedures and materials; incorporate electronic lockouts and tagouts that enhance traditional

MAXIMO's integrated safety module provides automated lockout/tagout/lineup features, as well as real-time documentation of hazardous materials. The system also offers sophisticated workflow capabilities, and the industry's most intuitive interface.

paper-oriented safety documents; and implement safety procedures that are completely integrated with your existing business processes.

MAXIMO has a proven reputation as the industry standard in asset maintenance. Shouldn't it be the logical choice for protecting the most valuable assets of them all? Call **014-837-187-33** or visit **www.maximo.com** to learn more. The life of your business could depend upon it.

MAXIMO®
Maintaining Your Enterprise
www.maximo.com

Chapter 5

The maintenance toolbox

In the insistent drama of the workplace it pays to remember not only the wide range of problems and duties facing the maintenance manager, but also the equally wide range of tools available to deal with them. This is the thinking behind this chapter. Each of the following sections considers a tool which can be usefully put to work.

Because the maintenance and repair of equipment concentrates in the period during and after operations it is natural to consider such work as being after the event, to think of maintenance as a response to previous incidents. This of course, is not the whole story: work in predictive and condition-based maintenance is devoted to identifying and preparing for the future, while the maintenance team frequently finds its attention focused on immediate problems, and concentrating on the present. There is a famous wry comment which captures this dilemma perfectly:

Don't tell me that the tap is dripping when the house is on fire.

As we have seen, however, the more warning the maintenance team has and the more preparation that can be completed, the more efficient and less costly maintenance work becomes. There is an understanding among maintenance personnel that the ratio between planned and unplanned work is important and generally the latter should be kept to a minimum. Many systems formally monitor the planned to unplanned ratio, and we look at this in more detail in Section 13.3. It is one of many tools or methods open to the maintenance manager and his team which can help balance the different demands on maintenance resources and make the transition from future to past more orderly and less painful.

In addition, the chapter discusses computer systems, the maintenance plan, the work order system, resources, data recording and other subjects besides. When considering the large range of such facilities the abundance and growing complexity are clear and in themselves introduce problems of knowledge, understanding and perhaps selection and introduction. It is easy for an individual manager to be overwhelmed by such variety and the gathering pace of change. There is an overriding constraint that is only touched upon in this manual: all methods or systems used *must serve the host organization*. Although a new system can usefully point the way forward and include fresh opportunities, it is not an independent entity: its very existence depends on useful improvement to the maintenance work management. In the world of fascination and opportunity it is very easy to be lost to the reasons for the whole exercise.

5.1 Computer-based systems

The initials CMMS refer to computer maintenance management systems and are increasingly applicable to maintenance systems in current use. The present surge in features and effectiveness of modern maintenance systems owes its momentum to the application of computers and attachment to this highly dynamic business sector.

Within the last five years personal computers in domestic use have progressed from hard disk memories of 20 Mbyte (20×10^6 bytes) or 20 million bytes, where each byte is equivalent to one letter or number to a figure of 2 Gbyte (2×10^9 bytes) on offer today (a factor of 100), with storage capacities rising and price and physical volume still falling.

Earlier computer systems were either based on large mainframes costing hundreds of thousands of pounds to buy and run, or networked terminals each with limited intelligence, linked to file server storage of both operational software and users' data. The present tendency is towards networked PCs each with considerable storage capability and processing power separately downloading processed data from a central file server and transmitting it to other users on the network when required. By this means the high cost and centralized nature of older systems has been broken and high local intelligence and power is supported by active databases. In addition, the computers in use reside within the organization structure that they serve without violating departmental boundaries, and place key facilities where they are needed.

Computers in maintenance use have increased in power, speed and flexibility, introducing features which until only recently were not possible. Systems are commercially available that can store colour pictures of maintenance activities and hardware for reprinting or display as part of maintenance history, work instructions, job training or preparation for repeated tasks. Drawings, schematics, photographs, manufacturers' instructions, illustrations and a host of relevant information may be connected via the computer system to allow immediate migration from an identified initial task to associated subject material and related company experience. Figure 5.1 shows an example.

Single job histories may be examined by different departments at the same time, allowing the preparation of associated work and the interactive contribution of different groups such as 'logistics' or 'safety' to current work in progress. Drawings and parts lists may be interrogated from the work site during actual work and video images of unexpected problems or opportunities transmitted to remote locations for authority or guidance. Maintenance planning engineers preparing shutdown work programmes or standby plans can refer to the growing volume and detail of relevant work history and performance experience.

Three-dimensional work site layouts can be examined using CAD (computer-aided design) system-generated images, and the space available for cumbersome parts

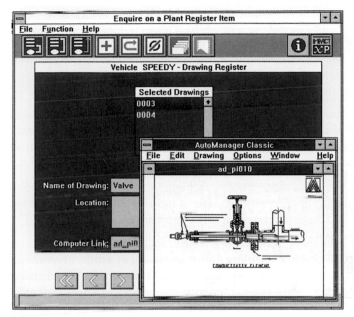

Figure 5.1
(By courtesy of Software Solutions Ltd, Wakefield.)

can be examined prior to movement and so called 'walk-through' features used by plant design engineers for layout modification and construction projects.

The possibilities are breathtaking and gathering pace all the time. The effect on maintenance work management is not yet fully understood, although it is already clear that coordination of work by the operating organization will improve as the isolated nature of maintenance work decreases and support is closer, more relevant and more immediate.

Computer system assets

With such a range of tantalizing opportunities it is easy to be swept into enthusiastic adventures which do not finally help maintenance management tasks, and it is necessary to examine those computer-based attributes which are of established benefit to the maintenance engineer. These, somewhat surprisingly, have changed very little, and it is essential to examine any proposal in terms of an effective maintenance system as opposed to an effective computer system. While the latter will aid the system operation, it is not the prime requisite, a consideration which is sometimes overlooked and can lead to disputes between computer and maintenance professionals[1].

---

1. In the following list it is assumed that the maintenance computer system is usually online, employs a single coordinated database and is networked to workstations or PCs located in contributing departments and close to site work locations.

(1) *Work history focus*
     The focusing of maintenance work history into a single database, providing random access to work records arranged in various selected ways, such as work history by plant item Tag No., grouped similar items, selection by date, location, job type and combinations of these.

(2) *Data discipline*
     Computer systems necessarily format the presentation, structure and processing of data. Their use concentrates maintenance work records to essential elements, and reporting variations are diminished and relevance improved.

(3) *Access to most recent data*
     Reference to work history, schedules, supporting documents, drawings, standards or other retained information will present that most recently known to the computer system at any enquiring location.

(4) *More effective work management*
     The selection, scheduling and prioritizing of work, together with discipline assignment, more effective use of personnel, coordination between them, clear delegation, work in progress and job history reporting, all aid maintenance work management, allowing greater concentration on the work itself.

(5) *More efficient system operation*
     The adjustments of plans, revisions to schedules, resource allocation generation of work orders, preparation of maintenance history abstracts and many similar tasks are only realistically possible with a computer-based system, because of rapid information changes and frequent user requirements.

(6) *Long-term work history*
     Manually maintained work records are bedevilled by variations in methods used by different personnel and changes in the formats used during the timespan of the equipment history. Use of a computer system imposes a consistency of data handling and work records while making retained history available over the years of operational life[2,3].

     The tag number, however, not only described the type of item and gives it a numerical identity, it also *defines its location*. So, if the item in question is changed out the *tag number stays where it is*. The old metalwork has been removed and the replacement unit adopts the tag number. For most computer system facilities this is no problem, but in work history confusion could arise if the tag number identity only is used. For this reason, *serial number tracking* of work order history is used for those items likely to be changed out.

---

2. There is a distinction between the history of a specific item of plant and the work history facility of the computer system. The former contain records of work conducted while the individual plant item was subject to maintenance care; the latter refers to the computer system capability.

3. Some computer systems use equipment tag numbers to identify items of plant in work planning, scheduling, work order preparation and (consequently) closed work orders, retained collectively as work history.

(7) *Numerical processing*

The use of formulae, arithmetical treatment and other numerical processes can be rapidly conducted by the computer within the short time often available. Calculations such as the MTBF or the 'hours run to calendar' conversion can be reliably and quickly available.

(8) *Selections from history*

Abstracts from the database can be selectively made and sorted into a variety of presentations. As the maintenance history covers a wider time span with system use, the number of records relevant to a selected sort grows, and the information provided is more extensive.

Logical selections of the sort criteria can be used to interpret the history records in various ways, such as[4]:

(a) The examination of failures, their numbers and frequency leading to forecasts to failure.

(b) The measurement of downtime by work period, critical tag numbered items, groups of items, location or operational system.

(c) Performance variations according to different locations, correlation with hours run or load duty.

(d) Examination of methods of failure discovery, by operators, by site inspection or specific instrumentation.

These and many other selections can be made, usually instigated by a maintenance engineer pursuing an enquiry before deciding on maintenance action to be taken.

Pitfalls and dangers

Maintenance systems have drawbacks whichever method is used, and they are often the consequences of not using the system properly. This is a serious point: maintenance engineers and technicians are not employed for their computer skills or their empathy with system operation. When work targets are difficult to meet, units are down and the pressure to restart is intense, the requirements of data recording and work management systems are sometimes left to be dealt with later, and while such situations are difficult to avoid, it is here that many subsequent computer problems arise[5].

---

4. Some work order systems include a 'How Found' code in the data recorded for a defect work order. When such work orders are closed out and held as part of the work order history the discovery sources can be examined by using the code in database evaluations.

5. When considering the introduction or extension of a maintenance computer system, the capital costs of system purchase and installation are usually the detailed subject of an AFE (application for expenditure). The costs of data acquisition and input, however, are often overlooked, and taken over the life of the system installation these will dwarf the initial investment. We should also remember that, when system installation is complete, without initial data input it cannot be used and a parallel programme of data preparation is needed. It is also usual to consider the maintenance computer system as part of plant operations, not as part of the original plant construction and installation project. As a result, the maintenance computer system is usually introduced after plant handover has been completed, and a valuable slice of equipment installation and commissioning work is lost to the equipment history. Because of the pressures on maintenance crews during early operation and consequent data recording, this section of maintenance work history is seldom recovered.

(1) *Data accuracy*

During hectic times, data is input to the computer after the job has been done, sometimes by a service clerk, sometimes by a different crew. System updating and data input are regrettably not seen as part of the job, and one result is data errors. These fall into two main groups:

(a) *Method errors*

Errors of method or technique, misuse of the system, recording correct data in the wrong fields, errors of omission, use of wrong codes and regarding data input in terms of a necessary immediate task without regard to the longer term consequences of error.

(b) *Data errors*

Introducing incorrect data to a correct record; for instance using the wrong tag number or work code number or assigning fictional manhours, symptoms investigated or corrections made.

(2) *Error transmission*

Once an error has been made and incorrect data accepted by the system, the database is corrupted; running totals, subtotals and statistics are wrong and corrections difficult, expensive or impossible to make. Even more costly are wrong decisions following wrong information, incorrect judgements made and healthy machines stripped down while others are left to fail, with consequential additional downtime and loss of maintenance credibility.

(3) *Change impediments*

Computer systems protect themselves from error where possible, using data formatting requirements and identifying obvious errors. This slows the data input process and makes correction more difficult when it does occur. Two major types of check are applied to the data input process:

(a) *Input format*

The checking of field sizes, number ranges, sequences and types of characters that will be accepted during input.

(b) *Logical check*

The checking of input data that meets the formal tests against a logical yardstick. For instance, the acceptable date range may be limited to the two weeks up to today only, or the running hours (total) figure read from the unit meter must be higher than the previous reading unless the meter is shown as replaced.

Revised data requirements

During the life of the plant conditions change: new items requiring maintenance attention are introduced, some are removed and the treatment of others is modified. Systems are designed to cater for such changes within the established framework, but revisions to the framework itself are far more difficult. For instance, the transfer of a unit such as a pump to a totally different duty will require a changed tag number, while the unit serial number remains the same. The previous maintenance work history referring to the earlier tag number and duty is not relevant to the new situation and the maintenance manager may decide to retain this history attached to the old tag. This means that the serial number of the unit will appear in two places and be barred by the computer.

Similarly, it may become evident that a key data item is missing from the work history because a needed evaluation cannot be done. If this is the recording of a frequently changing variable, say the engine rpm, it may simply not be possible. Alternatively, where future changes are contemplated or where a few values are needed and can be established from other reliable records, such as the equipment orientation, then an addition to both the record structure and data of every record in history can be made.

There are also occasions when existing records are simplified so that data input can be made more accurate or data errors reduced. If, for instance, it is found that the two codes

(1) Replacement item
(2) New item

have been repeatedly confused during data input, then the combination of these two into a single code in all work history records will eliminate the error. Such changes can be achieved by both data and structural changes in the work history database. As a simple rule for existing records in most cases:

Data reduction is far easier than data enlargement.

Limits of useful information

The previous item refers to the dilemma facing the system analyst at the time of the computer system creation. When the work records are designed there is little plant experience available and future data requirements have to be predicted. It is a 'chicken and egg' situation: if too much is required because of what might be needed, omissions and errors are inevitable and processing costs are higher, while overly simple records are less prone to error and easier to process, but offer less information from subsequent investigations.

A careful balance has to be made, and one factor is underlined by experience:

If the maintenance work team participation is lost, work history will be inaccurate and incomplete and computer-generated reports and analyses will be of no value.

In addition to the natural work preferences of the maintenance team discussed above:

the computer system is seen as an extra chore offering little to the maintenance technician in support of his work.

With this constraint in mind the system needs to limit the requirements of data processing while preserving useful history evaluation for the maintenance engineer and enhancing system reporting and support for the working technician.

It is essential for the maintenance technician to see and experience clear benefit to his work of frequent computer system use.

System ingredients

During the late 1970s and early 1980s, major operating companies found that, while the need for computerized maintenance systems was accepted, there were not suitable systems already developed. The result was customized computer systems written to meet these needs for individual companies. Although this was an expensive option, the resulting systems were directly compatible with their own operations and were not obliged to meet a wide range of commercial variety.

Since that time, confidence and experience has grown and several 'off the shelf' maintenance system packages have been developed and are now on the market at more modest prices. These developments are still going on, and much of the impetus arises from the developments in power and capacity of modern PCs. Similarly, there has been an explosion in the software field, with huge numbers of new software packages suitable for use in an 'open systems' environment and usable through a windowing operating system. The data files prepared by such systems are not part of an integrated approach, but they are usually compatible, allowing data prepared by one software package to be transferred for further use by another.

In spite of the widening numbers of system applications, and the increasing sophistication of proliferating new software packages, the basic elements of the maintenance computer system are unchanged, and these are described in more detail in Sections 5.2–5.4.

First, the overall disposition needs to be visited.

The general computer maintenance management system

In Figure 5.2, continuous lined boxes show features of the general computer system and broken line boxes show the major associated activities which produce or require data for the general system. Between blocks, links are numbered and described, while arrows indicate the general direction of information flow.

Link 1

Construction projects will often affect the nature of maintenance work and the equipment population to which it is applied. Additions and changes to the equipment database often result.

Link 2

The stock control module shown here refers to parts and materials held in support of maintenance activities. Changes in the equipment database will often require modification of stocks held.

Link 3

Changes in the equipment population requiring maintenance as part of the equipment database will require corresponding changes in the maintenance plan.

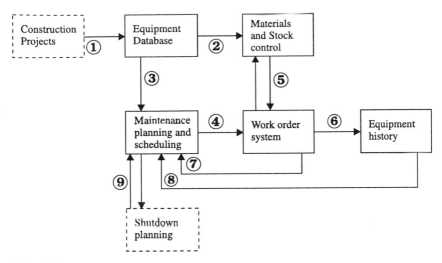

Figure 5.2
The general computer maintenance management system.

Link 4

The maintenance plan is regularly harvested to produce the schedule for a selected work period. Work orders are 'cut' prior to issue through the work order system.

Link 5

Materials required for completion of a work order are usually requisitioned from stock in the week prior to the scheduled work tour and prepared for site. Major long lead items are identified by the maintenance planning engineer and availability arranged with the warehouse in advance.

Link 6

Once work has been completed the work order is closed via the computer. The closed work order became an item of equipment history.

Link 7

Some work orders not attempted during the scheduled period will be rescheduled for subsequent attention by the same crew.

Link 8

Investigation of the equipment history and other information by the maintenance engineer will often result in changes to the maintenance content and frequency of work for selected items. Changes to the maintenance plan result.

Link 9

Shutdown plans may include construction and maintenance work. Of the latter, some planned jobs will be undertaken during the shutdown, retimed to avoid it or cancelled altogether.

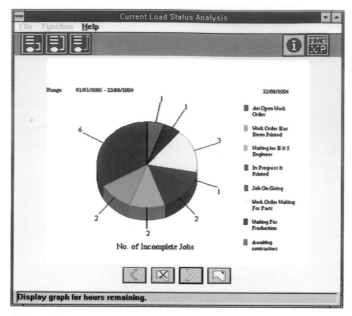

Figure 5.3
Maintenance backlog presented as a pie chart (by courtesy of
Software Solutions Ltd).

It is fatal to assume that knowledge of computers in general or the alternative use of computers elsewhere in your organization makes the selection of a maintenance computer system simpler. It is a solemn fact that the reverse is likely to be true. Maintenance departments are seldom the first part of an organization to assign computers to their work and it is very tempting to tap previous experience from other areas to ease the difficulty of system selection. As the pages of this handbook emphasize, and as in other areas of speciality, maintenance must be regarded as different from other types of work, which means in turn that the systems employed will be similarly specialized and regions of general operating similarity will be few. Although the principles of data handling and system operation are relevant in all areas, it is the

system as a maintenance application which drives both need and choice.

The selection of the system must remain firmly in the hands of the maintenance team, and the inclination to choose one for mainly computing reasons should be rigorously avoided. This is not to say that such influences should be ignored – far from it – but they are best regarded as one of a range of factors to be considered when the choice is made.

We should also remind ourselves that maintenance teams are not primarily computer motivated, and it is easy but wrong to assume that others know more about the specialized subject than they do. Herein lies a danger: the maintenance team may itself shrink at the difficulty of system choice and advocate the adoption of a computer 'professionals' approach. If this path is chosen the resulting selection will inevitably be biased, with a strong chance that the resultant installation will fail to meet its primary maintenance objectives, resulting in disaffection and lack of use. Such disaffection needs to be underlined: if significant investments are made without resultant support and improvement, then the situation is actually worse than before. Further development and financial support by the company will inevitably be refused and the whole maintenance management improvement enterprise will stall.

Such an approach is uncomplicated if the system is confined to the maintenance group alone and the database used only for maintenance purposes. Often, however, data is shared by several different departmental users employing an *integrated database*, or by the rapid communication between related elements of an *open system* With the huge proliferation of separate specialized software modules available on the open market, the latter method is becoming more and more widely adopted. In addition, the interconnection of PCs to form a company-wide network emphasizes the communication aspect, and it is in this area in particular that the computer professional's thinking is necessary.

Another major influence is of course the apparent complexity of choice and the huge proliferation of competing suppliers and different maintenance systems. There is a straightforward way in which this dilemma can be resolved and a final selection made. This is discussed below, but we should first emphasize that although this method is certain it is not quick. If, without preparation, it is imperative that you have a decision this afternoon – use a pin.

Another important factor is to consider the genealogy of the different systems which are available. There are systems on the market *specifically designed for maintenance management*, which are usually the ones meriting serious attention. Be wary of maintenance systems which are added on to major existing systems initially designed for something else. This method is often inevitable if overall system integration is intended and gains much overall data processing power, but the danger is of course that later developments are obliged to fit an existing framework, and the maintenance package is constrained as a result.

The following approach assumes that the purchasing company has an existing means of selecting suppliers suitable to its business operations, and this method of choice should be outlined at the beginning to bear on the initial choices. It should be addressed again in more detail once the shortlist has been established.

Selection guidelines

It is a fateful reality that, having completed a rigorous process of selection, installation and subsequent use, many of us would glumly prefer to begin the whole process all over again. This is an inevitable result of human experience. The original imperative to install or revise a maintenance system is often based on incomplete information, resulting from failures of existing methods, pressure of growing business, ageing machines or the simple realization that 'something must be done'.

Conducting the review, considering effects, targets and alternatives, is in itself a learning process and the team's knowledge is more current and extensive once the choice is made. Of course, it does not stop there: the decision to change introduces an era of accelerated learning which will only slowly diminish as the new methods become established and understood. Knowing of this effect at the beginning of the process is not immediately very helpful. The learning influence is certainly impossible to avoid and these guidelines emerge as useful.

(1) Appoint the team who will make and arrange the process of selection. Each user group should be included, together with someone with a computer background if possible. The chairman or team leader should be the maintenance representative.

(2) Be consciously aware of the effects above and manage the selection so that choices are made as far into the learning effect as the timetable will allow.

(3) Follow the selection framework (or one similar) laid out below. Be careful of changing the sequence of actions or of conducting key steps in parallel when the successful onset of one depends on the previous completion of another.

(4) Stimulate and include change. This is a step of initiation, not one of reaction. It is not sufficient merely to respond to, or to allow for change; it should be positively addressed during the programme.

(5) Be constantly aware of what each step is for. Be very careful of using data gathering, sorting and comparison merely to reinforce decisions that have already been taken.

(6) Find a red pen. The range of options is growing all the time, and some favourite options will have to be discarded. Make a distinction between personal preference and what the company can tolerate. *For instance, small companies have neither the wealth nor the time to experiment: failure can be fatal. Large organizations may wish to run different pilot studies before final selection, where apparently small differences influence large sums of money.*

The selection process

First, think about selection as a process to be followed step by step rather than early consideration of the competing systems themselves. If you regard the procedure as starting wide at the beginning and narrowing down to a point of final choice, then the opening stages should be as wide as possible during the gathering of relevant information. Human beings find it very difficult to reverse a decision or a choice once it has been made, and if this step is taken too early, apart from limiting the number of options considered, later effort is inevitably assigned to justifying a decision already made.

(1) *Define the purpose of the exercise*
     Consider, for example, whether you intend to extend the existing system or replace it. This is not as easy as it would initially appear, and in some cases this will not be settled until after knowledge and costs of current options are known. This is a powerful consideration. Maintenance computer systems

are changing very rapidly, and the important thing is to ensure that this is clearly settled before any choice is made.

(2) *Consider the scope of application*
Will the new system be confined to the maintenance department only or be used by others? There is a strong case for production control, stock control, purchasing, spare parts and goods inwards all to interface with each other and employ common data (see notes above).

(3) *Question the authority*
Which department will act as the operating authority, making the system available conducting system upkeep, revisions, backups, database reorganization and central operations if applicable? Separately, which group has authority over the data being processed and stored, permitting or refusing the destruction of 'old' information and interpreting that in recent use?[6]

(4) *Outline what you are buying*
The most attractive commercial purchase is a proprietary system already developed and ready to run without adaptation or modification on a wide range of inexpensive platforms. The price and features of such a product can be stated confidently, simplifying the choice at the outset. As soon as any adaptation or customizing is added the price rises and becomes uncertain. When some customers require a system totally designed to suit their own 'in-house' specification, not only does the resulting software become exceedingly expensive, but the creation of the system design specification itself becomes another source of cost.

Additionally, some customized systems, usually of advanced years, have been given a facelift and remarketed as proprietary systems, inevitably without the features and advantages of leaner and younger competition.

Finally, we should refer to the current emphasis on the windows-style operating systems. Around 80% of new users prefer this, and some of the better systems have been designed to cater for this method.

(5) *Sketch the intended application*
Although it will not be possible to define the final system specification at this point, a general understanding will be known at the outset and certain features will be known as you start.

(6) *Gather information about realistic options*
Even at this stage, some initial selection is necessary. There are simply too many systems on offer to consider detailed work on more than a few. The

---

6. Maintenance regularly attends to long life cycle machines, and overall operations of 25 years are not unusual. Machine histories collected over this whole period refer to infrequent events which can become relevant at any time. These are not the qualities which usually apply to stored data, and here are the ingredients of disagreement with other groups, who process for the short term and ruthlessly weed out apparently 'old and irrelevant' material.

'overall likely cost' is often used to make a first selection, coupled with general technical and operational features.[7]

(7) *Narrow down the choices*

Prepare a simple questionnaire expanding the details referred to in step 4. Use computer professional input to help prepare this document, but ensure that its strong maintenance foundation remains (some of the elements of proposals are listed below). Send these documents with a letter explaining the nature of the exercise and examine the replies carefully. Once this is done a second selection reducing the options further should be made. Advise suppliers that, if selected, there will be a further presentation (refer to steps 9 and 10 below), during which a formal sales presentation will be included.

Repeat this step using more gathered information to reduce the number of possible system suppliers to a manageable number. Six to eight should be sufficient.

The following list is to act as an *aide mémoire* as the final contents of the enquiry should be tailored to suit the customer's application.

Company name
Contact name
Address
Telephone
Fax
System purpose
Database management system
Number of current installations
Geographical region of installations (e.g. UK, USA, Europe, worldwide)
On which computer (mainframe/mini/micro)
Usable operating systems
Storage capacity on disk
Number of current users
Typical prices:
    Software
    Hardware

System facilities included

Work scheduling
Equipment register
Shutdown planning
Equipment/work history
Costing

---

7. It should be emphasized that high price and high performance are not the same. Given the present high rates of change the selected system's ability to handle change is vital, and the nominally modest systems are giving far better value for money.

Budgeting
Purchasing
Performance analysis
Hours run monitoring
Standard work instructions
Backlog monitoring
Fault coding/analysis
Condition-based triggers
History search/query facilities
Navigation
User special reporting
Graphical selectable reporting
Statutory diary (e.g. the flagging of statutory inspections or surveys at planned times)

Is system easily transferable to other different computers during upgrade?
Can system readily integrate with other software?
Is access security included?
Minimum RAM requirement
Hard disk storage required
In which country was the main software written?

(8) *Expand the questionnaire*
Having gone through the early selections the team should have a greater understanding of the different alternatives. Use this understanding to expand the questionnaire into an enquiry profile, detailing the requirements of the application, the features required and the supporting services needed. One way of doing this is to invite the various future user departments to submit an outline of their needs which can be merged under separated sections into a single 'enquiry profile' and linked to the attractive features now known to be available.

(9) *Obtain internal agreement*
The 'enquiry profile' to be used should be generally understood by its 'internal customers' and any emergent difficulties for local applications can be settled, Suggestions for other departmental users can be considered and, most importantly, the prospects of future disagreement can be diminished.

(10) *Present the profile*
Present the profile in direct face-to-face discussions, with each of the companies remaining on the short list, preferably on their premises, and record the responses.

Remember that the presentation is intended to formally establish how the requirements of the profile are to be met.

Still bearing in mind that certainties are likely to change with further knowledge, repeatedly compare the profile with the different options which will inevitably feature during the presentations.

(11) *Ask for a sales demonstration*

All successful sales departments will be fully capable of presenting their system in its most attractive manner, and it is important to both customer and supplier that this is done. It is best arranged at this point in the selection process, encouraging the supplier to emphasize standard features which address the customer's system profile and introducing alternative and perhaps superior options which may have been overlooked. Also, the customer is at his most receptive, understanding more clearly what he is being told and being more relevant and perceptive in his enquiries.

Notice that step 7 ensures that potential suppliers are advised at the outset that the sales demonstrations will be needed to ensure that their product will be fully considered and that the process of selection is a fair one. Also, the resulting sale must be based on agreement and understanding. The sale is the end of the selection process, but it is also the beginning of a successful application. It is in nobody's interest for a poor selection to lead to failure and future recrimination.

Regard the selected system supplier more as a future partner whose motivation and commercial well-being are important to your own company's future success.

(12) *Make the selection*

The preceding process will lead naturally to a final and well documented choice, though it may be necessary to repeat steps, especially as fresh information becomes available and understanding improves. Make sure that all members of the investigation team are included in the final selection. Any disagreement must be dealt with at this stage; proceeding with installation in the teeth of dissent will inevitably lead to future operational problems.

(13) *Visit existing installations*

The final act in the selection process should be to visit installations where the chosen system already operates successfully, preferably a user in similar operations to the customer. This step has three purposes:

(a) To demonstrate that the selected system operates successfully without user problems.

(b) To demonstrate that the software needed by the customer is already in use and not under significant development.

(c) To pinpoint any areas of difficulty experienced by the existing user, particularly during installation, which could be avoided with care by the new user.

The choice of the installation to visit is not easy. The system supplier would naturally prefer the best available, while the customer wants to avoid a single special case. The best approach is to ask the system supplier for a selection of five or six names that he is happy with, for which a visit may be requested.

The purpose of the visit is to gather confidence. It should be done with the support of the system supplier and clearly aimed at reassurance, not at system selection. This step should have been covered already.

(d) *Place a purchase order*
The preceding means of selection is aimed at system choice, not purchase order processing. However, many of the requirements of the purchase order will already feature in the selection, and the ordering should be considerably easier. Purchase orders are considered in general in Section 8.2.

5.2 The equipment database

Often referred to as the asset register or the plant list, information gathered in the equipment database is fundamental to any formal system of maintenance management. It may seem obvious, but maintenance work cannot be properly managed without defining what it is that has to be maintained.

There are four paramount purposes of the equipment database:

(1) To identify separately and clearly all equipment subject to maintenance management.
(2) To outline technically each separate item, its type and rated performance.
(3) To provide an audit trail for equipment replacement, maintenance and repair, together with access to more detailed and current data required during working life.
(4) To provide the foundation of the computer system database and information for planning, scheduling and the creation of work orders.

These four statements have an unashamed maintenance bias, and the equipment database described below is a principal foundation of the maintenance system, *whichever maintenance management philosophy is used*. It is also usual for the database to contain information for other purposes; for example, financial information for item and equipment population value, initial purchase price, insurance requirements and recorded depreciation, plus engineering information by cross-reference or transfer access to the original plant design database.

Similarly, for the purchasing and contracts departments, access to manufacturers' and service suppliers' company details, project contractors or plant operators purchase orders and details of existing or previous special contracts could be included, along with details of maintenance contracts, manufacturers' warranties and special operating provisions.

Some operating companies with less coordinated information systems employ more than one equipment list, each arising from a different history and purpose. For instance, the finance department may use a list which includes only those items of plant with a positive current value, while the operations group refers to listed equipment under operational duty. The equipment list for the maintenance department would also be different:

the maintenance equipment database defines all main items of equipment requiring maintenance attention.

Notice the word 'attention' is used, not the word 'work'. Whenever a machine is on site it requires maintenance attention, even if that attention results in no work. As we have observed elsewhere, equipment foundations and power supplies may remain and require attention long after the equipment has been removed. Also, the delivery of new equipment to the site may introduce a maintenance requirement merely by being there, before installation, commissioning or power. Equipment deteriorates from the moment of creation, not the moment of installation. It should therefore be clear that the list of maintenance equipment and the list of operational equipment *are not the same.*

The maintenance equipment database includes plant items whose operational duty is finished but are yet to be decommissioned, other items which have already been removed and new plant items awaiting installation, commissioning or operational use.

The broad type of maintenance attention assigned is sometimes referred to in the equipment database, using codes for safety inspection only, pre-use maintenance, normal maintenance, site checks and others.

It is also likely that the installed status is shown, usually as a simple single-character code, indicating whether each plant item is in normal operation, awaiting removal, or yet to be installed. It is then a simple matter to prepare reports of all main items awaiting commissioning or all items in operational use, and others, by using this code in the report selection criteria. An interesting outcome of this technique is that decommissioned items which have been removed from the site may be retained on the equipment database, providing historical data for contractual or purchasing reasons, or for comparison with technically similar items which remain in active use.

Some overall company systems use one central equipment list coded for each group putting it to work. A single alphabetical character 'Departmental Selection' code is used (such as O = operations, S = stock control, F = finance, M = maintenance). When considering the maintenance equipment database the phrase 'functional coherence' above underlines the nature of the items represented. So each of six identical 'main oil line pumps' would be separately entered, but their component parts would not. Alternatively, 1000 electrical junction boxes to the same technical specification, individually numbered, would be included in the equipment database in a few defined groups under the respective electrical distribution systems. So because many (junction boxes) are physically spaced apart, the coherence in this case is electrical and not mechanical, and maintenance treats them as coherent groups.

Contents of the equipment database

(1) *Plant item or main equipment identity and description*
    Assigned plant number or tag number and by alphanumeric character description field, e.g.

    GE 1401A gas turbine generator

(2) *Plant item purpose*
    Also an alphanumeric character field, which describes the functional purpose of the equipment item, for instance, a pressure vessel could have a listed purpose of:

    crude oil separation

(3) *Manufacturer's company code number*
The code number (generated from a separate list) is used with purchase and supply collated information to secure details of the original equipment manufacturer's company name and further details.

(4) *Supplier's company code number*
Referring to the same information as 3 above, this identification is used when the equipment has been provided by a third party, e.g. a stockist or local subsidiary of the OEM (original equipment manufacturer).

(5) *OEM's reference number*
These are usually confined to model number and type number recorded in the database using an unformatted alphanumeric field[8].

(6) *Technical specification reference*
Many operators create a separate technical specification for each major or critical item of plant, collating selected technical highlights from the manufacturer's detailed information. A cross-reference to this technical specification is provided in the equipment database.

(7) *Departmental selection*
Single character code used for the creation of various abstracts to suit different departments.

(8) *Maintenance responsibility*
Identification by company name and address or in-house maintenance group of those responsible for maintenance attention.

(9) *Maintenance contract reference*
In conjunction with item 8 above, the Y/N confirmation of an existing maintenance contract and its reference in the contracts department database.

(10) *Plant design reference*
Prepared as the originating contractor's database by the initial project design group, and held by the operating company as primary plant information.

In keeping with other features of the maintenance computer system, the equipment database is accessible via a system keyboard and provides information for a wide range of enquiries as well as forming a main system foundation.

5.3 The maintenance plan

Any serious examination of effective management systems and operations will quickly discover a vital working link to the future.

---

8. For selected major items of replaceable plant the serial number is also recorded. It is sometimes called a 'plated number' because it is stamped by the manufacturer on an identification plate which is riveted to the body of the machine. Although it may seem obvious, every machine which leaves the factory has its own unique serial number, and at times this number provides the most definite means of tracing the history of the machine.

Anticipation is a feature of good management.

For maintenance management this points to a problem: the natural character of maintenance actions is to respond to an event, and such an event may be the issue of a condition monitoring warning, the judgement of an experienced technician, or the result of an inspection. Whatever the initiating happening, it occurs before the maintenance action. We look elsewhere at the *ability to respond*, and in spite of its undeniable importance this approach addresses maintenance as a reactive rather than a proactive science. Many exciting and recent developments in maintenance improve work targeting and shorten reaction times, but do less for the ability to anticipate. There is one major area where this is not true, and its foundation is of course the maintenance plan. Although it carries the charge of 'over-maintenance', there are simply too many deteriorating items in the equipment population to merely wait for an occurrence. An effective maintenance management system needs a footprint in the future, and the maintenance plan provides such an imprint. If maintenance requirements did not change, the plan could tell us much of what to expect in ten years' time.

It is a feature of good planning that the process starts in an imprecise and distant future with gradually increasing accuracy and intent until detail and action finally occur. All managers, whether they formalize it or not, work in the future. Their activities are sometimes described as the process required to make things happen, and it is sobering to remember that forecasts of output targets and working life covering a twenty or thirty year period are solemnly made before any major site investments are made. The whole maintenance disposition follows the working of a predictive process.

A feature of most major systems, the maintenance plan is a formal and usually pictorial means of presenting the work forecast. Referring to both equipment and time, use of the plan allows the preparation of work schedules for specified periods for variable time spans of 1 to 52 weeks, commencing at specific dates or week numbers in the future.

A most convenient method employs a rolling plan, constantly moving to keep pace with the advancing calendar, spanning the 52 week period from the current week. The presentation is used when regular tasks are the main ingredients of the plan and forthcoming maintenance PPM routines are shown in the calendar week in which they are planned to occur.

So, in the screen display or printout of the plan, the listed PPM maintenance tasks for a specific tag number, say GE 1401B gas turbine generator on Ninian Central Platform, could show a 'tick' against Week No. 12 for the six-monthly PPM routine and another 'tick' in week 38 for the 12-monthly PPM. The whole plan is built up in this manner for each main tag numbered item and each definable maintenance job. Although there a few simple structural rules, the plan construction is straightforward, and in some cases is initially computer generated.

Main types of plan

Before looking at the rules which apply to the regular work plan it is also important to distinguish between two quite different types of plan and the ways that both are involved in the maintenance management process.

The first, which is the main subject of this section, is the work plan[9], which defines the time and resource requirements of largely independent jobs or 'work packages' and is the usual method applied to maintenance PPM routines.

Job duration is generally limited to one work tour, so where a 'two on, two off' (weeks) regime is in use, the maximum job would take two weeks[10]. Logical relationships between different jobs are weak, and they can be changed or rescheduled with little effect on others in the work plan provided the constraints of work association are not violated (see Section 8.3)[11].

So, if the six-monthly PPM on a water injection pump is postponed for three months it has little effect on plan activities applied to other equipment apart from the resource requirements which will change.

The second type of plan is the network plan, used mainly in project work, where the various activities in the plan are generally dissimilar and their logical interconnection is very strong. So if the design specification activity of a construction project is extended by three months, almost every other activity in the project is delayed and the network plan duration is increased by roughly the same amount. There is a strong dependence of each activity on the completion of its predecessors, and this is the foundation of the network created. Such networks are produced by PERT (programme evaluation and review technique) or CPM (critical path method), which are usually described as critical path methods and are individually prepared for the project and have little further use once the work has been completed.

Network planning is used very effectively when a maintenance job is treated as a separate project. For instance, the hot gas path inspection of a gas turbine requires treatment as a project involving the employment of a specialist maintenance team, tools, materials etc. and a work duration greater than other normal tasks and beyond the extent of a single work tour.

This, of course, is very simplistic. There are also logical effects in the work plan method, but they are not usually sequentially dependent, and flexibility among various maintenance tasks is a very strong feature of this method.

---

9. Some commentators refer to such plans as 'work schedules', and there can be confusion. In this handbook, 'schedules' refer to the collection of jobs arranged for a single work tour only and include unplanned tasks, repairs and rescheduled work previously issued but not done. Most work schedules facilities allow for the addition of unplanned work occurring during the work tour, although a fresh printout would be necessary (see Section 2.6).

10. This rule is not universal. Some sites operate three on, two off, giving a nominal maximum of three weeks. Also, selected jobs are accepted as exceeding the normal maximum.

11. When a work association facility is in use the replanning of a task into an invalid association is routinely prevented by the computer.

Names and abbreviations

Although names and abbreviations are referred to in both the text and the glossary this particular topic is rich in terms which deserve expansion.

(1) PPM: pre-planned maintenance. The method used to select and assign maintenance to specific machines and requisite week numbers.

(2) PPMs: used by maintenance personnel in the plural to refer to planned maintenance jobs in the PPM programme.

(3) Maintenance jobs: jobs which require the action of the maintenance team (either PPMs or defect repairs).

(4) Maintenance tasks: specific work elements of maintenance jobs defined in PPMs as sequential steps in work instruction (see Section 9.4).

(5) Defect work orders: initially unplanned maintenance jobs issued as work orders by the work order system which arise from equipment failures or other defects.

(6) PPM work orders: work orders issued by the work order system for work required on PPM maintenance jobs (see Section 5.4).

(7) Package number: numbers which apply to groups of PPM jobs pertaining to the same plant item or equipment group, responding to the same time datum and 'packaged together'.

(8) Package or PPM datum: the year and week number at which the lowest frequency PPM in a group occurs and about which other PPM jobs are timed and planned.

(9) PERT: programme evaluation and review technique.

Properties of the work plan

(1) Each future routine maintenance job for the subject item of equipment and the timing of intended work by week number.[12].

(2) The precise equipment identities of items to be maintained are abstracted from the equipment database and referred to by tag numbers[13].

(3) Maintenance groups, companies or technical disciplines are identified in the plan, showing where they are required to complete both single or multi-discipline tasks.

(4) Each individual discipline and overall resources allocated to the plan is calculated in manhours, subtotalled for each week of the plan and usually presented as a histogram (see Sections 2.6 and 3.7).

(5) Facilities for change in the plan itself, including the removal of jobs or equipment no longer needed, and the addition of jobs (standard or special) defined and included in the plan with extra resources required[14].

(6) Some computer-based planning tools will automatically add fresh recurrent

---

12. What is to be done is detailed in maintenance work instructions, standard procedures or manufacturer's selected routines (see Section 9.4).

13. Sometimes plant numbers, equipment groups or descriptions are used.

14. Although possible, it is not usual to add normally planned jobs to the plan for the current week.

jobs when generating a new plan as the period covered reaches new week numbers.

For instance, a twelve-monthly PPM currently displayed in week 24 of this year will automatically be added to week 24 of next year once the plan advances to include that week.

(7) The work plan is the focus of maintenance support and supervisory planning activity, acting as the vehicle for converting fault reports and failure warnings into scheduled maintenance tasks.

Rules and arithmetic

There are a few simple rules which apply to the way in which calendar-based PPMs are organized into working packages and assigned to the correct week number. Many PPM work instructions are prepared so that when all are taken together they constitute one full year's maintenance work. Some cycles are longer than this, taking up to five years, and a few exceed even this.

Under most circumstances the main (or lowest frequency) PPM acts as the datum which, when located in time, automatically defines the applicable times of all subordinate PPMs. So if the main PPM had a one year frequency planned for the first week in January, the corresponding six-monthly PPM will be planned for the first week in July and the two three-monthly tasks planned for the first week in April and the first week in October.

The full set of PPMs could look like this[15,16,17]:

1 off 12 monthly
1 off 6 monthly
2 off 3 monthlies
8 off 1 monthlies

It is immediately evident that in the preceding list there are only one six-monthly, two three-monthlies and eight monthlies, which seems to deny the requirements of the annual calendar. The reason is simple: the lowest frequency, or most senior PPM, overrides the smaller one because work specified in the latter is included in the former.

---

15. The set of PPMs is made up to suit the required frequency of certain items of work, and when all are combined they constitute a complete maintenance work set.

16. Daily and weekly routines are also included in the work done, but are regarded as part of the working background and not specifically defined in the work plan. This work is usually confined to first line maintenance such as greasing, lube oil replenishment, coolant level checks, indicator zeroing and other similar steps.

17. If the subject item of a PPM has been replaced, reconditioned or undergone major repair, the relevance of unchanged PPM routines is questionable in work content and timing. It would be expensive and unnecessary for a plant item to be replaced in week 21 by a defect or repair work order, only to face a two-yearly PPM in week 22.

If the repair or reconditioning work is modest, then a content modification of the subsequent PPM may be all that is required. If, however, the unit has been replaced, then the maintenance PPM 'clock' is effectively put back to zero. For instance, if the lowest frequency PPM is two-yearly, then the PPM maintenance datum should be set to two years after the time of replacement and a note of this change added to the repair work order at the time of closeout.

So in the preceding example, the twelve-monthly would include the PPM work specified in the six-monthly, which in turn includes work required by the three-monthly, which finally repeats the steps defined in the monthly (Figure 5.4).

This method of 'nesting' the PPM routines together can be applied to a variety of different frequency combinations, such as 2, 8 and 24 months or 1, 2 and 4 months, and is a powerful method provided the work required is cumulative. In some cases they are not. Quite apart from the wider constraints of 'work association' it is often convenient, more efficient or even not possible to connect certain actions in the same work package, and the instruction may be deliberately designed to see that they are kept apart. For instance, alignment measurements on an item of rotating equipment can only be conducted when it is running, while cam follower clearance checks require cold, stationary conditions. In such a case, different steps could be arranged serially, but the extra time required may make the one work tour limit difficult to meet. Similarly, acoustic leak checking of a valve requires fluid passing through the pipeline, while the valve's replacement is best achieved using isolation of the pipeline section and no fluid flow.

In other cases the positioning of tools for one task may use the space required for another; for example, the electrical charging unit used to recharge main d.c. standby batteries on an offshore platform weighs several cwt (over 200 kg) and is as big as a desk. It requires two men and some swearing to move it into position up steel stairways and along narrow walkways. While it is in use (for a period of many hours) nobody is inclined to move it, and local access to the connecting electrical circuitry is very restricted. It would be most unwise to schedule maintenance work on the local electrical distribution circuitry during this period.

There are many instances of this sort, requiring inclusion or exclusion of work steps and both the content (see Section 9.4) and coincidence of such steps is reflected

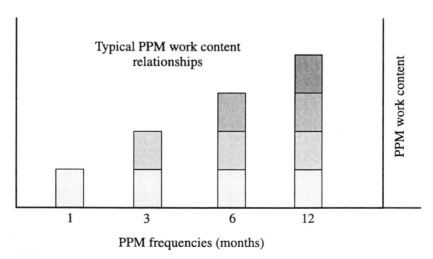

Figure 5.4
The content of minor PPMs is included in the makeup of the less frequent major PPMs.

in the way that PPMs relate to each other in the work plan. One of the great benefits of work plan creation is the recognition and removal of problems of this kind.

The work package

In some cases, maintenance work will be called for in the work plan at the same frequency as another PPM conducted at a different time, e.g. a pedestal south crane boom inspection is conducted annually in the summer, while crane operating controls are also checked annually, but in the winter. The two PPMs have the same frequency but they do not coincide. In these cases separate 'package' numbers are used as separate PPMs in the plan:

12.01   Crane boom mechanical inspection
12.02   Crane cabin instruments and controls check

Apart from the practical considerations, such as resources, work space and that it is the same host machine being maintained the work requirements of the different work packages are treated separately. There is no link, for instance between the two packages 01 and 02 above, each running to their own datum. So if the two packages and their datums are six months apart, so too are the six-monthlies which may arise, and the planning arrangement would look something like this:

Week No. 2   Week No. 28
12.01   12.02
06.02   06.01

In the case of the d.c. batteries, this technique is usefully employed for discharging and recharging work carried out on each battery every twelve months. For a complement of 24 defined batteries, 24 separate package numbers, 1–24, are used in the work plan, one of which is called up in each offshore work tour (every two weeks). Apart from this careful planning distinction, the same work instruction is used in each case.

In much of the text the word 'frequency' is used to describe how often something occurs, which is its correct meaning and is expressed in units per time. So the frequency of a PPM might be described as twice per year.

Maintenance planners, however, use the reciprocal of this for simple convenience and it is referred to as the 'periodicity'. So the twice per year PPM frequency referred to above would have a six-monthly periodicity.

The periodicity and the package number are conveniently stated together:

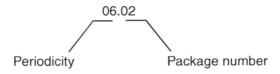

Periodicity   Package number

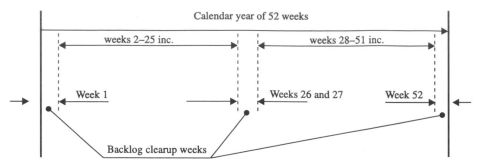

Figure 5.5
Typical annual work pattern for pre-planned maintenance.

Structure of the work plan

One site crew work pattern divides the calendar year into two similar 24 week blocks, with two two-week unplanned periods in weeks 1 and 52 (Christmas/New Year) and weeks 26 and 27 (mid-year) used for clearing outstanding work (Figure 5.5).

Although the work content of the 24 week blocks is deliberately similar, they are not identical. Twelve monthly PPMs, for example, will appear in one block and not the other, while frequencies of less than one year (two-yearly, five-yearly etc.) may not appear in the current year at all. If a maximum frequency of one month applied, the expected total PPMs for the year would of course be 12[18].

Most computer systems allow the maintenance planning engineers to reschedule or cancel PPMs, though there are formal limits on what is permitted. For example, rescheduling a PPM is usually restricted to prevent conflict with the next one due in the plan and shortly due for issue. If the job cannot be completed during the permitted reschedule period then they are closed out as 'not done' and the extent of the permitted reschedule will depend on the number and pattern of PPMs planned. For example, the PPM programme for a pump above may look like this:

PPM periodicity (Months)	Week number	Crew colour
12	33	Blue
6	9	Red
3	21	Red
	45	Blue
1	5,13,17,25	Red
	29,37,41,49	Blue

Any delay in the twelve-monthly PPM in week 33 would soon impact the monthly due in week 37.

---

18. In some cases this will be reduced to 11 if one falls unavoidably due at year end or mid-year, in which case one of the highest frequency PPMs would probably be cancelled.

Figure 5.6 gives an abstract of a typical work plan, shown for a gas turbine generator. Notice that 1, 3 and 6-monthly PPMs are planned in package 1, whilst the 12, 24, and 48-monthlies are planned for package 2 and that these latter PPMs are also 'hours run'-triggered: the figures shown in brackets are the estimated hours of equivalent running.

Maintenance planning

Although the work plan is a vital ingredient of the maintenance manager's toolbox, it is maintenance planning as an activity leading to work order creation which transforms the requirements of the plan into active reality and can be regarded in seven principal areas of action:

(1) The upkeep of the work plan by introducing new tasks, rescheduling backlogged work, deleting tasks no longer required and retiming maintenance cycles following equipment replacement (see note 17 above).

(2) Coordination with the offshore crews and maintenance engineers to agree the content of forthcoming work schedules, including the closing or rescheduling of work orders undone at the end of the current work tour.

(3) The regular issuing of plan abstracts in the form of work schedules for forthcoming work periods.

(4) The 'cutting' of work orders in preparation for the next scheduled PPM maintenance work tour (see Section 5.4 and item (3) above).

(5) The upkeep of maintenance work instructions, maintenance standards and test records, either on the computer itself or in an offline file.

(6) Preparation of maintenance system management reports defining the status and progress of the maintenance programme according to previously selected performance measurement criteria (see Section 13.4).

(7) Preparation of work in progress, work order backlogs and work orders closed 'not done' reports which summarize the work position and highlight outstanding jobs requiring urgent attention.

Using the work plan and work schedules, work selection and delegation are completed by the site maintenance supervisor directed at the next immediate crew shift or work tour, when tactical decisions about work sequencing coordination and priority are made[19].

In addition to the activities described above, maintenance planning activities are often conducted in parallel with the preparation of work history abstracts. These are produced to aid analyses into maintenance occurrences and the determination of expected times to failure, evaluation of PPM work frequencies and work instruction content.

Maintenance planning, like the maintenance system as a whole, operates as a regular office exercise week by week. Because of this, it can become routine, with the real

---

19. Test records are required for hard-copy recording of readings taken during the completion of some PPM routines. These are mainly required for instrument procedures, but all maintenance disciplines use this feature of work instructions.

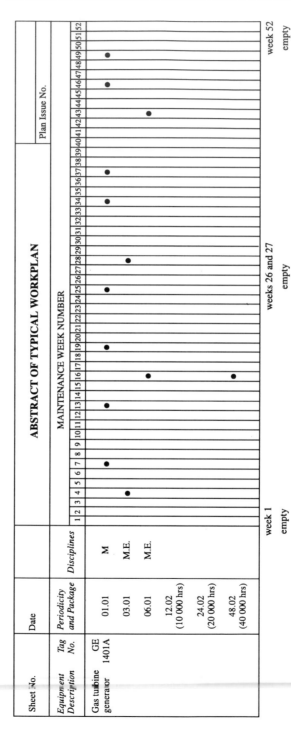

Figure 5.6
Abstract of a typical work plan.

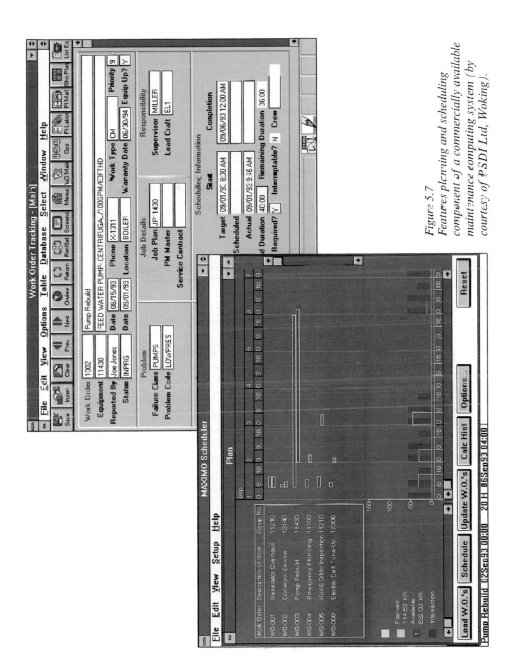

Figure 5.7
Features planning and scheduling
component of a commercially available
maintenance computing system (by
courtesy of PSDI Ltd, Woking).

danger that operation of the 'system' becomes the target rather than maintenance of the plant itself. For this reason, certain exercises conducted during maintenance planning have an extra significance for senior management and are highlighted as follows:

(1) Reports of backlog and work orders closed out 'not done' (item (7) above) indicate whether:
   (a) The work is being done.
   (b) The resources are adequate for the combination of planned and unplanned work required.
(2) The number and type of defects arising give an insight into plant condition, both overall and by selecting data, for specific groups of equipment.
(3) The ratio of planned to unplanned work helps to quantify the effectiveness of the maintenance programme, both in general and for selected groups of equipment (see Section 13.3).

The strategic planning of maintenance work is usually conducted by the maintenance manager and has a wide variety of sources, including company senior management, other departmental managers and the maintenance teams and supervisors themselves. Inputs to be considered would include:

(1) Statutory influences: perhaps the introduction of a Christmas tree changeout programme or a Department of Transport lifeboat inspection schedule.
(2) Influences of planned activity on adjacent or connected sites, such as the installation of subsea gas pipeline shutoff valves or shutdown maintenance at 'receiving' oil refineries requiring the suspension of production and oil export activities.
(3) Effects of operator company-wide targets or objectives, such as the restarting of an underwater structural inspection programme, or the structural modification of equipment to reroute the fluid flow in production processes.

These and many other strategic influences are translated by the maintenance manager into various effects, one of which is to require changes to the maintenance work plan affecting medium to long-term activities, which are processed by the maintenance planning activity.

5.4 The work order system

Maintenance management systems normally include a work order facility which produces hard copy orders for work to be carried out, and work orders created by different maintenance systems are very similar. Each work order presents information required for the maintenance job to be done and provides data input fields as both job and record are completed. This is not a glamorous part of maintenance management: it has little to catch the imagination, no neat graphical solutions or intricate analyses, and each work order merely collects job data in a formal manner.

This is a great pity, because *the work order is the key initiator of maintenance action*, and subsequently all work history, maintenance analysis and management understanding are based on the successful completion of this step. Regrettably, work orders are skimped, bypassed and overlooked; lacking the interest or rewards attached to other work, they are the subject of inaccuracy and evasion. It takes unremitting and accurate work to get them right, and though such effort may pass unnoticed, if it is not done the whole maintenance management system is degraded or worthless.

The work order system translates jobs in the work schedule or defects identified on site into hard-copy work orders authorizing work to be done.

For maintenance purposes there are two types of work order, often given numbers which become part of the work order number:

(1)   Preplanned or PPM work orders (Type 1).
(2)   Defect or repair work orders (Type 2).

Bearing in mind the planned to unplanned ratio (see Section 13.3), Type 1 work orders are more numerous than Type 2 and usually require less work. The value of the ratio, however, is calculated from the total manhours spent in each category rather than the number of work orders issued. Refer also to 'defect' and 'defect work order' in the glossary.

The relationship between PPM and defect work orders

PPM work orders are produced from the maintenance schedule and 'cut' in computer memory by the maintenance planner, prior to call-off by the site maintenance supervisor, while defect work orders, because they are less predictable, cannot be included in the maintenance work plan. The latter may be raised on the platform or by the planning engineer on the beach according to a defect discovered at the work site. They are often raised and acted upon very quickly, sometimes missing the formal

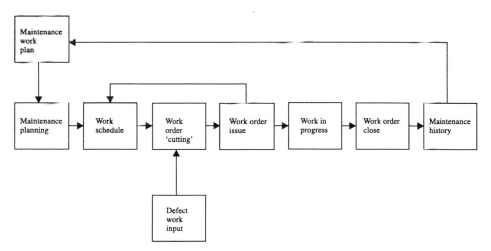

Figure 5.8
Stages of work order creation and issue.

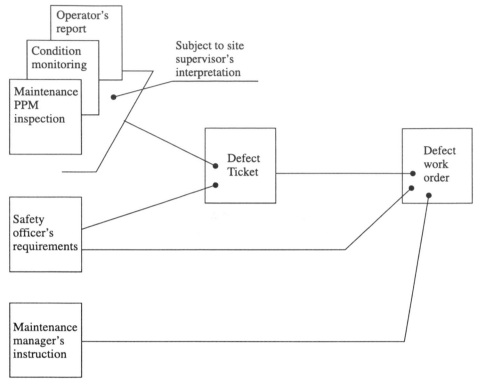

Figure 5.9
Sources of defect work orders.

hard-copy issue of the schedule, being raised and completed before the following schedule is produced. The computer enquiry screen will of course show any additional work orders included (see Section 5.3)[20].

Defect work orders are raised when a fault requiring rectification has been reported, and they can arise from several sources (Figure 5.9).

(1) Technician's report following a PPM routine.
(2) Condition monitoring: equipment running problem detected.
(3) Production operator's fault report.
(4) Safety officer's requirement.
(5) Maintenance manager's instruction.
(6) Statutory authority/third party inspection requirement.

Items (1), (2) and (3) require the site maintenance supervisor's interpretation and decision before the work orders are raised.

---

20. Some systems exclude the maintenance planning function in the work creation mechanism, instituting automatic or library work orders which repeat at preset interval defined at the outset. Under these circumstances, maintenance routines are formalized but not so readily apparent.

Some maintenance commentators refer to PPM inspections as 'fault-finding tasks' or 'fault location routines' where defect work is identified and defect work orders raised. This of course is partly reflected in group (1) above, but we should also note that PPMs are more than this ideal inspection alone:

(1) In many cases, bringing the technician and his tools to the machine is a key job purpose. Significant non-invasive or adjustment maintenance can be carried out without the need for a defect work order and a return visit.

(2) Much preplanned work will be done irrespective of previous PPM inspection results. It will be dictated by maintenance frequencies, expected times of failure or a condition monitoring requirement.

(3) PPM jobs can be regarded as segments or elements of the total maintenance programme content, where the full cycle may take a year or several years to complete (see Section 5.3).

PPM work is driven by the maintenance plan, not by fault or failure.

(4) PPM work orders refer to specific *work instructions*, which define in detail each component task in the job that has to be done. Additional tasks can often be added to a PPM work order, modifying the work required for the single occasion.

The condition monitoring result referred to above triggers or retimes a preplanned routine rather than requiring a defect work order. Unfortunately, not all condition monitoring reports require action which is so convenient. Similarly, not all faults located during PPM routines will result in a defect work order, some being noted for checking at subsequent PPMs with no immediate action taken. Others will result in tasks added to the next PPM work order as a one-off additional task.

Creation of a PPM work order

Because the Type 1 work order results from the planning and scheduling process, creation of the work order is completed by the computer system. PPM jobs pertinent to the work tour are 'cut' by the maintenance planner and held in the computer memory. These, plus defect jobs already input, are included in the issued schedule and 'called off' by the site maintenance team during the work programme.

Referring to the maintenance plan, and in some cases the equipment database, much of the data required for the creation of a PPM work order[21] will be generated by the computer.

Maintenance code number
Equipment tag number
Maintenance package and periodicity
Selected maintenance week number
Date of work order issue

---

21. A PPM work order calls up the maintenance package and periodicity, which by reference to the standard work instruction defines the work to be done.

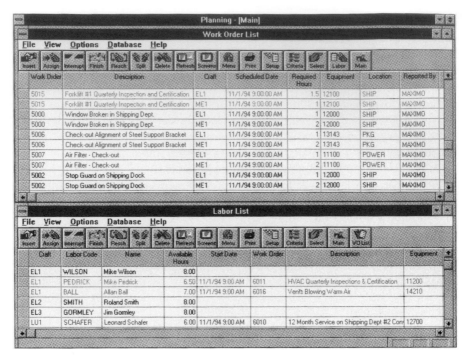

Figure 5.10
Labour available for completion of work orders listed in a current maintenance computer
system (by courtesy of PSDI Ltd, Woking).

Resource disciplines included
Estimated manhours per discipline
Issue number of applicable work instructions

Creation of a defect work order

When preparing a defect work order, a preprinted defect ticket completed on site would
be passed from the operator or technician with details of equipment identity and defect
symptoms to the maintenance supervisor for the addition of estimated times, tools and
materials required, plus the personnel assigned and other details listed below.

Completed data is input through the keyboard for work order creation by the
computer system. Not all data referred to below will be relevant or known during
initial input and may be added later or permanently left blank. However, some data
fields, such as the equipment identity, estimated production downtime and assigned crew/discipline are regarded as mandatory, and a work order will not be
produced unless this is included[22].

---

22. Work orders may be 'called off' (i.e. printed and issued) by instruction from the site maintenance
supervisor according to his judgement of the immediate situation.

Initial data input:

Maintenance code number (alphanumeric)
Equipment tag number (alphanumeric)
Equipment description (alpha description)
Input date and time (Numeric)
Likely defect cause (by code)
Means of discovery (by code)
Defect symptoms (alpha description)
Estimated manhours per discipline to repair (numeric hours)
Estimated job duration (numeric hours)
Estimated machine downtime (numeric hours)
Estimated production downtime (numeric hours)
Crew/discipline assigned (alpha description or numeric code)
Specific safety requirements (alpha description)
Relevant company standards and work instructions (alpha description)
Required interface with 'permit to work' system (alpha description)

Work in progress

Data input during work in progress – PPM only:

Progress status
Unusual work commentary
Supervisory comments
Additions to work narrative

Data input during work in progress – defect only:

Symptoms confirmation and other work narrative
Progress status
Revised estimate of duration or downtime if applicable
Revision of discipline manhours needed

Work order closeout – defect and PPM

Job completion date and time
Final equipment status
Equipment downtime hours
Production downtime hours
Man-hours used for each discipline
Work completed (by code)
Work completed (by alpha description)
Parts and materials used (by stock numbers and quantities)
Likely cause of defect (by code – defect only)
Technician's name

Once the work tour is over, all work orders issued will have been processed by one of the following methods:

(1) *Work orders completed and closed out*
    Data has been processed as described above and the closed work order attached to equipment work history

(2) *Work orders not started*
    Shown as remaining undone from the original schedule, these are processed by the maintenance planner in one of the following ways:
    (a) Closed out as 'not done', usually where the subject is a high-frequency PPM and previous work history is favourable.
    (b) Rescheduled forward to be included in a future work schedule of the same crew.
    (c) Rescheduled forward into the next schedule of the same crew.
    (d) Rescheduled forward to the next immediate crew.

Groups (b), (c) and (d), will be added to listed backlog.

(3) *Jobs still in progress*

Depending on the site supervisor's decision, jobs which have been started but which remain unfinished at the end of the tour may continue to be worked by the next crew on board, or may be deferred until the initiating crew returns.

5.5 Experts and disciplines

Assigning the right discipline can make a huge difference in the number of hours required to complete a piece of work, but making the choice is more difficult than it would initially appear. Selecting from a known and necessarily narrow in-house range has traditionally been simple, such as choosing a mechanic or an electrician. Unfortunately, such simple choices are a thing of the past, following a steady increase in the range of equipment used, its design complexity and the variety of competing alternatives. Team members have usually been selected for their ability in multiple technical sectors rather than their specialized knowledge in a few. This means that maintenance teams can respond quickly to most situations by wide experience and prompt action, keeping downtime losses to a minimum. There is, of course, a downside: because of the wide variety and infrequency of many equipment failure tasks the acquisition of specific skills within the team is very difficult and can only be seriously contemplated at the expense of the wide technical experience which is so vital.

Increasing the number of maintenance disciplines increases the skills available and reduces the time required for the completion of some jobs. The workload need, however, has the annoying habit of overwhelming a particular speciality while leaving another underemployed.

Remember also that the fluctuations in required skills are worse for breakdown or emergency work orders, where rapidity of response is most important. For PPM or preplanned work orders the discipline requirements are predictable and resource smoothing more effective. Even in these cases, however, the resource requirements of the maintenance plan are often not referred to the availability of key staff for the planned times. The high incidence of training, holidays, stand-ins, special assignments and sickness means that with the most careful planning, wide technical experience remains an essential qualification of key staff.

There is a further important consideration. Of all the technical regions of activity, maintenance is one of the least popular. It is seen by some engineers as reactive and unskilled. While this is clearly and increasingly untrue, it partly explains why good maintenance practitioners are so hard to find. There are three key maintenance motivators (the reasons why a vital few regard maintenance as the most important part of engineering and the key ingredient of reliable quality operations):

(1)   The important nature of the work itself.
(2)   The emergency adrenalin in many situations.
(3)   The wide variety of different tasks

It is this last item which specifically refers to the discipline questions and underlines the need to have the necessary staff to get the job done at all.

The following is a list of typical maintenance disciplines and some associated work. It is a feature of the modern technical scene that new disciplines are arising all the time, and while many items of installed plant survive for twenty years, the older ones will remain.

Mechanical (M)	Boilers, heaters, cranes, lifting equipment, dehydrators, de-aerators, driers, filters, generators, pumps, turbines, engines, gearboxes, compressors, blowers, tanks, pressure vessels, mooring turrets, manifolds, separation trains, piping, fluid and gas transmission systems, air compression, mechanical handling, drilling equipment, hydraulic power.
	Water, sewage, fuel, essential services, doors, gangways.
Electrical (E)	Batteries, chargers and rectifiers, distribution boards, inverters, panels, switches, switchboards, transformers, lighting and heating systems, high and low power generation and distribution systems, navigation aids, public address systems, emergency lighting and power, electrical displays, motors, instrumentation.
Instrumentation (I), Fire and gas (F&G)	Plant air, instrument air, fire detection, gas detection, fire protection, gas protection, instrument panels, emergency shutdown system, local instruments, transducers, control room displays, transmission cabling, condition monitoring, termination cabinets.
Heating, ventilating, air conditioning (HVAC)	Air handling units, refrigeration, condensers, humidifiers, fans, controllers, dampers, ducting, space heaters, calorifiers, control panels, cooling coils, filter washers, heat exchangers.
Safety (S)	Hydrant cabinets, valves, hoses, hose reels and cabinets, couplings, firemen's cabinets, helideck crash cabinets,

firewater/foam monitors, life rafts, life jackets/cabinets, breathing apparatus, survival suits, portable ladders, safety shuts, lifeboats, lifeboat winches, boat maintenance, engines, radios.

Telecommunications   Telecommunications equipment, personal radios, site or platform radios, public address system, telephones, emergency communications system, line of sight transmission/reception system, antennae.

The external expert

Choosing the correct external expert is even more difficult, given that competing contractors will sometimes exaggerate their capabilities without admitting their limitations, and it is frequently the case that true abilities are only known after the work has been completed.

As we have already observed, the acquisition of a working skill requires high work volume and repetition, contrary to wide labour variety. The use of expertise, frequently the means of reducing prime costs, is born of familiarity and practice. Maintenance work, particularly when done by the 'in-house' team, is more varied than almost any other form of work. Maintenance technicians face many types of parent machines and many different devices, manufacturers, installations and methods of use; opportunities to specialize are lacking.

There is a further ingredient to consider – knowledge – applied to the equipment, the maintenance job and the working environment. As complexity increases, so does the knowledge required, and when a machine, or job, is visited infrequently, the acquisition of knowledge before the job can be started adds to the overall manhours.

A small but infrequent job will show a high proportion of learning time, with less time to recover during the task itself. Also, according to the number of men assigned, the working duration and consequently the equipment downtime are affected. The hours lost are multiplied when a whole team is delayed while knowledge is collected by a single member.

It would be simple if this was all that had to be considered before a choice was made but using experts introduces penalties of work management. 'In-house' teams gain sharply through knowledge of the site, logistics and working methods. In addition, other and sometimes more powerful influences apply, such as downtime costs or safety implications (see Chapters 11 and 12) and in these two cases speed of response would be the paramount characteristic.

External expert or 'in-house' team – a choice

Assuming that the maintenance manager is not bound to a certain course by existing contracts or company directives, the matter of choice is not as obvious as it may at first appear. Ironically, the difficulties are sometimes compounded by the coincidence of many such choices and the conflicting need to give each the attention it requires. This problem is offset by evaluating different classes of work in advance and detailing the response to be followed in the majority of cases. In making such arrangements or subsequent choices the following ingredients will have an effect.

Type of failure

Many failures, such as a drop in operating efficiency or difficult starting, will not affect the principal functions of the machine or only introduce a gradual deterioration. In such circumstances, time pressures are subdued and work required may be scheduled in a forthcoming shutdown or PPM to await the arrival of an expert.

Safety implication

Failures, such as arcing contacts in a gas present (zone 0) environment or the centrifugal explosion of high-revving machinery, demand immediate and paramount attention before any external expert can arrive. Knowledge of the site's emergency shutdown system, supporting safety and emergency procedures, parent machine detection and protection systems, plus shutdown and isolation procedures is necessary for correct judgements and actions to be made.

Safety work, especially where a rapid response is necessary, favours locally trained and based safety personnel.

Machine criticality

Production- or safety-critical items with severe safety implications, which affect serial main elements of production output, or which have immediate effect on life support systems will be functionally critical and usually designated with a numerically valued *criticality factor*. Operator instructions usually insist on rapid and immediate return to full function in the event of failure (see Chapter 12).

Production downtime

The assignment of high criticality factors arises in many cases because of the severe financial effects of equipment failure and the resulting loss of production caused by *downtime*. Severe opportunity costs are sometimes of such magnitude that they swamp the applicable maintenance or repair costs, and subsequent actions are totally dominated by achieving the swiftest possible return to full production.

The callout response

The time between callout and work start is affected by response times agreed in the maintenance contract, travel time taken, personnel processing time on site arrival (such as safety officer's briefing), on board reporting and maintenance supervisor's Instruction and kit collection.

Other work on site

Because of access or accommodation restrictions, other ongoing work on site may have to be suspended to make room for incoming teams. Choices may require 'bumping' less essential personnel by transferring them off site or returning them to the beach.

Parts availability

Many maintenance or repair tasks simply cannot be done without correct replacement parts and materials. Operator stocks may have been held for several years, increasing the risks of shelf damage, or recently acquired, increasing the risk of misfit. The supply of correct parts on time at the correct location is a major preoccupation, compounded by the fact that the onset of repairs often identifies unexpected requirements. On a few occasions an incoming expert team may deliver key parts on arrival, identified during callout and by previous experience.

Summary

'In-house' teams

Much more rapid response, both time to site and time to commence work.
Knowledge of the working environment.
Familiarity with working systems and methods of reporting.
Superior use of logistical services.
Greater appreciation of influences on other processes and equipment.
Improved methods of communication.

Expert teams

Specialist and detailed machine knowledge.
Possessed of relevant work skills through practice.
Greater knowledge of other sites and related problems or opportunities.
More specific but short-term support from site supervision.
Shorter job duration times.

Questions of choice

The complexity of the work.
The preparation time.
Logistical support required.
The number of expert teams already on site.
The availability of other work for in-house teams released.
Selection of the correct expert.

The preceding commentary describes the attributes of in-house and external personnel and the circumstances which favour one group or the other. We should consider the consequences of making the wrong choice, which in some circumstances can introduce severe penalties. The downside of each decision should be remembered as the choice is made:

(1) *Wrongly choosing the in-house discipline*
    The wrong choice, either from among site disciplines or in failing to call in the expert, will usually be detected quickly and the decision changed. The main penalty is the time lost if external personnel have to be awaited.

(2) *Wrongly selecting the wrong specialist*
External experts are both reluctant to admit their limitations and narrow in their expertise. This error is more serious: it takes longer to detect, will cause unwanted delays and will be expensive to put right.

Black trades

This is the name given to experienced personnel who are often brought in to carry out certain types of work, such as cutting and welding, pipe fitting, scaffold erection and the whole raft of construction and installation work. When present on site they are often called upon to give specific working assistance to maintenance teams.

5.6 Resources

It is a regrettable feature of business life that maintenance, while recognized as an area of important and growing management activity, is almost certain to be under-resourced at the worksite. This state of affairs applies to both in-house or externally supplied manpower and arises for several reasons.

(1) The current trend in company organization is for headcount reduction, offsetting reduced manpower availability with increased working efficiency. Unfortunately, while the opportunity exists in some branches of manpower, it is far more difficult to apply to maintenance.
(2) The twin effects of more and more new designs of equipment to be maintained and older long-service equipment needing wear-out replacement and repair, leads to uncertainty in actual maintenance work to be done.
(3) The definition of resource requirements for uncertain tasks is more difficult. Needs often change when symptoms are analysed more fully or covers are removed.
(4) There is frequent interaction with other departments and groups in the organization, leading to delays and changing priorities, reducing maintenance crews to periods of waiting or strenuous activity.
(5) Because of the uncertainties involved, some companies do not prepare a formal assessment of maintenance resource requirements for a specified work period. Revisions to short-term maintenance plans do not always result in changes to defined resources.
(6) Where resource requirements are quantified within the maintenance plan it can prove difficult to coordinate with the actual resource numbers available. Absences arising from holidays, sickness, stand-in duties, special assignments etc. reduce or destroy resource levels expected.
(7) Difficulties or delays in early task activities cause timing problems as the work programme proceeds. When changes in disciplines or resource levels are needed, previous delays can cause waiting delays or under-resourcing at critical times.

The advent of such resource limitations gives rise to these unpalatable effects:

(1) Growth in the maintenance work backlog.
(2) Less efficient working, higher work cost and reduced results from given expenditure.

(3) Greater difficulty in estimating the consequences of equipment failures and downtime effects.

(4) Unnecessary inaccuracy of maintenance plans and difficulty in estimating activity durations, date milestones, job costs and timing of subsequent work.

(5) More difficult to trigger staff commitment and assign the correct skill or discipline required for the task.

The maintenance programme needs to cater quickly for both planned and unplanned work with correct timing, interlocking and balancing of many disciplines and specialities, resources. By providing higher resource levels and flexibility at the outset, programmed action can be achieved and efficiency improved, and fewer resources are finally required.

Figure 5.11 shows typical variations of discipline resource requirements by week number, and similar features are included in most maintenance computer systems (see Figure 5.10).

5.7 Status reporting

It is usual practice for site management and supervision to prepare various different reports for the company's general management, which come under the loose title of 'status reports' and refer to the following.

Production status

The reported output of the different product lines and predicted totals over the selected work period. Reports sometimes include figures for the generation of waste material, current levels of scrap and through flow of parts and bulk in respect of materials handling, transport and short-term storage.

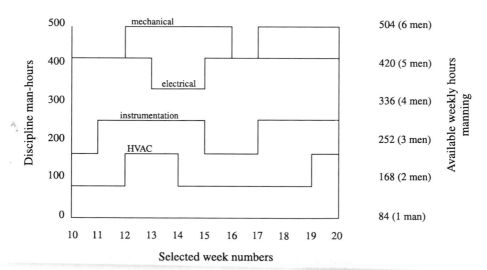

Figure 5.11
Discipline resource histogram.

For example, oil platform production requires water injection to maintain reservoir pressures and maintain oil outflow. The relationship between oil recovered and total fluid processed (oil plus water) varies during field life and production. Status reports usually quote total fluids processed as a measure of operational effectiveness, as well as the figure for actual oil recovered.

Project status

Construction, maintenance, production safety, drilling, and logistics will all be involved in site projects and all affect operations. Such projects are planned and managed individually and have two main considerations for the maintenance manager:

(1) The effect on other maintenance work affected by project access, proximity of work, and availability of materials.
(2) Changes in future maintenance work arising from equipment and installation changes completed by the project.

The progress positions of major projects under way are often summarized in the status report.

Job status

The status of major customer orders forming part of work in progress is often included in the status report in a batch manufacturing environment. Planned maintenance activity affecting production work on selected high-profile orders is a usual inclusion.

Equipment status

It is this section of the status report which has most significance to the maintenance group, as changes in machine status can have immediate effects on the maintenance work programme. An obvious example is when a planned strip-down of a standby machine is interrupted by the sudden failure of the parent machine. It is likely that the maintenance team would immediately box up the standby, switch on and transfer to production, when attention would be diverted to the parent machine. The status of both machines has changed. There are several machine states usually contained in the status report:

(1)	Normal operation	Switched on
(2)	Restricted operation	Switched on
(3)	Available – normal standby	Switched off
(4)	Available – but taken out of service	Switched off
(5)	Equipment failed – awaiting repair	Switched off
(6)	Equipment failed – under repair	Switched off
(7)	Equipment available – awaiting other equipment return to operation	Switched off
(8)	Equipment not available – installation and commissioning	Not connected
(9)	Equipment not available – awaiting removal	Switched off Power supply terminated

Only the first two conditions refer to equipment switched on. Item 2 is in operational use, but is subject to a defined load constraint (e.g. reduced lift tonnage on the pedestal crane) or is subject to shorter periods of inspection

Items 3 and 4 refer to equipment not in actual use but which will operate normally when switched on. Item 4 refers usually to switching off when production outputs are low and quantities can be processed by fewer machines.

Some status reports include general load conditions applicable to the machine in current operations.

There is a strong temptation to regard status reporting as a daily or half-daily exercise of reporting to management. In truth, it is more important than this, as it is the information foundation reporting status to all working groups currently on the site. The advent of computers and electronic communication provides the means of meeting the principal status reporting need:

current status should be reported all the time with any changes shown at once

BS 4778, Section 3.2, Figure 191–6 shows the classification of item states, which offers an alternative way of considering the status, it is repeated below as Figure 5.12. One interesting aspect is the clear distinction between 'up state' and 'down state'.

5.8 Shutdowns and projects

We should distinguish at once between a plant shutdown, a production shutdown and a machine or equipment shutdown, none of which is total. Life support and emergency services will be operational whenever men are on site, even on Christmas day, when many factories are limited to security personnel. When work is under way, power generation, distribution, detection, protection and safety systems will all be in operation.

During production shutdowns, platforms or sites are usually totally manned and accommodation is full, so safety and emergency services are essential. It is usual to conduct full maintenance and testing of such safety systems prior to the main shutdown.

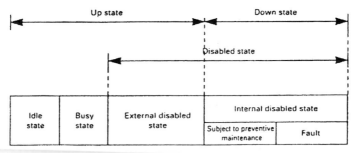

Figure 5.12
Classification of item states.

Plant shutdowns usually apply when major works construction projects are to be done or special-purpose maintenance teams are used to flood the site and carry out major maintenance tasks on operational equipment not usually available. Such programmes are sometimes conducted during the company employees' annual holiday, and the maintained plant is restored to full operation before their return.

Machinery or equipment shutdowns are a normal part of the maintenance and repair programme defined in the work order or work instructions pertaining to the job, and often specified in the maintenance plan.

It is the *production shutdown* which concerns us here, which allows simultaneous and valuable access to critical production systems at the expense of zero output. Key jobs included in the shutdown cannot be undertaken any other way and are specific elements of shutdown classification A – 'Instigating essential' see below.

The phrase 'partial shutdown' is also used and refers to production-restricted output. For example, a spirit bottling plant operating two bottle filling lines would suffer a 50% partial shutdown if one line was brought to a standstill for maintenance work.

BS 4778, Section 3.1, 1991 describes shutdown maintenance as: 'Maintenance which can only be carried out when the item is out of service'. This refers to machinery or equipment shutdown maintenance work, while the treatment of shutdown here also discusses it as a state or situation.

Some maintenance jobs, by their nature or speciality, are best treated as part of a shutdown programme rather than the maintenance plan itself, the latter's role being confined to triggering and identifying those standard maintenance tasks to be included. Non-standard constituent jobs are prepared in advance, often using a network-style planning tool (Figure 5.13) by an assigned supervisor or project manager. Jobs would naturally centre on those best or only done when production is at a standstill and would usually be graded according to their importance.

Three job shutdown classifications are often used:

(A) *Instigating essential*
    These jobs instigate the shutdown in the first place. The plant cannot be returned to production after shutdown until these jobs have been completed.
    Typical jobs –
    (a) Main production vessel five-yearly internal inspection.
    (b) Ball valve replacement in main production piping header.
    (c) Performance testing of production train pressure release valves.

(B) *Positive inclusion*
    Once the shutdown decision has been made there are many jobs which benefit if they are included without the necessary imperative to cause a shutdown themselves. Benefits would arise from improved access, the non-operation of in-circuit machinery, reduced safety risk and additional personnel on site.

(C) *Prepared option*
    Production shutdown opportunities do not occur very often, and it is good

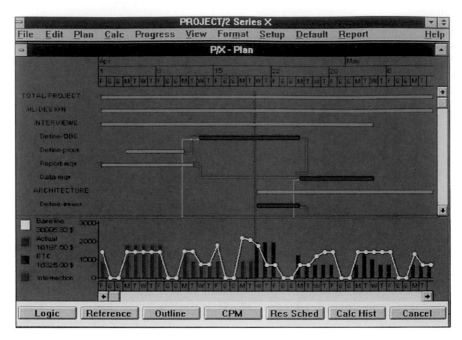

Figure 5.13
A commercially available software tool used for network planning of important
documents (by courtesy of PSDI Ltd, Woking).

practice to complete preparations for more jobs than can be realistically attempted, Category (C) fulfils this role, and such jobs take advantage of unexpected opportunities occurring in the programme through delays in categories (A) or (B).

Jobs are prepared for immediate attention and injected into the work programme on the instructions of the on-site shutdown controller. It is important to be prepared for delays or opportunities arising in categories (A) and (B); however, it is not unknown for shutdowns to be satisfactorily completed without any category (C) jobs being attempted. The reverse is more pernicious: if a delay occurs in the main shutdown programme, the inclusion of category (C) work provides some mitigation of the increased downtime cost.

Steps to take

Maintenance work in general is characterized by job variety, founded on infrequency and some uncertainty. Shutdowns extend this feature to the plant as a whole, with the introduction of unusual site conditions, job variation, specific project work, fresh staff unknown to the safety group and unfamiliar with the site. The scope for errors and accidents is immense, often made clear before the shutdown, during the programme and at the time of start-up. The role of the safety officer during the shutdown is paramount, and is often the only influence offsetting impatient shortcuts and dangerous actions.

Such concerns are met by:

(1) Agreed and defined shutdown objectives, referring to the job categories below and each of the headings listed here.

(2) Exhaustive preparation of individual job programmes, including tools, materials, manhours, discipline requirements and job duration.

(3) Identification and induction of working teams, including team members, authority and reporting, means of communication, contact names, domestic arrangements, responsible supervisor and safety officer.

(4) A refined and agreed overall shutdown plan, defining jobs and their categories, discipline assignments, estimated durations, logical interdependencies, fallback actions and alternatives.

(5) Ever-present safety representation, referred to in all job actions, work site preparation, provision of safety services, shutdown plans, labour induction and safety briefing.

(6) Clear work instructions, by previous reference to on-site experience, equipment manufacturer's maintenance repair and retest instructions, equipment maintenance and repair history, standard work instructions and procedures where applicable, team technical job preparation.

(7) Equipment documentation, drawings, parts/component lists, circuit diagrams, recalibration or certifying requirements if applicable.[23]

(8) Exact shutdown procedure, including correct sequence of events, controls and equipment required, perceived results of each step, procedure supervision, disposition of manpower in process communication, stage by stage monitoring and reporting of completion handover to maintenance teams.

(9) Efficient work management, with identified work in progress, clear job objectives and procedures, non-ambiguous task delegation and on-site shutdown coordination and control.

(10) Regular progress reporting: job progress, special actions, jobs completed, options selected, problems encountered and achievement statistics.

(11) Run-up and handback procedure: definition of maintenance work complete, accounting work complete, accounting for personnel, detectors and safety devices, contact with shutdown supervision and production, and overall position handback isolation reversal.

These observations are not exhaustive. They are intended to act as an outline, prompting similar but more elaborate preparation individual to the plant.

---

23. In some cases the repair of an item of equipment (particularly one used for measurement) or even its repositioning may violate certificaton or pretesting applicable to its use.

Shutdown management

Because many different jobs, companies, departments, disciplines and crews are affected by a shutdown, management of the whole programme, particularly on-site control coordination, requires careful and senior attention. Also, company management, supervisory and technical personnel will probably be assigned to special short-term roles, with their normal functions attended to by a stand-in. There may be a consequent reluctance to vacate a known situation for an uncertain one, coupled with problems to be encountered on the return to normal duty. In addition, some production personnel may be absent from the site altogether and warily return to the different behaviour of apparently familiar equipment at the end of the programme. For these reasons and those detailed above, the management of the whole programme presents unusual requirements and must be managed with great care. The following organization steps should be considered:

(1) Assign a director or senior manager to act as the company's strategic focus of fears, problems and the wider picture for his peers and shutdown management. His role should include contact with statutory bodies, such as the fire brigade, the Department of Energy and the DTI, access to relevant regulations, company safety department, and shutdown management. He should have senior management presence for programme shutdown and start-up, monitoring of handover and handback processes, and the recipient of progress reporting and job options from site.

> *The role and duties of the shutdown director should be clearly defined prior to planning and preparation.*

In addition, the shutdown director will define the shutdown organization to be employed for its likely duration, appointment of officers and programme management procedures.

(2) Appoint a shutdown controller. The shutdown controller will conduct tactical management of the programme on-site, managing the efforts of team leaders, supervisors and engineers, and reporting to the off-site shutdown director.

Both the shutdown director and the shutdown controller should be involved in the limited objective setting of the overall programme.

A question of timing

The adoption of the three job categories introduces a timing question. Job category (A) would normally define the duration of the programme, and during planning category (B) jobs would be chosen to be completed prior to the finish of category (A). Category (C) jobs would only be chosen if they did not affect the (A) or (B) schedules. However, sometimes things go wrong, and in the *presence of very strong pressure to return to production operations* the programme may be delayed by a category (B) or (C) job. For this reason, some additional information is very useful.

(1) For all category (B) and (C) jobs: *define the consequences of them not being done* – particularly the likelihood of causing subsequent downtime.

(2.) For these jobs, estimate the *abandon and return to production time*. High figures should identify the greatest overrun and should have no noticeable relationship to the overall duration.

5.9 Data recording

The question marks attached to any computer system and its installation refer to a host of topics, such as the hardware specification, the selection or design of software and the steadily changing requirements of different users. Such topics tend to quieten down as the system is established, and initial problems are ironed out as general understanding improves.

Data, however, is another matter. It is, firstly, the most likely requirement of a new system to be overlooked, and unlike many other questions it is a potential problem present throughout the life of the system. We should distinguish between data recording as part of system installation and data recording in the chain of normal work. It is the latter that concerns us here, and whether data is entered on cards or input through a computer keyboard, all maintenance management systems require job data for the update of working records, the extension of equipment history, the preparation of job costs and the provision of a basis for further action. There is a clear succession of steps, starting with data input, which (taken with other information) is interpreted to supply material for decision-making and subsequent action.

Careful data gathering and input is regrettably often regarded as a chore and delegated to the most junior member of the team. Inaccurate or incomplete data may have little apparent effect, affecting future jobs only after a long delay and reinforcing the temptation to do the data input task badly.

(1) Poor and inaccurate equipment histories mean incomplete job knowledge and data gathering before a new job can start.

   Data gathering has to be done somewhere.

(2) Poor job knowledge means lessons are not learned, previous experience has no value, extra tasks have to be done and work is inefficient.

   Poor information means higher job cost.

(3) Inefficient and extended work times demand more technical resources and mean that

   other jobs do not get done.

(4) Poor data means poor job knowledge, job times are extended, machine availability is diminished and equipment and production downtime are increased.

   Company production and hence revenues go down.

(5) As work becomes less efficient, more hours are spent on repairs and opportunities for work improvement are reduced.

   Valuable lessons and improvements are lost.

(6) Without effective data input and resulting knowledge,

expensive human resources have to be increased.

(7) Data gathering, input and processing are the vital but frequently forgotten

Lifeblood of the computer system.

(8) Input is conducted by system users, not computer specialists

User motivation lies elsewhere.

(9) Quality of data input work varies markedly with the effort applied.

Input quality varies with the individual.

(10) The retraining, transfer or promotion of personnel quickly affects work quality.

Repeated training improves task performance.

(11) Personnel who do this and other work well are more likely to be promoted.

New staff need immediate training.

Some data errors can be anticipated and checks included in software to identify and prevent them, such as days in a month greater than 31 or acceptable numbers in a defined range. The use of such checks, however, can be counter-productive: they slow down the data input process and confuse the system operator, and are often seen as an impediment to be circumvented as quickly as possible. Similarly, the data fields themselves may not be understood by an inexperienced user, adding reluctance to use the system. This can lead to annoyance and the introduction of fictional data merely to bypass the uncertainty and get the job finished. For example, some work order jobs can justify several dates:

Date of defect occurrence
Date of work start
Date of completion
Date of return to operation
Date of data input

The resulting confusion for the unskilled is certain, and perhaps worse is the discovery half-way through that the early inputs were wrong.

Unchecked data is a liability while data checks can induce hostility and error.

Types of data error

(1) *Errors of technique*
Errors of method or technique, misuse of the system, entering correct data in the wrong field, using fictional inputs to bypass checks.

(2) *Incorrect primary data*
Use of wrong tag numbers, discipline hours, maintenance code numbers, data times etc.

(3) *Errors of interpretation*
Omitting significant facts, assigning incorrect codes, fictional reporting, faulty interpretation.

(4) *Errors of logic*
Assumed causes and consequences, failure to record contributory facts or incidents, underlining of unconnected events[24].

(5) *False input*
On rare occasions, incorrect records can be introduced to be called upon later to justify an action or support a decision. Alternatively, incorrect records can be used to disguise or divert attention from an error of judgement, operational blunder, or failure by personnel. The tragedy is that the falsehood can lead to subsequent failures that could have been prevented.

The failings listed above can be attributed to many aspects of systems failure, but poor data input is the most insidious and pervasive. Occasional mistakes can be forgiven and rectified, but poor standards cannot. Clear requirements with constant audits and reminders are the best approach[25].

5.10 Supervisor's information[26]

The classical requirements of good supervision have been known for many years and apply in widely different fields, maintenance included. These are usually presented as:

Planning
Decision-making
Delegating
Coordinating
Controlling
Organizing

All are exercised by effective *communication*. Apart from observing that these headings represent supervisory similarity applicable in most fields, there is little need to expand, as they are well treated elsewhere. However, we should look harder at supervisory differences in any attempt to improve our understanding of the maintenance supervisor's role and situation.

The first item to note is that the classical model assumes two things:

(1) Tasks required of the group are known in advance.

---

24. Item (3) is most difficult, as without subsequent knowledge and examination the correct interpretation may not be known. Perhaps the greater danger is to assign a certainty of observation or interpretation where certainty does not exist.

25. Planed or PPM work using previously input work orders usually has the advantage of being computer-generated by most maintenance systems.

26. This section is not a treatise on maintenance supervision, but is intended as a reminder of actions already endorsed as good practice. If on reading this section it becomes clear that your present methods are wanting, then take preventative action at once. Men, machines and output targets could be in danger.

(2) Information and decisions flow down the command chain to the point of action.

Thus good supervision is proactive and good supervisors hold the initiative. Sadly, most would recognize these conditions as ideal, applying (particularly in the maintenance field) less often than most. Item (1) returns us to the maintenance preoccupation with the balance between *planned* and *unplanned* work (see Section 13.3), and many maintenance systems monitor this effect as a work or planning ratio, where 70–80% planned to total work is regarded as ideal. The imperative nature of many maintenance tasks can regrettably turn maintenance work programmes into scrap paper, destroying supervisory opportunity and reducing decision-making to action by default. This brings us to the second point: because of the nature of maintenance work, information about failures or requests for work often reach the maintenance team before the supervisor, diminishing his influence on workplace activity and frequently requiring him to react to situations rather than initiate them.

The supervisor's need for information therefore varies with planned and unplanned tasks. In the former case, the disciplines applicable and the work required, together with the exact definitions of affected equipment materials required, shutdown and isolation procedures etc. are included in the task definition issued through the planning and work order modules. In the unplanned case, no such luxuries apply. Tasks may arise from equipment breakdown, failure to start, operational damage, adverse local reports, gas leak or fire alert in the region, or safety officer's instruction. Reports are often incomplete, inexact, contradictory, confusing and apply to the wrong item of equipment. Decisions taken, rely on the rapid interpretation of reported symptoms, and need a wide knowledge of the environmental status and of any secondary effects.

Even though the need for rapidity is recognized, it understates the compelling pressure on maintenance decisions and actions when critical failures introduce production downtime and pressure to restart from every quarter.

Referring to the maintenance box job classification (Section 2.3), these comments apply mainly to category B2 ('Timing unknown, Content unknown'), and it should be clear that under such circumstances the need for good supervision is paramount and most difficult. The supervisor's tasks, in addition to his classical role, will include:

(1) Gather and interpret symptoms of failure.
(2) Clearly identify affected items and equipment criticality.
(3) Determine the effect on production output and associated equipment and processes, assigning work priority and working with the safety officer.
(4) Assess risk and special safety equipment, life support, relevant dangers, protective actions, work programme and all steps and effects needed.
(5) Identify disciplines and assign work details.
(6) Update the work programme sequence work to be done, from a range of competing priorities.

(7) Estimate manhours required, task duration, start and finish times, production downtime and restart time.

(8) Initiate check of materials, tools and special skills required to assess availability, and acquire where needed.

(9) Identify special equipment shut-off and isolation requirements, integrate with work programme. (Refer to Section 9.3).

(10) Institute a work programme, selecting the start time and other milestones. Inform participants.

(11) Mobilize incoming personnel. Assign to work programme; book transport and accommodation where necessary.

(12) Inform transportation, safety and production officers and all on-site supervisors affected.

(13) Assign a work coordinator to focus activities of work participants and report to the shutdown controller.

(14) Establish and promote clear work targets and objectives.

(15) Discuss with operations the completion time targets, support personnel attendance or absence required, handback procedures and secondary effects.

(16) Identify isolation steps, sequence and shut-off requirements for equipment under repair and associated secondary items or processes.

(17) Specifically advise operational management, supervisors and site control rooms of work programmes in advance, of personnel presence on site, and of shutdown and isolation provisions. Repeat this step repeatedly during the work programme, particularly when fresh activities are instigated and personnel are exposed to hazards.

(18) Check all requirements of the permit to work procedure and raise permits when requisite preparatory actions are complete. (Refer to Section 11.9).

(19) Ensure that interpersonnel communications are established and fully understood by all participating and affected personnel.

(20) Thoroughly brief all team and affected personnel; ensure that work steps, actions and sequences are fully understood. Underline targets and objectives together with the influences of different actions, possible or likely effects, all hidden or visible dangers, and responses to emergencies.

(21) Check that the safety group has specifically advised or reminded all work participants, particularly visiting personnel, of company safety requirements and procedures.

(22) Research equipment history and knowledge of similar or related work, together with the currency of work and technical information to be used.

(23) Ensure that work data is properly collected and entered to the work order system, both during and after task completion.

It should be clear from this list that the maintenance supervisor's role is central to a wide variety of activities, and his relationships with peers and staff in all groups are vital.

Chapter 6

Professional organizations

Many organizations offer services to operations and maintenance departments of different businesses and industries. The following are some of those to be approached when searching for further information.

USA

American Association of Engineering
Societies
1111 19th Street, NW, Suite 608
Washington DC 20036
USA
Tel: 1 202 296 2237
Fax: 1 202 296 1151

American Institute of Mining,
Metallurgical and Petroleum Engineers
(AIME)
345 East 47th Street
New York NY 10017
USA
Tel: 1 212 705 7695
Fax: 1 212 371 9622

American National Standards Institute
11 West 42nd Street
New York NY 10036
USA
Tel: 1 212 642 4900
Fax: 1 212 398 0023

American Petroleum Institute (API)
1200 L Street Northwest
Washington DC 20005–4070
USA
Tel: 1 202 682 8000

American Society of Civil Engineers
345 East 47th Street
New York NY 10017
USA
Tel: 1 212 705 7010
Fax: 1 212 705 7712

American Society of Heating,
Refrigeration and Air-Conditioning
Engineers
1791 Tullie Circle NE
Atlanta GA 30329
USA
Tel: 1 404 636 8400
Fax: 1 404 321 5478

American Society of Mechanical
Engineers
22 Law Dr, PO Box 2900
Fairfield NJ 07007–2900
USA
Tel: 1 201 882 1167
Fax: 1 201 882 5155

American Society of Petroleum
Operating Engineers
PO Box 6174
Arlington VA 22206
USA

Association for Manufacturing
Technology
7901 Westpark Drive
McLean VA 22102
USA
Tel: 1 703 893 2900
Fax: 1 703 893 1151

Association of Energy Engineers
4025 Pleasantdale Road, Suite 420
Atlanta GA 30340
USA
Tel: 1 770 447 5083
Fax: 1 770 446 3969

Institute of Electrical and Electronics
Engineers (IEEE)
445 Hoes Lane, PO Box 1331
Piscataway New Jersey 08855 1331
USA
Tel: 1 908 562 3800
Fax: 1 908 562 1571

Institute of Gas Technology
1700 S Mt Prospect Rd
Des Plaines IL 60018
USA
Tel: 1 847 768 0500
Fax: 1 847 768 0501

International Gas Turbine Institute
PO Box 422029
Atlanta GA 30342
USA
Tel: 1 404 847 0072
Fax: 1 404 847 0151

Marine Technology Society
1828 L Street NW, Suite 906
Washington DC 20036
USA
Tel: 1 202 775 5966
Fax: 1 202 429 9417

Society of Manufacturing Engineers
1 SME Drive, PO Box 930
Dearborn MI 48121–0930
USA
Tel: 1 313 271 1500

Society of Naval Architects and
Marine Engineers
601 Pavonia Avenue
Jersey City NJ 07306
USA
Tel: 1 201 798 4800
Fax: 1 201 798 4975

Society of Petroleum Engineers (SPE)
PO Box 833836, 222 Palisades Creek
Drive
Richardson TX 75803–3836
USA
Tel: 1 214 952 9393
Fax: 1 214 952 9435

American Chemical Society
1155, 16th Street NW
Washington DC 20036
USA
Tel: 1 202 872 4600
Fax: 1 202 872 4615

Department of Energy (DOE)
Headquarters, Forrestal Building
1000 Independence Avenue SW
Washington DC 20585
USA
Tel: 1 202 586 5575
Fax: 1 202 586 3045

Department of the Interior
Minerals Management Service
1849 C Street NW
Washington DC 20240
USA
Tel: 1 202 208 3985
Fax: 1 202 208 3968

Gas Industry Standards Board
1100 Louisiana, Suite 4925
Houston, TX 77002
USA
Tel: 1 713 757 4175
Fax: 1 713 757 2491

National Association of
Environmental Professionals
5165 Macarthur Blvd NW

Washington DC 20016–3315
USA
Tel: 1 202 966 1500

National Fire Protection Association
1 Butterymarch Park
Quincy MA 02269–9101
USA
Tel: 1 617 770 3000
Fax: 1 617 770 0700

National Institute of Standards &
Technology (NIST)
Administration Bldg A903
Gaithersburg MD 20899–0001
USA
Tel: 1 301 975 3058

NIST Boulder, Colorado Laboratories
325 Broadway
Boulder CO 80303
Tel: 1 303 497 3017
Fax: 1 303 497 3371

Germany

Berliner Energie-Agentur
Energy Agency of Berlin
Rudower Chaussee 5
D-12489 Berlin
Germany
Tel: 49 30 695495 0
Fax: 49 30 695495 99

Det Norske Veritas
Neue Große Bergstr 18
D-22767 Hamburg
Germany
Tel: 49 40 389 1600
Fax: 49 40 389 43045

Germanischer Lloyd
Vorsetzen 32
PO Box 111606
D-20459 Hamburg
Germany
Tel: 49 40 361 490
Fax: 49 40 361 49250

Gesellschaft Deutscher Chemiker
German Chemical Society
Postfach 90 04 40
D-60444 Frankfurt am Main
Germany

Lloyd's Register
Monckebergstr 27
D-20095 Hamburg
Germany
Tel: 49 40 328 1070
Fax: 49 40 335 710

Uniti Bundesverband Mitteiständischer
Mineralolunternehmen
(National Association of Mineral Oil
Enterprises)
Buchtstrasse 10
22087 Hamburg
Germany
Tel: 49 40 227 0030
Fax: 49 40 227 00338

VDMA – Verband Deutscher
Maschinen-und Anlagenbau eV
Postfach 71 08 64
60498 Frankfurt

Germany
Tel: 49 69 660 30
Fax: 49 69 660 31511

Verband für Schiffbau und
Meerestechnik
(German Shipbuilding & Oceanic
Industries Association)
An Der Alster 1, D-20099 Hamburg
Germany
Tel: 49 40 246 205
Fax: 49 40 246 287

Wirtschaftsverband Erdol und
Erdgasgewinnung
(Association of Crude Oil & Gas
Producers)
Bruhlstraße 9
30169 Hannover
Germany
Tel: 49 511 131 9555
Fax: 49 511 131 6739

United Kingdom

Association of British Offshore
Industries
4th Floor, 30 Great Guildford Street
London SE1 0HS
UK
Tel: 44 171 928 9199
Fax: 44 171 928 6599

British Constructional Steelwork
Association
4 Whitehall Court
Westminster
London SW1A 2ES
UK
Tel: 44 171 839 8566
Fax: 44 171 976 1634

British Marine Industries
Federation
Meadlake Place, Thorpe Lea Road
Egham Surrey TW20 8HE
UK
Tel: 44 1784 473377
Fax: 44 1784 439678

British Standards Institute
389 Chiswick High Road
Chiswick
London W4 4AL
UK
Tel: 44 181 996 9000
Fax: 44 181 996 7400

Bureau Veritas
Capital House, 42 Weston Street
London SE1 3QL
UK
Tel: 44 171 403 6266
Fax: 44 171 403 1590

Lloyd's Register of Shipping
100 Leadenhall Street
London EC3A 3BP
UK
Tel: 44 171 709 9166
Fax: 44 171 488 4796

Marine Accident Investigation Branch
5/7 Brunswick Place
Southampton
Hants SO15 2AN
UK

Norwegian Trade Council
Charles House, 5–11 Lower Regent
Street
London SW1Y 4LR
UK
Tel: 44 171 973 0188
Fax: 44 171 973 0189

Offshore Certification Bureau
61 Southwark Street
London SE1 1SA
UK
Tel: 44 171 620 0802
Fax: 44 171 261 9497

Ports Safety Organisation
Room 220, Africa House
64–78 Kingsway
London WC2B 6AH
UK
Tel: 44 171 242 3538
Fax: 44 171 404 6806

Pressure Gauge & Dial Thermometer
Association of Europe
136 Hagley Road
Edgbaston
Birmingham B16 9PN
UK
Tel: 44 121 454 4141
Fax: 44 121 454 4949

Sira Test and Certification
Saighton Lane
Saighton
Chester
Cheshire CH3 6EG
UK
Tel: 44 1244 442200
Fax: 44 1244 332112

United Kingdom Offshore Operators
Association (UKOOA)
3 Hans Crescent
London SW1X 0LN
UK
Tel: 44 171 589 5255
Fax: 44 171 589 8961

United Kingdom Onshore Operators
Group (UKOOG)
63 Duke Street
London W1M 5DH
UK
Tel: 44 171 355 3393
Fax: 44 171 355 3704

University of Dundee Centre for
Petroleum & Mineral Law and Policy
Park Place
Dundee DD1 4HN
Scotland
UK
Tel: 44 1382 344300
Fax: 44 1382 322578

Department of Transport, Marine
Safety Agency
Spring Place, 105 Commercial Road
Southampton
Hampshire SO15 1EG
UK
Tel: 44 1703 329100
Fax: 44 1703 329298

Det Norske Veritas Classification AS
Palace House
3 Cathedral Street
London SE1 9DE
UK
Tel: 44 171 357 6080
Fax: 44 171 357 6048

Energy Industries Council
Newcombe House
45 Notting Hill Gate
London W11 3LQ
UK
Tel: 44 171 221 2043
Fax: 44 171 221 8813

Engineering Council
10 Maltravers Street
London WC2R 3ER
UK
Tel: 44 171 240 7891
Fax: 44 171 379 5586

Health and Safety Executive
Rose Court
2 Southwark Bridge
London SE1 9HS
UK
Tel: 44 171 717 6000
Fax: 44 171 717 6907

Meteorological Office
Offshore & Marine Services
Seaforth Centre
Lime Street
Aberdeen AB11 5FJ
Scotland
UK
Tel: 44 1224 211840
Fax: 44 1224 210575

National Corrosion Service
Queens Road
Teddington
Middlesex TW11 0LW
UK
Tel: 44 181 977 3222
Fax: 44 181 943 2989

North Sea Medical Centre
3 Lowestoft Road
Gorleston
Great Yarmouth
Norfolk NR31 6SG
UK
Tel: 44 1493 600011
Fax: 44 1493 656253

Institute of Oceanographic Sciences
Marine Information & Advisory
Services
Deacon Laboratory
Wormley
Godalming
Surrey GU8 5UB
UK
Tel: 44 1702 595000
Fax: 44 1703 6833066

Institute of Petroleum
61 New Cavendish Street
London W1M 8AR
UK
Tel: 44 1017 467 7100
Fax: 44 1017 255 1472

Institute of Physics
76–78 Portland Place
London W1N 4AA
UK
Tel: 44 171 470 4800
Fax: 44 171 470 4848

Institution of Chemical Engineers
165–189 Railway Terrace
Rugby CV21 3HQ
UK
Tel: 44 1788 578214
Fax: 44 1788 560833

Institution of Civil Engineers
1 Great George Street
London SW1P 3AA
UK
Tel: 44 171 222 7722
Fax: 44 171 222 7500

Institution of Electrical Engineers
Savoy Place
London WC2R 0BL
UK
Tel: 44 171 240 1871
Fax: 44 171 240 7735

Institution of Electronics & Electrical
Incorporated Engineers
Savoy Hill House
Savoy Hill
London WC2R 0BS
UK
Tel: 44 171 836 3357
Fax: 44 171 497 9006

Institution of Fire Engineers
148 New Walk
Leicester
Leicestershire LE1 7QB

UK
Tel: 44 116 255 3654
Fax: 44 116 247 1231

Institution of Gas Engineers
21 Portland Place
London W1N 3AF
UK
Tel: 44 171 636 6603
Fax: 44 171 636 6602

Institution of Mechanical Engineers
1 Birdcage Walk
London SW1H 9JJ
UK
Tel: 44 171 222 7899
Fax: 44 171 222 4557

Institution of Mechanical Incorporated
Engineers
3 Birdcage Walk
London SW1H 9JN
UK
Tel: 44 171 799 1808
Fax: 44 171 799 2243

Institution of Mining & Metallurgy
44 Portland Place
London W1N 4BR
UK
Tel: 44 171 580 3802
Fax: 44 171 436 5388

International Institute of Risk and
Safety Management
National Safety Centre
Chancellor's Road
London W6 9RS
UK
Tel: 44 181 741 1231
Fax: 44 181 741 4555

Chapter 7

Maintenance and finance

One of the attractive features of maintenance work is its tangible nature. Once a machine has been stripped down, there may be arguments about causes but fewer about what needs to be done. Also, when maintenance work is finished the machine can be given a new lease of life: the transformation is both apparent and satisfying.

Regrettably, these attractions are less evident with figurative or abstract work. Cost and financial affairs suffer from this aspect, and when we work in this area satisfaction comes from sound preparation, careful data processing and from the numbers themselves. Practical people often groan when they are obliged to concentrate their efforts on work of this kind and it is tempting to skimp or rush to get on with something 'more important'!

In almost all companies, maintenance work will be identified by its own section of company financial affairs, and we concentrate below on those matters which are direct and limited maintenance effects, or, as in the case of budgets (Section 7.3) refer to an area of duty facing the maintenance manager in respect of departmental management.

Purpose

There are a variety of methods and philosophies in use, and the purpose here is to discuss the arguments and effects of various features and ideas, without attempting to duplicate or replace formal financial control systems. The main themes are *labour cost estimating* and *labour cost monitoring*, which are the key techniques available to keep track of principal maintenance costs and provide useful links to progress calculations against a plan.

Methods described constitute a form of 'ready reckoning' based directly on maintenance work orders intending to allow rapid response to identifiable trends or changes and giving maintenance supervision more immediate knowledge of costs.

There is one particular difficulty regarding the treatment of planned and unplanned work: as the latter is more difficult to forecast, some cost and progress monitoring techniques refer to planned work only, particularly in manufacturing, making comparison with planned figures more meaningful. However, unplanned work is a part of maintenance, and the following notes include an unplanned allowance in the early forecasts to keep overall calculated costs closer to the budget and progress achieved closer to the total maintenance work being done (see Section 7.3).

Questions of over- and under-maintenance abound in maintenance affairs, and the existence of the two states indicates the third or 'optimum' opportunity. Making the right decisions to achieve optimum cost performance is only one driving force at work when choices are made. The results of such choices often appear as unplanned work, a usually more expensive feature of 'economic risk', defined in BS4778, Section 3.1, as: 'The risk of financial loss associated with a product, system or plant due to potential hazards causing loss of production, or other financial consequence'.

However the collection of prime cost information is used in the company financial management system. It is used here in support of direct maintenance for these additional reasons:

(1) To allow comparison with forecast costs by individual job, groups of jobs or overall.
(2) To provide an aid to decision-making, perhaps where limited funds are best applied, or evaluating the impact on costs of different actions or methods[1].

The chapter cannot and must not be regarded as a treatise on financial management, which is accurately and comprehensively presented elsewhere by the accountancy profession. Within the text, the term financial management is intended to refer to the formal treatment of company finances rather than the narrower matters faced by the maintenance department itself.

Although it is necessary to clarify these distinctions and to underline what is attempted, there are some qualities which apply to financial management systems in general which are worth repeating.

(1) In most organizations, the measurement of profit and sales and the principal measures of performance are prepared in financial terms.
(2) Financial measures particularly present a uniform means of reporting and comparing the wide and disparate range of company activities.
(3) Use of financial measurements and comparisons presents the manager with tasks of reporting, assessing and communicating.
(4) Financial reports will present the manager with problems and opportunities, both including direct reference to departmental work.

Each of these influences the business environment in which the teams do their work and provides a means of motivation, control and understanding.

7.1 Work order costs

The basic work unit of most maintenance work management systems is the 'work order', which acts as the focus for the maintenance team's actions in carrying out each specified piece of work. The work order system plus the processing and requirements of each work order are described more fully in Section 5.4. In this chapter

---

1. Item 2 surfaces in more detail in Section 7.4, when we consider optimum cost.

we turn again to the work order as the vehicle for collating and quantifying maintenance costs. It is assumed here that the maintenance management computer system includes the features described here, although regrettably this is not always the case. Although these mechanisms are straightforward, maintenance management may work in parallel with a separate financial management system, where cost information is gathered through time-sheets, purchases and stock issues, collecting them under broader categories than the individual work order. For the maintenance manager, identification of cost to each work order is more direct, but can produce a greater number of cost elements than the company-wide system presents.

The measure of cost described below is:

$$\text{Prime cost} = \text{Labour cost} + \text{Material cost}$$

where both labour and material apply to the requirements of the work order itself. If we examine the 'work order closeout – defect and PPM' subsection in Section 5.4 (p. 119), the following items are noted for entry during work order closeout:

(1) Man-hours used for each discipline.
(2) Parts and materials used (by stock numbers and quantities).

For comparing different methods of work, where different local cost effects lead to decisions about maintenance work, the prime cost is very useful, and it should be possible to refer to the work order records for this information.

The tables that follow in Section 7.2 refer to labour costs, as this is the area where maintenance management has more scope and can have more effect. When full cost implications are being considered, however (perhaps replacement alternatives are compared or budget targets revisited), the whole range of cost effects must be included, with the labour figures and the entire package treated under a formal framework[2].

(1) Costs of external services, such as laboratory analysis, drawing preparation and clerical support.
(2) Contractor costs, for work charged directly to the job such as specialist technical services, on-site refurbishment and third party inspection.
(3) General on-site contractor costs: work in overall support such as cleaning, blasting, painting and descaling.
(4) Transport costs: special shipping and handling costs are sometimes assigned to the job itself.
(5) Off-site repair costs: cost demarcation is sometimes confusing, with site removal and replacement costs borne by maintenance, while refurbishment costs are charged to stock. Repair costs applicable to an item returned to the site for reinstallation is usually regarded as a job cost.
(6) Internal cost transfer: other departments, such as the drawing office or the computer department, may charge internally for their services.

---

2. At the time of normal work order closeout it is often convenient to use a check flag in the computer record giving 'yes' or 'no' to the question 'will subcontractor invoices apply to this work order?' In the event of a 'yes' check, the work order can be retained in the history, awaiting further data input. Any work order cost summary can indicate those work orders which await the input of contractor data.

TABLE 7.1

Discipline	Man-hours used	Applicable Labour rate	Discipline cost
Mechanical	20	12	240
Electrician	6	14	84
Instrument technician	8	14	112
Computer technician	0	15	0
Work order labour cost (in-house)			436

This is the point where the fundamentals are trapped, leading to an automatic conversion into prime cost.

Labour costs

The work order closeout routine requires the man-hours used for each discipline, so the cost figure recorded is the sum of each discipline figure extended by the applicable labour rate. Table 7.1[3] shows an example.

A question of rate changes

Because prime costs are presented as financial figures, the labour man-hours required for a job are not readily apparent. This is particularly annoying in maintenance affairs when comparisons are to be made between repeated but identical jobs carried out when different labour rates apply. It is sometimes the case that improvements in work performance are smaller than the increase caused by a changing rate, which leads to an apparent deterioration in performance, which is not the case.

This is more serious than it would appear. Small improvements can refer to large sums of money and yet be disregarded because they are invisible. Similarly, improvements which could be extended to wide use can be reversed or abandoned for the same reason. The answer is to recall this problem when making comparisons where small variations in prime cost are evident, and refer to the manhours figure on the work order record.

Material costs

When materials or spare parts are issued to a specified work order the transaction will be held in the materials management system as a stock issue. Many major systems, particularly ERP systems (enterprise resource planning) integrate materials management and maintenance activities which can allow such costs to be added to the material cost field attached to the correct work order. In cases where there is no such integration, the material cost can be taken from the appropriate stock

3. Not all disciplines are represented here and the rates shown are fictional. The labour rates used will be held as accessible constants in the computer system and referred to during the calculation. The values used are probably input by the maintenance team, but they should be agreed with the accounts department.

issue report and re-entered to the materials cost field in the work order. It is essential to formalize such a process, as more than one material issue may apply to the work order and there is a danger that they can be issued twice or overlooked altogether. Because the maintenance and materials operations are usually separate, there is scope here for error or misunderstanding. If the maintenance group requires software changes it must make its requirements clear.

There is also the matter of bulk supplies, which will apply to many maintenance jobs but not be the subject of a separate material issue. Fortunately, materials such as oil, grease and cleaning fluids apply mainly to routine PPMs, and their assignment can be predicted. For specific groups of high-frequency work orders this is the one area that a material allocation can be justified. Remember that although the amounts used on the specified PPMs can be predicted, the amounts between the various jobs will be different.

Contractor costs

When an external contractor is used the maintenance manager will be unaware of the costs and their make-up unless

(1) He approves (or at least sees) the contractor's invoice.
(2) The invoice breaks down into labour and material components.

When a contractor's invoice is approved by the maintenance manager it is usually after the work has been completed. The work order will be completed in work order history and be shown as closed. The contractor's invoice figures in their two components of labour and material, using access to the work order, can then be added to, or changed in, the work order record by entry through the maintenance manager's keyboard.

It is sometimes argued that from a maintenance point of view the definition of prime cost should be labour + materials + contractor's charges, with the opportunity to input the latter without breaking them down. The company accounts department would almost certainly have a view on this, but the author recommends against it. Material costs will be open to pressure from the company purchasing group and the labour charges can be subject to comparison with alternative 'in-house' charges (see Section 5.5), and such scrutiny cannot be retained if a single figure only is used.

It is possible that more than one contractor is used, leading to invoices from different companies which will arrive at different times and all require the treatment outlined above. When the facility to input contractor charges is a feature of the computer system the options to remain open or finally close the work order cost record should be part of the maintenance manager's input routine.

7.2 Maintenance cost targets

One of the features of budget cost preparation is the estimate of planned costs at different times during the year according to forecasts based on the maintenance plan. In addition, overall costs of unplanned work are estimated using previous history, and these costs can be assigned as an allowance to different periods of the year.

This means that total estimated costs can simply be plotted against the year's time base as a curve of forecast costs. The total cost figure used[4] must agree with all jobs in the maintenance plan plus the estimate for unplanned work.

Total expected cost = Estimated costs of planned jobs + Costs of
estimated allowance for unplanned jobs

Useful as this information is, it is even more effective when presented in graphical form. At the time of budget preparation forecasts are prepared, and a task that faces the maintenance manager is the monitoring and reporting of performance against these forecasts for the year. Because of the wide variety of maintenance jobs and cost-bearing effects, the financial presentation is common to them all and is the most useful means of comparison.

The tables and graphs that follow look at the financial effects of all maintenance work conducted over the full year. It is clear that the maintenance manager will look in more detail in many areas, but the initial consideration is how to keep a general understanding of the wide scope and variety which are characteristic of maintenance work. The figures in Table 7.2 are for a mythical maintenance year, *for a single mechanical discipline labour only.*

Figures are assigned to 13 equal four-weekly periods, each representing two two-week work tours[5]. The numbers are deliberately kept simple and do not relate to any real installation; their purpose is to illustrate the method. Man-hours included in planned jobs in column 2 reflect reduced workloads at mid-year and at the beginning and end of the year, which is the structure of the plan. The estimate for unplanned work is approximately 21% of the total and is amortized evenly except at the year end, where the load on machinery is expected to be lower. The even nature of the unplanned estimates assumes that unplanned work can arise at any time[6].

The total expected manhours shown in column 4 is the sum of columns 2 and 3. Column 6 shows the total estimated accumulated amount as it rises during the year. Notice that the final cumulative figure for work period 13 is the sum calculated for the whole year, shown as the column 4 sum.

Column 5, the total expected man-hour costs, is the product of column 4 with the £12 per hour labour rate (see Section 7.1), while column 7 is the cumulative costs for the whole year.

---

4. The total figure used in the budget may be slightly different if financial allowance is made for different options.

5. There are many terms used to describe this subject, and their meanings can easily overlap and cause confusion. The following have been used here:

Estimated costs = Number of manhours × Applicable labour rate
Expected costs = Estimated planned costs + Estimated unplanned costs
Forecast costs = All costs which are expected to apply to future work
Actual costs = Costs calculated from recorded job hours input to the system multiplied by the appropriate labour rate

6. When dealing with different disciplines, different ratios of planned to unplanned work apply. Control of safety equipment, maintained by instrumentation or safety personnel, respectively, are often over-maintained to ensure that unplanned breakdowns do not occur. In addition, some groups of equipment are less prone to unexpected failure than others.

TABLE 7.2 ESTIMATED PLANNED AND UNPLANNED MAN-HOURS FOR MECHANICAL DISCIPLINE – MAINTENANCE YEAR 1997

Maintenance period	Sum of jobs maintenance plan man-hours ($\times 100$)	Allowance for unplanned work ($\times 100$)	Total expected man-hours ($\times 100$)	Total expected man-hours costs ($\times 100$)	Cumulative total expected ($\times 100$)	Cumulative total expected costs ($\times 100$)
(1)	(2)	(3)	(4)	(5)	(6)	(7)
1	24	4	28	336	28	336
2	30	8	38	456	66	792
3	34	8	42	504	108	1296
4	28	8	36	432	144	1728
5	30	8	38	456	182	2184
6	38	8	46	552	228	2736
7	16	8	24	288	252	3024
8	32	8	40	480	292	3504
9	30	8	38	456	330	3960
10	30	8	38	456	368	4416
11	36	8	44	528	412	4944
12	32	8	40	480	452	5424
13	24	4	28	336	480	5760
	384	96	480	5760		

Similar tables would usually be prepared for each of the disciplines concerned, and a summary table produced for all disciplines together (Table 7.3). Although the figures would naturally be much higher the graphical presentation is the same as one for a single discipline.

The expected or target figures from columns 6 and 7 of Table 7.3 are plotted against the maintenance period time base.

In Figure 7.1 the forecast cumulative cost or target line is shown rising period by period through the year to reach a total of £12,494 × 100 at the end of period 13. The same figures may be expressed as a percentage of the total amount, and these figures are shown in column 7[1].

Actual costs

At the end of each working period the *actual cost incurred* is plotted on the same graph, with actual costs shown on a separate rising curve. The figure is calculated as *the actual labour cost sum* (labour hours × labour rate for each discipline, summed) of all maintenance job work orders closed out in the period.

---

7. The forecast cumulative cost line is shown as essentially linear and assumes that resources in the maintenance plan are all used and steadily assigned throughout. This is unlikely: practical effects will modify the shape of the line. Also, for network planned projects the line will follow the shape of an 'S', as opening stages of a project have to build up to maximum costs before tailing off as the work is completed and the costs assigned to the work decline.

**TABLE 7.3 TOTAL FORECAST MAN-HOUR COSTS –
MAINTENANCE YEAR 1997**

Maintenance period	Total expected mechanical cost (× 100)	Total expected electrical cost (× 100)	Total expected instrumentation cost (× 100)	Total expected cost, all disciplines (× 100)	Total expected cumulative cost (× 100)	Expected cumulative cost (× %)
(1)	(2)	(3)	(4)	(5)	(6)	(7)
1	336	294	98	728	728	5.83
2	456	392	140	988	1716	13.73
3	504	448	154	1106	2822	22.59
4	432	378	126	936	3758	30.08
5	456	392	140	988	4746	37.98
6	552	490	154	1196	5942	47.56
7	288	252	84	624	6566	52.55
8	480	420	140	1040	7606	60.88
9	456	392	140	988	8594	68.78
10	456	392	140	988	9582	76.69
11	528	462	154	1144	10726	85.85
12	480	420	140	1040	11766	94.17
13	336	294	98	728	12494	100

The use of Figure 7.1

The total cumulative cost curve shown in Figure 7.1 shows the total expected costs over the year and is mainly based on the maintenance plan. The actual cost curve is the plot of total costs which have occurred. Their relationship allows the maintenance manager to monitor the financial effects of the whole range of activity and take action where necessary. 'Actual' figures used in Fig. 7.1 are taken from column 4 of Table 13.3 page 320.

7.3 Budgets

Founded on a departmental organization, budgets, their preparation and control are a formal part of the company's financial management system, and the rules for the budgeting exercise will be set by the finance department.

Budgets are prepared to assign resources and responsibilities for a specified period, usually of one year, to the different sections or departments of the organization concerned. Expressed in financial terms, they are founded on future needs, and employ estimates of all ingredients of the work and its timing to be concluded within the budgeted period.

By carefully identifying the company's expectations, budgets act as targets for contributing departments, provide a basis for in-progress financial monitoring and are accompanied by a transfer of authority to the specified departments.

The maintenance budget is a key part of the company budget and is a supporting pillar of the maintenance management operation. Although preparation and use follow financial mechanisms, their content, scope, timing and estimated amounts derive directly from the maintenance teams' estimates of future events. Because accounting and general management are also part of maintenance budget preparation, it sometimes leads to the mistaken belief that they are also responsible for the forecast. So let us look again at the different contributing responsibilities to budget preparation.

(1) It is general management's responsibility to define the overall targets that the maintenance work programme will attempt to achieve. Here is the first point of departure. The 'automatic' maintenance response would be 'to maintain'; to keep, or perhaps improve the existing plant's functional capability and undertake the maintenance needs of new installations and old equipment removal.

This is true only in an unchanged business environment, whereas general management may require a change of stance. The site may be the subject of a major refurbishment, or perhaps plant items are to be mothballed or deliberately allowed to deteriorate. (See Sections 3.1, 3.2 and 3.7.) In any of these cases, the maintenance work programme will be affected. The company may also change the assignment of work, using more subcontractors and a

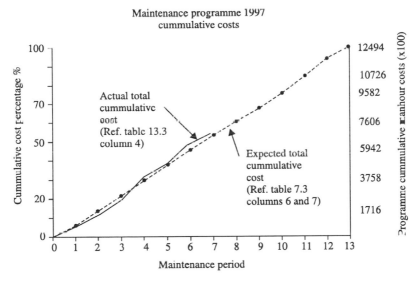

Figure 7.1
1 *The right hand 'y' axis shows cumulative costs based on column 6 of Table 7.3 page 152*
2 *The left hand 'y' axis shows cumulative costs expressed as a percentage based on column ? of Table 7.3*
3 *Cost figures are plotted against maintenance periods 1–13 ie column 1 of Table 7.3*
4 *The full curve with figures for all periods 1–13 is the 'expected'*
5 *Actual figures plotted on the same periods are shown in column 4 of Table 13.3 page 320*
6 *Not all figures from the table have been reproduced on the outline graph but they ???*

smaller in-house team (see Section 5.5), or it may, by switching from one contractor to another, change the way the work is done, its cost, timing or duration. Although aware of departments' preferences and submissions, it is the general management's responsibility to make their decisions and requirements clear.

(2) It is the accounting department's responsibility to apply the financial methods and mechanisms, ensuring that the effects of options and choices are highlighted while imposed financial limitations are met.

(3) It is the maintenance team's responsibility to forecast forthcoming work needed to meet the general objectives outlined in point 1 above, estimate the financial cost of each job and emphasize the needs and consequences of various options.

Preparation of the budget

Although the budget preparation exercise, apart from ongoing adjustments, takes place at certain times of the year, the maintenance plan (Section 5.3) will already exist. The one weakness of this rolling plan approach is that, as they progress from year to year, it is assumed that previously planned maintenance jobs are still required, whereas a definite change to the plan is needed to cater for changed circumstances. It is sometimes the preparation of the budget which crystallizes future changes, so while budget preparation depends on the plan, it will need to be interpreted during the budget preparation process.

When completed, the maintenance budget should make provision for action in respect of planned and unplanned work. The preparation of estimates, including an allowance for the unexpected, should still confirm the planned/unplanned ratio discussed in Section 13.3, but we should remember that, as the plan is changed during budget creation, this ratio may also change and a revised calculation will be necessary.

(1) Take a printout from the computer system of the maintenance plan which refers to the whole of the budget period and ensure that the periods match.

(2) Be sure that the man-hour resource total for each discipline is noted for each week of the plan and as a total for the complete budget period.

(3) Check that the current planned/unplanned ratio figure for the preceding twelve months (to match the budget period) is available[8].

(4) Take the maintenance plan printout and strike out all jobs where maintenance is no longer needed.

(5) Similarly, on separate sheets, prepare maintenance plan insertions for items

---

8. This ratio is based on historical information and will be different for each discipline; changes will have an effect. For example, the number of unplanned jobs rises as the plant gets older or drops if troublesome equipment is taken out of service. A check of different previous periods will show how volatile the ratio is. Some adjustments will be necessary.

to be newly installed and check items referred to in pre-use maintenance (Section 2.5) which may be absent from the plan. Adjust the discipline resource totals of the maintenance plan.

(6) Once all planned maintenance jobs have been identified, review the scope and nature of work planned. The duties of some machines may be changing or equipment may be the fresh subject of a replacement changeout programme for statutory reasons (see Section 8.9).

Modify the maintenance if necessary, adjusting the forecast resource figure for each discipline again.

(7) Separately, list those so-far-unspecified jobs which you would like to be done. Prioritize them according to their urgency and estimate resource manhours required for each job and each discipline. Put the list to one side.

(8) Taking the newly estimated planned/unplanned ratio (item 3 above) apply it to the forecast resource figure for each discipline, producing estimated planned and unplanned figures for each discipline. For example:

Discipline	Estimated planned man-hours	Estimated unplanned %	Estimated unplanned hours	Total estimated man-hours
Electricians	1100	18	198	1298
Instrument technicians	750	16	120	870
Mechanics	1460	21	307	1767

and similarly for all disciplines. These are the estimated figures for the unplanned work in total over the budget period. Note that the planned/unplanned ratio is used here as a percentage in column 3.

Return to the separate list of unspecified jobs (item 7 above), selecting those which would arise as unplanned work in the budget period and could be considered as already covered by the figure above.

Estimate man-hour figures for those of the remaining jobs which you decide to include and revise the unplanned figures.

Review the different maintenance tasks outlined above and adjust and sum the planned and unplanned man-hour figures for each discipline.

Factor each discipline's total man-hour figure by the applicable labour rate, aggregating into time period figures according to the requirements of the budget.

This process is summarized in Table 7.4. In addition to the steps listed, 'specials' which have been included should be identified in a footnote.

Materials

The material content of some jobs can swamp the prime cost, such as exchanging the rotor on a major electric generator, where the component cost is very high. Conversely, for other jobs, such as routine PPM inspections, the material cost is trivial.

TABLE 7.4

Step No.	Action	Applied to	Planned	Unplanned
1	Acquire	Maintenance plan printout	–	–
2	Check	Discipline man-hour figures available	–	–
3	Check	Current planned/unplanned ration available from work history	–	–
4	Add or delete	Add new jobs; remove jobs no longer needed	Adjusts planned figures	–
5	Add	Insert new items yet to be included	Adjusts planned figures	–
6	Review	Scope and nature of planned work	Adjusts planned figures	–
7	List	Prepare list of specials	–	–
8	Calculate	Applies planned/unplanned ratio to revise discipline man-hour figures	–	Produces estimates of unplanned manhours for each discipline
9	Identify	Form list of special jobs likely to be already included	–	No change
10	Select	Specials not covered but to be included	–	Revises unplanned figures
11	Summarize	Figures for each discipline	Adjusted planned totals	Adjusted unplanned totals
12	Aggregate	Factor each discipline: 1. Man-hours total by the applicable labour rate sum both planned 2. Unplanned figures 3. Aggregate into monthly or quarterly totals to suit the budget format		

For many jobs in the maintenance plan, material costs are modest and predictable, and the same PPM jobs previously completed and recorded in the work history should hold figures for the materials used, which, with adjustment for price changes, can be used as material cost estimates for the job concerned.

In this way, material estimates for planned jobs can be acquired, and as so many are repeated throughout the plan they can be summarized and added to the budget estimates.

Unplanned materials are more difficult to estimate because requirements are obviously unknown. Ironically, they may be easier to prepare by referring to previous totals in the accounts or work history and after adjustment for plant ageing and price change merely calculate the ratio against the unplanned resource figures and add the resulting estimate to budgeted materials.

It is the 'specials' which are the problem, and they require additional effort. Because there is so much potential variety the first option is to do it the long way. For each job included in the budget and featured on the specials list (item 7 above) the material necessities of the job have to be identified and costs prepared. However, there may be a way out. For each job concerned:

(1) Check work history for previous records of this job conducted on the same type of machine.
(2) If (1) fails, check for similarities in other applications who can provide a broad guide.
(3) Check other operating companies in similar businesses to your own who may employ the same machine.
(4) Check with the manufacturer, who may have estimates for known major jobs.

If all these fail, the first option remains.

Retained estimates

One of the options discussed in Section 13.3 is the comparison of grouped or individual job costs with the estimated costs included in the budget. For such work the estimates prepared for the budget need to be available in a form for ready comparison with actual prime costs, which will attach to each work order following closeout. The first place for such a comparison is the work order itself. Fields are provided in both PPM and defect work orders for the input of estimated manhours per discipline (see Section 5.4), and following creation of the maintenance budget these figures can be input directly through the keyboard as each work order is raised, or more efficiently held on the computer for automatic insertion to each planned work order as it is raised for issue.

Estimates for defect work orders cannot be treated this way, and they are usually input to the individual work order after it has been raised and the scope of the job is known.

However, the estimated unplanned manhour totals for each discipline are included in the budget and the actual prime costs attaching to each work order after closeout can be added to a rising total in the computer and compared with total or group estimates provided in the budget.

This allows the maintenance manager to monitor:

(1) Budget estimates with actual figures for PPM work orders.
(2) Aggregated totals for unplanned jobs to compare with allowed for totals in the budget.

In the latter case the maintenance manager will be aware of rising unplanned costs and can apply for increased funding before a breach of the budget figure takes place.

7.4 Optimum cost

Somewhere between doing nothing at all and repeating activity without further benefit is an optimum level of action where, for the maintenance man, work is effective without being overdone and the correct balance between under- and over-maintenance is achieved. Judgements are made about departmental optimum levels by different organs of the business following strategies to meet their delegated business objectives. Regrettably, such judgements do not necessarily consider the effects on the business as a whole, and cannot ensure that the enterprise operates at optimum levels. Stock control, for example, may use the number of stockouts in a selected time period as a measure of performance; similarly, production will measure output in barrels of oil produced or total volume of fluids processed, while a manufacturing enterprise would monitor the number of items produced and maintenance will measure plant 'availability' production 'downtime', or the number of equipment failures. There are a whole range of matters like these where the selection of individual departmental measures needs to be considered within the context of the business as a whole. Sometimes the over-rigorous pursuit of one target can make another impossible to achieve.

Active improvements in one area can make other areas worse.

As a further complication, the use of chosen techniques will vary. 'Run to destruction', for example would not be employed on an item of critical equipment, or a zero stock level would never be accepted for a long lead spare required for a critical safety system. A department like maintenance will use a combination of methods, whose use will vary with time or circumstance. Also, the attendant risk will change, the risk of failure will rise as planned maintenance frequencies decrease, the risk of stockouts will rise as the spare parts inventory falls, and the risk of product failure will rise as the design or processing complexity increases; with so many possible variations, selection of the most effective level is difficult.

Fortunately, the effects of any contributory strategy can be judged for their financial impact, and cost/risk evaluation is available to help clarify the various options. Not only is the approach becoming more widely understood as an effective method, it is also available as a range of commercial products which allow an enterprise to examine the effects of following any option. These are computer-based systems which can conduct numerical evaluation of cost and risk in key areas of business operation and explore the dependencies of specific variables on different sets of data and upon each other. Such methods consider the effects of changes, analysing different options and defining best and worst cases and optimum combinations.

Sources of information

One of the downfalls of some methods of analysis is the requirement for extensive and detailed information, which can only be gathered from the company's financial accounts, a process which is time-consuming and expensive, disrupting normal financial business and failing to enjoy the direct benefit of company officers' experience of regular operations. Conversely, and perhaps surprisingly, modern cost/risk analysis systems use estimated information only, garnered from the company participants in the evaluation exercise, to arrive at rapid and concise results without the interruption of normal data processing. Data is collected during analysis using ranged estimates, and although broad estimate bands are used, close results arise from the interaction of the variables considered.

Sensitivity analysis

Revealing evaluations are produced when optimum values of the analysis are compared one with another and when changes are made in the estimated data used to prepare the initial results. A large change in one or other of the data ranges used may produce a surprisingly small change in the optimum results achieved, revealing a low-sensitivity data source. Conversely, small changes in the data estimated for another parameter may produce large changes in the optimum value, indicating just the opposite. This technique is very useful. It uses the evaluation methods themselves to identify where extra data accuracy is necessary and where wide approximations may be used.

Present methods available fall into three main areas of optimum analysis:

(1) *Projects and changes*
   Cost/risk evaluation can be successfully employed to analyse project ideas or proposed changes. Uneconomic alternatives can be identified and removed, and the direct benefits can be clarified and demonstrated. In addition, less tangible costs or advantages can be considered and their effects included, such as external constraints, legal obligations, environmental results and safety implications. Use of this approach can help rank project proposals in order of benefit and demonstrate those which are not worthwhile. Evaluations can be prepared to:
   (a) Rank projects from the best and those not worthwhile.
   (b) Identify which data is necessary for the evaluation.
   (c) Calculate the costs and effects of intangible effects (above).
   (d) Engage company experience and skill to make choices in an efficient manner.
   (e) Clarify and present options considered and choices made.
   (f) Record assumptions made and monitor results.

(2) *Holding spares*
   Deciding which spares to hold and how many of them is difficult to judge using usual management tools, especially for an almost new machine where little operational experience has been accumulated or for the required levels of slow-moving 'insurance' spares. The question is made more uncertain because it refers to improbable events (low probability). Cost/risk evaluation can identify those areas where stocks are too low and corresponding areas where stock levels can be safely reduced. This feature of optimum analysis can:
   (a) Calculate requisite and optimum spares, determining which to hold and how many.
   (b) Consider the effects of spares unavailability on downtime costs and additional machine damage incurred through overrunning.
   (c) Examine the viability of the present stockholding strategies employed.

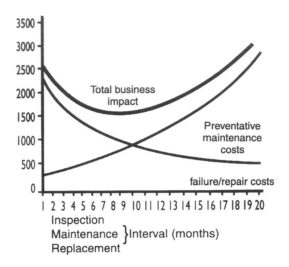

Figure 7.2
(reproduced by courtesy of Asset Performance Tools Ltd).

(d)　Determine the best methods of measurement for the evaluation and regular stock management.

(e)　Decide which data to gather and which to disregard.

(3)　*Maintenance and risk*

Maintenance concerns uncertainty; managers are involved with risk and judgements about risk, whether they formalize it or not. They are frequently questioned on the matters of over- or under-maintenance and preventive work must be shown to reduce subsequent failures, overall work required, downtime costs and preserve the business operation in a productive state. Cost/risk evaluation can produce optimum answers to the following questions.

(a)　What is the most effective combination of philosophies (see Chapter 3) to suit a particular business?

(b)　How do we know that preventive maintenance is more economic?

(c)　Where calendar-based maintenance applies, what maintenance interval should be used?

Figure 7.2 refers to defined items, either singly or as a group, but considers maintenance costs only. The optimum interval X indicates a value Y of failure costs.

The line B, however, could be plotted to include the cost of lost production raising the curve and shifting the optimum to the left.

The company referred to in Figure 7.3 employs combinations of hard and soft data in the estimates, typified by best and worst case scenarios and giving a surprisingly narrow band of results.

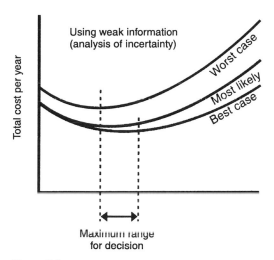

Total cost per year

Using weak information
(analysis of incertainty)

Worst case

Most likely

Best case

Maximum range
for decision

Figure 7.3
(reproduced by courtesy of Asset Performance Tools
Ltd).

The results of such optimum analysis can be stunning, with least expensive options achievable with adjustments to operational arrangements. By referring to the interaction between different variables, broadband estimates give more accurate results than the tolerance applicable to any single estimate would suggest[9].

9. BS 5760, 'Reliability of systems equipment and components', part 0, Introductory guide to reliability, item 4, The cost aspect of reliability.

Chapter 8

Preparation and response

Maintenance, like all areas of business, is not conducted in a vacuum. Its effectiveness is affected by the quality of a wide variety of general organizational facilities, and support from these sources can be noticeably improved when they are conducted with a maintenance flavour.

When regarding downtime arising from equipment maintenance or repair work, maintenance professionals naturally devote their attention to the identification of the problem, gathering of resources, delegation of the task and completion of the work. These may be regarded as the main time-consuming elements of the job, and the time between 'work reported' and 'work started' is often referred to as the 'response'. That response is shorter and less costly when the maintenance department is prepared, and preparation can be generally improved by the activities discussed below.

Good preparation means a more rapid response.

Spare capacity, made up of extra resources such as parts materials and trained personnel, can be quickly assigned to meet a sudden need when coupled with access to affected items of plant and available opportunity in repair workshops.

The case for holding additional resources in the form of spare parts and material is well understood, while additional resources in other areas, particularly personnel, is not. Unfortunately, the gains made by improved responses are offset by the costs of holding surplus resources, and the optimum balance is both difficult to establish, and is also changing all the time (see Section 7.4)[1,2].

8.1 Replacement parts

It is usual to buy additional spares at the time a new item of equipment is purchased and to hold these as maintenance stock until they are needed for breakdown or routine

---

1. The manual often refers to materials and spare parts when discussing stocks held or job resources required. The distinction meant here is that materials will require further processing, possibly at the work site, whereas spare parts can be used without change.

2. Unlike most maintenance philosophies, campaign maintenance actively reduces some resources available to conduct maintenance work (see Section 3.7).

maintenance. Unfortunately, during purchasing the latest designs are naturally favoured, and those components most likely to fail are not always known with certainty. The purchase of replacements for machines that are in current production is not usually a problem, provided the money is available, and well-established designs may be supported by more than one source of supply. Designers and plant operators, however, are usually working to increase the life of the equipment concerned, and major failures will occur late in the operational life cycle, sometimes after the parent machine is no longer in production. Many manufacturers will maintain the supply of spare parts for several years after production of the main equipment has ceased, and will advise all known users when the availability of replacements is finally to be discontinued.

Although even this is not the end, the red flag is now flying and any future failure can be corrected mainly by the operator's resources alone. It becomes a maintenance problem when a generally sound piece of plant is kept idle because one broken part cannot be replaced. This leads to a *maintenance man's golden rule* with regard to parts availability.

The life of operating machinery depends on the availability of spare parts.

There are some simple steps which can mitigate these effects, although as equipment gets older and the pace of redesign quickens such situations become more likely.

(1) *Consider the content of the original purchase contract*
    If the supply of replacement parts is made part of the original purchase order, effective over a defined period, the original equipment manufacturer (OEM) or the supplier can be bound to advise when the supported period is to end or the production of spares will cease. Remember also that some items will be sourced from overseas, adding to difficulties of obtaining future replacements[3].

(2) *Constantly refer to equipment history*
    Maintenance stocks should also refer to the use of replacements in equipment history. Although movements are recorded in the stock control system, equipment history will refer to use and operational effects, allowing more accurate failure estimates of both timing and material.

(3) *Carefully define the part*
    Alternative parts are frequently considered with questions of price or availability. It is not unknown for key parts to be 'hot shotted' from distant parts of the world, which, coupled with the effects of machine downtime, can be exceedingly expensive. This is not the time to discover that the wrong part has been ordered or that the offered alternative is not suitable.

(4) *Maintain other sources of supply*
    It makes sense to *constantly* and formally examine other sources of supply in detail, including trial purchases, before the supply becomes critical or main sources suddenly fail.

---

3. A review of equipment purchase orders, starting with information in the equipment database, will yield useful information even when the items are several years old and the supply of supporting spares is not referred to.

(5) *Consider technical options*

Some replacement items will be inferior in performance to the items they replace, but insistence on the initial specification may prevent supply. It is important for the maintenance engineers to confirm the suitability of any changes in item specification before replacement part purchases are confirmed. Also, some changes will require revisions in equipment loading or duty, and operations personnel need to be consulted.

(6) *Act on suppliers' warnings*

Do not let suppliers' advice of impending supply termination pass unheeded merely because current operations are satisfactory. Follow up all warnings of this kind and order additional parts to lift stocks held where there is a danger of future shortages.

(7) *Keep in contact with other users*

Other users have the same general problem if supply is discontinued, but their experience may apply to a different part or a different supplier. There is scope for cooperation.

(8) *Consider assembly redesign*

When it is clear that replacement supplies will eventually dry up, it may be possible to redesign a housing or an assembly to allow manufacture, well in advance of the expected failure, from parts of current manufacture.

(9) *Consider manufacture*

If other options are closed, it is sometimes worth manufacturing the replacement parts directly. Drawings and technical specifications will often be provided by the OEM, who may be persuaded to subcontract the work or to inspect the finished article.

(10) *Work with stock control*

Organize regular and frequent contact with the stock control manager and formally advise of expected failure or replacement problems. Also be sure that maintenance are fully aware of changes in stock levels and can introduce a maintenance 'veto' if key items are to be removed from the umbrella of company materials management[4]

---

4. The maintenance response to this and indeed other situations will be fundamentally different for two main reasons:

    (a)   Unlike many other disciplines, the maintenance concept of time is based on a 20 or 25 year life cycle. Most items of equipment will require attention and materials through changing item condition and its environment throughout this period.

       Maintenance responsibility does not end when the design team has gone away.

    (b)   When considering stock control the essential difference between 'revenue support stocks', i.e. regularly used and sold to a customer, usually after a manufacturing or processing treatment, and 'maintenance support stocks', requiring low stock quantities held for long periods

Continued

(11) *Remember new equipment and new designs*

Some items are newly required but of an existing design, while others are newly required but of a new design, and the criteria to emphasize for both are operational and maintenance experience.

In the first case, refer to company maintenance history or information from other users, while in the second use the manufacturer's estimates and the history of closely similar equipment.

(12) *Look at the outside world*

The availability of replacement material is affected by both the host equipment lifespan, i.e. the existence of a replacement market and the behavior of the market into which it is sold.

Equipment which has generated low sales or has proved unreliable in service will remain in use for a shorter time and will be supported for a shorter period. Conversely some products (such as personal computer systems) have been so successful that competition is feverish, the product sales life is measured in months and replacement parts are equally difficult to find.

For both 11 and 12 modify estimates according to the item itself and the way it has behaved.

8.2 Purchase order data

Purchase orders are such a normal and numerous part of business life that it is easy to forget just how important they are. Raised by the customer, they are the main initiating vehicle for the transfer of goods or services from the supplier to the purchaser. We also tend to forget that purchase orders are the opening 'half' of a legal contact between buyer and seller, carrying all the constraints and advantages of a legally enforceable procedure. Purchase orders define the customer's business identity, the extent and content of the proposed purchase, the acceptable time and location of delivery or application and all additional conditions, including methods of shipment, terms of payment, contact names, telephone numbers and fax or telex numbers, plus terms considered under contract warranty.

The maintenance manager is interested in the content of many purchase orders placed in support or anticipation of maintenance or repairs:

(1) Purchase orders for items of machinery or equipment which are part of the operational process or direct replacements for them.

---

4. Continued

of materials which may never be used, is the reason for different objectives in different areas of stock.

For stock control, revenue support and maintenance support require different stocks and different decisions.

For some maintenance parts, in the early years *maintenance requires the purchase of materials which are available but of uncertain need*, and towards the end of working life, maintenance requires *the purchase of materials of certain need but uncertain availability.*

(2) Purchases of spare parts and assemblies for insurance stock for which there is a perceived future maintenance need.

(3) Purchases for services, equipment or machinery or their replacements which directly support the environment in which the plant operates, such as fork-lift trucks, cranes and winches.

(4) Orders for safety equipment defined by the safety group for the support of on-site personnel, such as breathing apparatus, lifeboats, fire control equipment, life jackets, protective clothing and escape equipment.

(5) On-site communications equipment, hand-held radios, public address systems, site telephones, aircraft and marine communications equipment, hazard beacons and warning lights.

(6) Purchases of measuring and monitoring equipment, reference instruments, calibration devices or services and workshop standard tools.

(7) Orders for site-mounted monitors for instrumentation and control, for smoke, flame, explosive and inert gas, gas detectors, halon/dry chemical bottles, water and air filters and many others.

(8) The purchasing of repair maintenance and consumable materials, such as lubricants, paint, sealants, adhesives, coolants, washdown solvents and degreasers.

(9) The purchasing of fuel oil, gas or electricity for heating offices or accommodation and energy for process or facilities equipment.

(10) The purchase of materials which will affect maintenance personnel in the conduct of their work, such as floor coverings, door handles and hinges, warning and directional signs, equipment trolleys, lifting hooks, ropes, eyes, new or replacement doors, stairway guard rails and manhole covers.

(11) The purchase of specialist services, including specific maintenance teams, for work on items like cranes or generators, teams for replacement or refurbishing of a group of items, such as valves or piping, and third party inspectors or engineers for work on lifting equipment, helicopter refuelling and pressure vessel inspection.

(12) The purchase of tools, equipment, materials, working aids and a host of items required to aid the maintenance technician in general, together with special devices needed to make some work possible.

The preceding list cannot be regarded as complete, but it emphasizes the wide variety of purchase orders which will affect the maintenance function and shows a huge source of information for the maintenance operation, both before and after the event, as a warning of things to come and as a record of key events.

The purchase order is a focal document. It must be carefully identified in the equipment database and retained for ready use throughout the life of the installation.

When we regard a major plant construction or modification project, the purchase order is often the first formal legal step in the developing relationship between buyer and seller. In some cases it follows long processes of technical evaluation, the development of the specific application within initial designs, product reviews, modifications, installation, operational and performance questions and a series of detailed negotiations.

Under these circumstances the process affects the nature and timing of a host of following activities which cannot be started until goods or services defined in the purchase order have been raised by the purchaser and acknowledged by the supplier.

In other cases, purchase orders are issued for items like a box of screws or a can of paint which are part of a well-established routine and require little preparation or development.

There are some general maintenance observations regarding the initial construction purchases:

(1) Construction purchase orders are raised by the project team, not the operator. These are often separate companies, and maintenance may feature in design decisions but rarely in the purchasing process.

(2) The placing of POs lies squarely on the construction project's critical path. Any delay extends the programme. Future maintenance activities are not often seen as project problems.

(3) Orders are placed by a 'buyer' working to a detailed design specification produced by the relevant project engineer. Maintenance needs are often featured, but they are not specific elements of the design and are difficult to quantify or measure.

(4) New project designs often include the latest equipment, whose operational and maintenance history is hardly known.

(5) The maintenance group is vitally interested in the content of the original purchase order because it has a founding effect on so much that follows.

Maintenance and purchasing

The following elements refer mainly to routine operational activity, but most also apply to an initial project stage, and it is wise to consider them for both situations.

(1) Arrange for maintenance group participation in major equipment purchase order preparation.

(2) Ensure that the originating purchase order requires the supplier to specify associated spare parts (see item 7).

(3) Check that equipment's working life span and expected maintenance timings are defined.

(4) Consider the need to conclude a maintenance contract separately with the OEM and cross-refer in the Equipment PO.

(5) New equipment will have a maintenance requirement affected by rated performance, operational loading and suitability of the equipment specification. Consider the relationship of these features over the period of working life: wide differences will inevitably lead to enhanced maintenance needs.

(6) Does the PO include a requirement for the supply of replacement parts over the full working life of the equipment?

(7) Does the PO specify the maintenance (and operational) information required? This includes:

> Assembly and disassembly routines
> Maintenance procedures

Fault finding
Symptoms of poor performance and impending failure
Definitions of materials and component parts
Lubricants and coolants used
Shutdown and start-up procedures
Operational procedures
First line or operator maintenance
Pre-use maintenance requirements
Installation and commissioning information
Power supplied to the assembly
Machine isolation and safety features
Machine stored energy and other sources of danger

(8) Check that packing, marking, inspection, product release and local storage instructions (when applicable) have been specified (see Section 8.4).

After the plant has been constructed and installed, the flow of stock information reflects the condition and maintenance of the plant and the manner in which it is being used. Interpreting such information is difficult, but previously identified links through the stock control system can be useful; for example, selected components under stock control can be flagged to advise the maintenance engineer when either an item is issued or a purchase order is raised.

Many purchase orders are raised following maintenance intervention after an item has failed. They reflect work already outstanding, and the delivery time awaited will affect the content of the work schedule. There are also purchase orders for identifiable work in the future where failure should be avoided, so for maintenance the PO ledger partly reflects organizational anticipation and partly what has already happened.

There are some types of regular purchase which refer to indirect problems:

(1) For construction or modification project equipment, introducing an additional maintenance workload.
(2) Purchase orders for replacement parts and materials held for long periods in case of failure (so-called 'insurance items'), where subsequent reordering may be a problem.
(3) Low-value short-delivery maintenance 'consumable items', e.g. O-rings or gaskets, which may give warning of a maintenance problem not identified by other sources.
(4) Purchases of equipment where a new specification is used for a fresh requirement or the introduction of an alternative for which the company has no comparable maintenance experience.

It is a common observation that in all forms of human communication what is said and what is meant are rarely the same. This is especially true for plant maintenance, as it is apparent that many maintenance or repair problems can be alleviated by careful work at all stages of the plant development and construction project, but *particularly during the preparation of the purchase order*[5,6,7].

8.3 Work association

It should be clear that the work schedule acts as the focus of all maintenance work, and its existence is vital whichever method of work selection is in use. Most sites are complex in both design and work content, with a wide variety of basic jobs and tasks active at different times. These are serious considerations: there is a natural tendency to consider work activity in isolation or as a series of sequential steps, but this is misleading and dangerous. Accidents rarely follow a single incident, but more often result from the coincidence of unlikely events. In usual operations there are many different actions simultaneously under way and all will have an effect on the work environment. Each job will include a wide number and variety of tasks, different materials, tools, chemicals, supporting equipment and personnel. Different processes, work activities and skills will be in use and the internals of process vessels or other equipment will be exposed to the environment when they are opened for inspection or repair. The effects of such combined variety are very difficult to predict.

The following notes describe the features and advantages of work association and how its use can help counter the influence of dangerous combinations. For all its apparent benefits, work association is not widely used, and the author knows of only two systems where provision for its use is included in system software. The principal reason for this is that it has to be constrained. It is not a method that can be applied too widely, and unless practical limits are applied it is possible to discern an association applicable to most items of work which can make the maintenance system unworkable. So, when applied, it is restricted to situations of major importance, some of which are mentioned below.

In the majority of cases, association is not applied, but in a few cases it is serious, and five things require definite attention.

(1) *It is essential for the operator to know what is going on.*
     Unknown or illicit activity in the workplace can have unpredictable consequences and we should reflect that there are plenty of uncertainties in the operation of complex plant already.

(2) *Control of work and the workplace is vital*
     Not only as a requirement for good supervision, but as a means of reducing the unexpected.

---

5. Many purchase orders will be backed by a design or procurement specification, and the preparation of such documents will involve a selection of the steps outlined above. BS 4773, 1991, 'Guide to the preparation of specifications' is a useful aid in such work.

6. Suggested reliability inclusions in a specification are described in Section 3.2.1, 'Reliability clauses in a specification' of BS 5760, Part 4.

7. The 'material specification' is similarly described with respect to reliability in BS 5760, Part 4, Section 4.4.1.

(3) *Fresh work or action should not be introduced to the workplace without first consulting the current schedule.*
This last is not always possible. Some actions result directly from a failure or an unexpected incident, and even with this proviso the presence of a fresh work variable may create the coincidence of danger, and a check with the schedule, even after starting, is always worthwhile.

(4) *There are predictable combinations of work which must be avoided.*
This introduces the subject of negative association: that combination of work circumstances which are to be prevented.

(5) *The supervisor's assent is needed when the work schedule is changed, particularly after it has been issued.*
The supervisor's involvement is required, partly as a matter of authority and partly as a matter of security. By focusing the work information in one office, resources are realistically assigned, safety matters can be properly addressed and problems of work association are more likely to be avoided.

Negative association

As we would expect, jobs are said to be negatively associated when there is good reason to ensure that they do not occur together. There are some immediate examples:

(1) (a) Helicopter fuel storage and delivery system inspection and tank replacement.
     (b) Helideck fire monitor six-monthly strip-down and examination.

Helidecks have at least two fire monitors to reduce the risk of multiple failures, but they are used only at times of test or in an emergency, which means that they can suffer from invisible failure.
Helicopter fuel system maintenance should be added to the work schedule only immediately after the fire monitor test routine on both pumps has been successfully completed.
Note the words 'immediately after': this requires an order of tests and the fact that they are not conducted together. The latter requirement is the next example of negative association.

(2) (a) Helideck Fire Monitor A – repair or strip-down
     (b) Helideck Fire Monitor B – six monthly examination

The subject of helideck maintenance contains several jobs which are negatively associated, together with other influences covered by standing or supervisory instructions at the time, like the presence or absence of the helicopter itself, the prevailing weather conditions, wind strength and visibility, and the possible movement of a pedestal crane boom during helicopter landing.
Negative association is best applied as a logical feature of both the work schedule and the work plan. In the former case, predicted combinations are avoided, with both planned and unplanned work being covered, in the

latter, the plan does not allow combinations which would subsequently be rejected by the scheduling computer run.

(3) *The twin machine facility*

The emphasis here is on critical machinery or emergency equipment: where a particular function is met by two (often identical) machines they are negatively associated, for example:

(a) Pedestal crane north: main hoist rope changeout.
(b) Pedestal crane south: boom foot pins replacement.

There is of course, a further penalty arising when both cranes are down[8].

Positive association

Jobs are said to be positively associated when there is good reason for them to be done simultaneously. Such relationships are natural components of work management. The jobs concerned may be the same, such as the stripdown of two identical pieces of plant, or different but sharing similar work ingredients, such as location, specialist equipment or expertise.

Example 1

(1) Six-monthly replacement of overhead lights and routine inspection in Module C.
(2) Six-monthly gas detector head local examination in Modules C and D.

In Module C, both overhead lights and gas detector heads are mounted in the module ceiling, 35 ft above floor level, and require scaffolds to be built for local work access. Jobs occur at the same frequency and are positively associated.

Example 2

(1) 18-inch spherical header valve No. 28 in platform crude import line – replacement.
(2) 18-inch spherical header valve No. 21 input line to separator train 2 – replacement.

The replacement of either valve can only be done during full production shutdown. Minimum downtime can be achieved when both jobs are tackled simultaneously.

The valves are both passing and each job is categorized as a repair; they are positively associated.

Example 3

(1) D.C. standby battery No. 27 – 12-monthly discharge/recharge and clean-up.

---

8. With the principal exceptions of mail, personnel and hand-carried baggage, materials are supplied to an offshore platform by supply boat and platform crane. When the latter is out of action, material cannot be transferred; food, drinking water, fuel, drill pipes and replacement parts all are included in this.

(2)   D.C. standby battery No. 28 – 12-monthly discharge/recharge and clean-up.

Unlike numbers 1–26 and 29–51, batteries 27 and 28 are mounted on a separate platform accessed by ladder. Recharging equipment which can output simultaneously to both batteries weighs 7 cwt, is 4 ft long x 3 ft high and has to be manhandled into its work position.

The jobs have the same frequency and are positively associated.

Imperative and advisory

Negative association by its nature is more imperative, and where it is a feature of the system the computer does not await selection: it imposes constraint.

Coincidence of defined tasks is prevented.

In many other cases, often those positively associated, the association is advisory only, and a decision is required during planning and scheduling. There are many ingredients of association, such as commercial consequences, work efficiency and convenience, and work access operators will often assign imperative association for strong commercial reasons.

Many tasks, such as weekly visual inspections, are unlikely to have association of either sort, while others, such as the testing of the fire main valves, will surface several times.

Plant work or status

It is tempting to assume that association concerns the relationships between items of plant and that they can be defined simply in the plant register. Second thoughts will identify *items of work* as the basis of association, which refers to jobs in the maintenance work schedule. Unfortunately, the question does not stop there: other influences, as we have seen, such as

> High gas alert
> Emergency generator switched off
> High winds
> Low ambient temperature

introduce associations beyond the scope of the maintenance work schedule. Some will be recognized as inputs to the platform's emergency shutdown system, whereby operations will cease when certain emergency signals are received. However, such systems themselves require maintenance, and ensuing work may refer to work association when they are shut down (see Chapter 12).

8.4 Packing, marking and inspection

The expensive and frantic acquisition of the wrong part is a classic maintenance problem. Having given confident and repeated assurances to production that all is under control, to be faced with the wrong item at the eleventh hour gives little chance of recovery. What is almost as bad is the stricken realization that the part supplied will fit the manufacturer's current machines but not the one installed. There is a further frequent disaster: the part supplied is correct for the right machine, on time, and supplied from stock, but its long sojourn on the stock shelf has left it in such poor condition that it is unfit for use.

If these were not enough, there are other dismal scenarios, such as the correct item of supply hot shotted across half the globe to arrive in unrepairable pieces, or those items that are correct and in good condition but incomplete, right-handed instead of left-handed, metric instead of imperial, with mounting holes drilled in the wrong place, or the right assembly with the wrong coupling.

One refuge for the maintenance supervisor in such circumstances is the reuse of second-hand parts. However, instead of the opportunity to replace non-faulty items which will become a problem later, he is faced with the certainty that reused parts will fail and the whole exercise will have to be repeated. It can also be the case that the stripdown has taken place in anticipation of the correct supply and the vital replacement is the item missing, which means an unexpected additional delay or work done without a solution being earned. This is most definitely not a practice to be recommended, but it survives in spite of all efforts to prevent it.

Some, but not all, of these problems can be traced back to the spare part purchase order, and it is a source of constant surprise that so many purchases do not attend to three basic requirements:

Packing, marking and inspection

It is understandable that when these subjects apply to normal business stocks the frequency of use is such that any problems surface and are corrected quickly. Parts for maintenance and repair, however, have an uncertain future, and it is often years before such parts are suddenly needed. If an omission was part of the original purchase, it could be the occasion of first failure before it becomes evident.

The purchase of replacement parts should be conducted with maintenance in mind (see Sections 8.2 and 10.2), and it is not unusual for major companies to employ a buying officer dedicated to this purpose. The following sectors will be addressed in his purchase orders in addition to their usual content. They are frequently defined by code additions to the PO and refer to a company's general purchase specification already issued to main suppliers.

Packing

The intentions are:

(1) To ensure that the purchased item completes the journey from the factory to the purchaser's warehouse or construction site undamaged, and fit for purpose.

(2) That the warehouse or site 'shelf packing' of the part required to provide

extended local storage or shelf life is a clearly identified part of the delivered package and marked for retention[9].

Inspection

The customer's purchase order refers to the following inspection codes or qualities:

(1) Requirement to ensure that all relevant inspection and test procedures performed by the manufacturer have been satisfactorily applied.
(2) That customer-appointed independent inspectors are shown to be part of the purchase order, processing when required.
(3) That the PO defines the goods inwards inspection criteria to be followed at the point of warehouse entry by the customer himself.
(4) That manufacturing and testing standards or procedures, test reports, required certificates of calibration or conformity and traceability to national or NAMAS standards are specified.

Marking

The customer's purchase order will define the marking to be applied to the outside of the packing crate, which will include:

(1) The customer's order number.
(2) Number of items included and purchase order items fulfilled.
(3) Overall weight, location of lifting points, base standing, top and bottom, loading and unloading instructions.
(4) Treatment of different packages contained, how they should be opened (or not) and the whole assembly retained intact.
(5) Danger signs for radiation, corrosive fluids and other hazards.

For construction projects it is sometimes required that goods are completed, inspected and tested to the requirements of the purchase order, but cannot be delivered to site until needed for the installation programme. Under these circumstances goods are often stored at the manufacturer's premises as finally packed, having been released by the customer's own inspectors or those of appointed third parties. It is important to remember that, in most cases, payment will be required by the manufacturer upon completion and that the goods have become the customer's property, making them subject to the company's maintenance regime. Pre-use maintenance procedures will apply (see Section 2.5), and conflict with the requirements of environmental packing may be inevitable.

8.5 Refurbished equipment

When an item of plant or machinery is refurbished it is usual to remove it from its existing operational position and immediately replace it with an identical item which

---

9. The goods receiving section of the warehouse is frequently obliged to open a package because no external marking is visible. By so doing they damage the careful component packing and can reduce the shelf life of the item.

is either new or has itself been previously reconditioned, whereupon the whole process is returned to immediate production. There are of course some instances where an item of occasional use, or one with a built in standby, is removed, refurbished and returned to its original position, while the production operation continues without it. Such situations are not as critical, although both methods are in use.

There are several reasons why refurbishing is such a useful part of the maintenance operation and is used so often:

(1) Workshop examination, maintenance and repair can be more thorough and effective, often extending equipment working life.

(2) Many of the usual replacement considerations are not relevant. Equipment performance, operation and maintenance are known and changed routines or retraining are not required.

(3) Statutory requirements, such as the testing of pressure-bearing items, may be conducted with more convenience and safety in the workshop, where the final performance may be more readily demonstrated and certified.

(4) Where production-critical machines are involved, rapid replacement of the item and the despatch of the reconditioned unit for refurbishing may be quicker than a site repair and retest, significantly reducing production downtime.

(5) The actual repair and reconditioning work is usually cheaper when conducted in a workshop with all facilities and tools to hand, unconstrained by access or other on site restrictions.

(6) It is occasionally the case that there is no choice, and the work required is sufficiently difficult that it cannot be done on-site.

(7) The reconditioning work is often more closely monitored and controlled in the workshop than could be realistically achieved on site.

(8) More effective inspection is possible with ease of access and use of workshop measuring equipment.

In the perverse manner of things, the major problem arising with refurbished equipment is that it is often seen as *less* of a problem than equipment replacements or major repairs. It is true that, usually, considerations of changed weight, shape, installation, processes, routines and retraining do not appear, but unfortunately others do instead.

(9) Things on site may have changed since the original installation or similar work was done. Other equipment may have been introduced, restricting access, doorways may have been lowered, narrowed or removed, and gangways rerouted or replaced.

(10) The preliminary description of plant items is usually by their function, e.g. Water Injection Pump A. Notice that at this level of definition it is not always clear which manufacturer or equipment type is in use. Replacement items come from an equipment population, and on being removed from one position may be reinstalled somewhere else. The time to discover that machines are functionally the same and physically different should be *prior* to the original equipment removal.

(11) Once an item is deinstalled it is either refurbished on-site or despatched to a previously selected repairer or the original equipment manufacturer. Obviously, the further the item is removed from its next operational location the more materials handling becomes an issue.

(12) Some items will require preparatory work, such as hosing down, before they can be safely handled for despatch to the repairer. Units may be corrosive, radioactive or subject to mechanical damage, which will make disassembly and despatch more difficult.

(13) In some situations, the condition of the equipment can be so poor that refurbished equipment is essential. Deinstallation become an irreversible step for the installed item, and in the check before removal some supervisors require the replacement alongside before deinstallation takes place.

(14) For production-critical items, the availability of the replacement is vital. If it is absent the travelling time to place of refurbishment adds to the production downtime.

8.6 Obsolete components

The term 'obsolete components' emerges from the language of stock control, not maintenance, and as the phrase is clear it is often puzzling to admit that it means different things to different people. This corner of stock management is most important to the maintenance group, and machine breakdown without the necessary replacement parts can easily become permanent. It is most alarming when the component has been used for the maintenance of similar machines before, but unless identified as a maintenance part, low stock movement will lead to an obsolete grading, and the maintenance manager's confident assumption about stock availability will be rudely broken at the time of failure.

There is another similar phrase to be borne in mind: 'Obsolescent stock'. Precise meanings of both phrases will vary according to the system in use, but the usual definitions are clear.

Obsolescent stock

Those items of stock which will no longer be supported once current stocks are exhausted. Replacement purchase orders will not be raised.

Obsolete components

Those items of stock which are no longer required. Where existing stocks exist, they will be sold or scrapped. Replacement purchase orders will not be raised.

The first point to emphasize is that most stores are dedicated to providing material for production and hence revenue-earning operations. Components and materials are held in stock until issued for processing and eventual sale to a customer. Certain characteristics apply:

(1) Materials or components are usually held in large quantities and stock movement numbers are significant.

(2) The component 'shelf resting' time is usually short; turnover of parts is rapid and component damage in storage is unusual.

(3) Part specifications are well-known, warehouses and inspectors recognize components and understand quality, handling and storage requirements.
(4) Material and components held are likely to be technically current, conforming to recent legislation and marketing needs.
(5) Stock reorder levels are usually high and current stock quantities are seldom allowed to drop to zero.
(6) The financial value of the stocks themselves is high; recent replacement prices are high and apply to large numbers.

For maintenance support stocks, in nearly all cases,

none of the preceding points apply.

Replacement parts are required and held in small quantities, stock levels of one or zero are not uncommon, and stock movements are measured as the statistics of small numbers. Parts are often old (repairs being more frequent in machinery approaching the end of its working life), reorder levels are tiny, shelf damage is a recurrent problem, and as stock control personnel may have only ever seen the box rather than the item itself, misidentification is more common.

In these circumstances it is vital for the maintenance position to be clear, and the following points and procedures apply:

(1) Make sure that existing replacement parts, components and materials which have a maintenance role are known to stock control managers and are properly identified in the materials management system.
(2) Remember that the stock manager's objectives are different, centring on movement and value. Decisions to regrade stocks will usually follow a review.
(3) The maintenance manager's objective is not the same: he aims for low stock usage and positive stock insurance.
(4) Ensure that the maintenance manager is circulated whenever stock reviews and regradings take place, with the opportunity to intervene.
(5) Low stock movement is not an argument to justify making a maintenance support component obsolete.

   Items must never be scrapped while the maintenance purpose remains.

(6) Even when the stock balance is zero the replacement part must remain on the computer as an identified and controlled part in the system

   The stock identity, definition and source must remain on the computer, even if the physical stocks do not exist.

8.7 Working life

The expected working life of a piece of equipment is often required during the original plant design, at the time of the item's replacement and for calculation of asset depreciation. Estimated figures lead to investment totals required and help define replacement times and costs. Regrettably, it is often the evidence of early operations and maintenance that show when the working life of an item is likely to be short.

There are at least two occasions when the question of a machine's replacement will justify review:

(1) When the unit ceases to achieve its working functions.
(2) When the financial breakeven has been reached, i.e. when combined running, maintenance and repair costs, referred to the cost of a new item make the replacement option worth taking.

This does not mean that the equipment has necessarily ceased to work; in fact, it may continue to operate usefully, but not to the standard originally specified, as with increasing age the performance of the unit gradually declines. Alternatively, it may be the duty of the item that has changed, whereby the functional requirements are increased and the existing item is unable to make the required change.

It is often the case that the item remains fully operational after points 1 and 2 above are over.

Also, the operational circumstances in which a replacement is included affects the case for the replacement itself. If the plant has reached an advanced age the arguments in favour of a single item's replacement are diminished. (Items are unlikely to be replaced if the whole plant is about to be shutdown.) One of the figures included in the financial judgement will be the second-hand value of the machine, and there are instances of items being resold to meet the primary function of a different owner and by its sale contribute to the cost of its own replacement.

We should distinguish between 'second-hand value', i.e. that which can be achieved from an equipment sale, and the 'depreciated value', which is the written-down item value in the asset register. Specialized equipment is often very expensive to purchase, but as it was designed and built for a unique installation the depreciated value may be high but the second-hand value equal to zero.

Estimates of working life are not easy to formalize, and several factors have to be taken into account.

(1) Definition of item duty can be uncertain when the original purchase order is placed. Initial functions can change and the manufacturer's specifications will quote the most optimistic performance based on ideal use.
(2) Equipment degradation starts from the moment of manufacture and becomes more severe following delivery to the site. Damage during shipment, craneage, unpacking and installation may be invisible or unnoticed by non-specialist work crews.
(3) Initial commissioning will require the fundamental operation of equipment to confirm correct coupling, connection and operation. There is a danger of temporary load conditions outside the duty envelope causing unforeseen and unnoticed damage.
(4) Operations may be more demanding than expected at the time of purchase. Provided that the operational duty required lies within the performance envelope equipment will usually perform satisfactorily, but regular severe duty will reduce the working life.

(5)   Variations in steady state conditions, such as frequent stop/starts, operation of switches, corrosive processes, environmental damage, fuel or lubricant contamination, incorrect use, severe local vibration, distortion of bed plates or couplings due to fluid flow or level changes, heavy material weight impact or transfer, severe stress or distortion due to thermal cycling, daytime/nighttime temperature changes, or water table or tidal effects on foundations, will all adversely affect the working life of the equipment.

(6)   Periods of equipment idleness due to output reductions on some production facilities, a long period awaiting installation or an item assigned to be a permanent standby. As described in pre-use maintenance (Section 2.5), machinery will degrade when not in use.

(7)   The production process itself may induce unexpected influences, such as changes in production fluid temperature viscosity, working pressure or corrosive effects. For example, the presence of hydrogen sulphide may increase, affecting the corrosion resistance of certain hardnesses of steel, and require replacement of pipes, valves and vessels etc. which lie on the wetted path. This latter effect will cause sharp reductions in affected equipment working life unless previously specified to cater with this form of contamination.

When attempting to estimate the working life, establish three things right at the outset:

(1)   The manufacturing 'start' date.
(2)   The equipment installation date.
(3)   The OEM's ideal working life figure.

Qualify these by referring to the effects outlined above under the headings:

(4)   History of installation, commissioning and pre-use testing.
(5)   Pre-use maintenance activity.
(6)   Comparison of expected performance and purchase specifications.
(7)   Nature of equipment duty, expected type of operation, stop/start and standby.
(8)   Known effects of local operations and adjacent machinery.
(9)   Expected future changes, loading and material movement over time, changes in production processes, geological and environmental effects and plant extensions.
(10) Known figures for previously replaced equipment of the same type.

Once the initial estimate is made, compare with experience of similar installations or machines, sister plants or other industrial experience; talk to the OEM, who will refer you to existing users as well as providing his own estimate.

Use the following points in your assessment

(1)   Diminish the 'ideal' working life figure for each damaging influence.
(2)   Remember that not all new designs will have a longer working life.
(3)   Where the designs are so new that ideal figures are suspect, look for those features that improve or penalize the working life.

8.8 Standing plans

Standing plans are prepared in advance of those situations which bring them into use, and it is military models which supply the best examples of how they work, providing the phrase 'constant readiness' to underline their purpose. Although they apply widely, we consider them here in terms of maintenance. Like all forms of work preparation, the intention is for a rapid organizational response, which, combined with improved work processing, will reduce production downtime, third party effects and, of course, costs.

The first point to consider is 'Which jobs should be given this treatment?' After all, prepared work instructions and standard procedures could also be regarded in the context of standing plans. The answer is simple: standard procedures apply to all maintenance work. All of those jobs in work box A1 (and to a lesser extent A2) should have work instructions already prepared and in regular use, while those in work box B1 will be candidates for standing plans. (Refer to pages 20 and 21). The word 'candidate' is used advisedly, because not all B1 jobs will merit a standing plan and some will be prepared for situations that will never occur. B1 jobs, like changing the slew ring on a pedestal crane, are known for content and that they lie in the future, though when will depend on equipment performance and duty, plus planning decisions for the work to take place.

This raises an interesting aspect of when to do the work. Some jobs, like the one mentioned above, arise because of a gradual deterioration of the machine in question, giving a wide band of suitable timing choices for careful preparation. It may in such circumstances be argued that, provided the in-house planning resources are always available and that sufficient time exists to include it in job preparation, then there is no need for a standing plan: all work necessary can be completed after the decision to act has been made.

Similarly, some obvious choices for the standing plan approach are already protected by built in 'standbys' where the rapid organizational response is created by immediate switchover in the event of sudden problems. The existence of the standby does not preclude the use of a standing plan, but it does reduce the need for this treatment.

In addition, there are many jobs which are simply not serious enough to merit this type of attention, perhaps members of a large equipment population where each apparently qualifying task is only represented by a small reduction in overall capability or perhaps a unit providing a service function whose absence can be tolerated for a few weeks. Neither of such groups would receive the standing plan treatment, though there are many others that will, and we must return to the original question, 'Which jobs should be given this treatment?'. It should already be clear that there are several points to consider.

Predictability

Many tasks, like the wall thickness inspection of a production pressure vessel or a Christmas tree changeout, will definitely occur, and will do so in a time window of two or three months. Although warnings will occur there may be insufficient time available to prepare the job completely, and a standing plan would help considerably.

Severity

Not all critical equipment is protected by standbys, and the severity of a job may involve financial consequences, danger to personnel, destruction of machinery, or contractual imperatives where the penalties are so great that a standing plan would be regarded as essential.

Uncertain timing

The timing of many qualifying jobs is difficult to predict, although some help can be gathered from the equipment's age, its treatment and its history. Older and heavily used equipment is more likely to give trouble and require rapid return to duty.

Immediacy

Many jobs, such as that caused by the loss of an impeller on a centrifugal pump, can erupt with no warning and apply to a critical item in constant use. Such cases are important enough to justify replacement stocks in hand and full standing plans.

For the maintenance engineer, preparation of standing plans should hopefully be done during the early installation 'honeymoon' period before their existence is vital. Even so, selection of which sequence of preparation to follow will require knowledge of the site situation and some care. The engineer may decide to delay preparing plans for the most immediate in favour of detailed attention to something more severe.

Prepare standing plans for qualifying jobs in the following sequence:

(1) Tasks of severe consequences with no built in standby.
(2) Tasks of immediate likelihood not already covered by (1).
(3) Predictable tasks expected to occur shortly.

Such a selection will not cover all jobs eventually treated by standing plans, but can be used as the initial foundation and followed by less stringent selection criteria later. Choices should be made by a maintenance engineer familiar with the plant status and operational processes.

As a first step before detailed planning, place on purchase order major long delivery items required for selected jobs and not already held in stock or on order.

Where the selection of purchases is impaired by financial constraints, then concentrate on materials required for the three groups which are on extended delivery.

When the standing plan is prepared, it should include the following items, which are treated in more detail in the normal maintenance planning process (Section 5.3).

(1) Applicable plant tag numbers or equipment identity numbers.
(2) Description of the job.
(3) Standing plan number.

(4)   Relevant company standard procedures and work instructions.
(5)   Work management procedures including contact names and numbers, relevant service or maintenance contracts, name of the associated maintenance engineer, housekeeping and travel arrangements.
(6)   Safety procedures: safety engineering requirements, on-site briefings, safety officers' presence, in work instructions, permit to work arrangements (Section 11.9).
(7)   Materials and parts required, sources of supply, company stocks, lead times and quantities required.
(8)   Sequence of work steps, description of actions, effects, adjustments, applicable maintenance work instructions (Section 9.4), other work procedures and previous relevant reports.
(9)   Sources of supporting information, equipment history, drawings, purchase and supplier specifications, other similar equipment histories and previous job records.

8.9 The replacement machine

When a machine approaches the end of its useful life, or for various other reasons, it may be decided that the simplest course is to purchase a replacement, although it does not always follow that an old or discontinued product can be readily replaced by an identical item or a new design of similar or superior specification.

Although the production team may complain about the poor performance of a plant item, they are unlikely to request a replacement for these reasons without the maintenance manager's agreement, and it often falls to him to activate a purchase order. Before taking this step there are several points for him to consider.

(1)   The introduction of a new machine will usually cure short-term maintenance problems and initiate short-term bedding down problems. Once the new item is running smoothly, maintenance interventions should be less frequent and failures postponed into the future. However, the maintenance of the replacement will change if a different machine is used and longer term effects will have to be considered. Speak to the manufacturer and existing users about the future maintenance requirements, the use of unusual tools, special disciplines, access requirements, special components or subassemblies and safe handling. There will come a time when the new machine will be the focus of immediate action and it pays to be prepared. The maintenance staff's well-worn observation about constant firefighting is sometimes partly of their own making!
(2)   The first questions arising from the AFE (application for expenditure) will be financial, but will lead on to other things:
   (a)   How much does it cost?
   (b)   Will it improve output and hence revenue?
   (c)   Will it affect the quality of the function it provides?
   (d)   Is it safe to operate? Are there special issues which have to be addressed?

(e) What are the maintenance costs to date of the existing machine?

(f) How much will the existing unit cost in future maintenance and repair?

(g) What will the new machine cost in future maintenance and repair?

(h) Does the existing machine have a residual value?

(i) What will the change do to the total asset value?

The accounts department will provide general financial evidence, but it may not be broken down to the level of the individual machine. Operations and maintenance will provide production costs, current repair and maintenance costs and estimates of future expenditure for both old and new machines. Estimates can be created by checking the equipment history of the replacement candidate and similar machines in a more advanced state while examining previous replacements of such items. Normally the replacement will be timed to achieve minimum cost overall, and there is a present emphasis on 'life costs', i.e. purchase price plus operation and maintenance costs, as opposed to 'lowest purchase price', where higher priced equipment may cost less to use, producing a lower total cost over its lifetime.

(3) One argument in support of a replacement is that the machine will have to be replaced eventually anyway. The main question here is the life of the plant compared to the life of the machine, and conventional cost comparisons often do not consider that it may all stop tomorrow.

When the plant is nominally permanent this is not a consideration, but if it is dedicated to an extraction operation, then it depends on the life of the main resource, such as a gravel bed or an oil field. When a facility will last for forty years, two expensive machines of twenty-year life would probably cost less than four cheaper machines of twelve years life with eight useful years wasted. One question the maintenance manager must address is:

How much longer can we expect to remain in operation?

(4) When a replacement machine is to be purchased, make sure that the AFE is placed well in advance of expected final failure. It would be tragic to undergo an expensive repair merely to keep the plant running until the replacement item arrives. Allow time for purchase order processing, the supplier's delivery period, shipment, installation and commissioning times, together with the manufacturer's lead and manufacturing times if the replacement has to be built to order. Remember also that manufacturers' lead times may extend sharply if expansion of their order position follows purchases placed for alternative projects, especially if major construction programmes are under way.

(5) Remember that a replacement machine may be different from the original,

Even when an identical item has been ordered.

Equipment to the original specification may not be available, product designs are changing all the time, companies go out of business and product lines may simply be discontinued.

(6) Take detailed care with the unit specification. It is imperative that the item ordered is precisely what is wanted. For construction projects the design engineers will be very exact about the technical specification. For replacement purchases, however, there is a strong temptation to repeat the specification that was used in the original purchase. For many reasons the requirement may have changed and the exercise needs to be done again! The penalties in money, time and production loss of a bad purchase are severe.

(7) It is always worth obtaining the shipping, handling and installation details well in advance of delivery so that preparations may be conducted during the purchase delivery period. It may also be sensible to despatch maintenance, construction and operational personnel to the originating factory for training in its installation and use.

(8) It is wise to obtain input from the operational group before a maintenance-initiated AFE is tabled (and vice versa). It is possible that they may take the opportunity to improve plant performance or replace several smaller machines with a larger, single unit or reconfigure the plant in a variety of ways.

(9) Non-identical machines will often have different mountings, couplings and power supplies. Machines from the USA will have American as opposed to metric threads, quote pressure and weight in pounds, use different electrical power measures and often respond to different statutory regulations. Manufacturers are aware of such differences, but it is essential to check carefully before the order is placed.

(10) The statutory requirements affecting both equipment and work are becoming increasingly stringent, particularly in the areas affecting quality and safety. Check the equipment itself and the work required in both installation and operation with associated regulations (see Chapters 11 and 12).

(11) A fresh installation may require rerouted pipework, strengthened floors, resited lifting equipment, repositioned ladders or access doors and associated purchase of special tools, higher rated support equipment or revised designs.

(12) A construction project will sometimes result from the fresh installation, requiring formal project management, defined resources, durations and activities portrayed in a network programme.

(13) The shipment and installation of the replacement item is sometimes part of the construction project and will require separate treatment,

   Especially if the weight, shape or size of the unit is greater than the item it replaces.

Size will affect access and may lead to the use of different gangways, hatches, doorways etc. Weight will impact lifting capability of cranes, deck and roadway compressive strengths and the size of hardstandings. The height of beams and entry doorways is sometimes overlooked, and in such circumstances work is needed before machinery movement.

Remember to consider the shape and size of other machines which have to be moved to provide access for the 'new' machine (refer to Section 9.2).

(14) The replacement of a major item may sometimes become the initiating task of a shutdown and become the focus of many different items of work (see Section 5.8).

(15) The operation and maintenance of the new machine is likely to be different: controls and displays will have been redesigned and access and operation changed. During the commissioning process, trips, indicators, alarms and safety devices will require setting to the new process variables. Both maintenance and operation personnel need to be familiar with the new situation.

(16) In all of these considerations, the important duty of the maintenance manager is

to act in advance of events

to maintain a careful and constant watch over the existing installation, ensuring that

necessary choices are tabled for decision before various options are lost by default.

To be driven by events is not a satisfactory technique. It is vital for the

maintenance manager to retain the working initiative.

Chapter 9

Maintenance and production

There are many situations where maintenance is required but production is not involved, the machinery and equipment being maintained to provide a service or preserve plant and equipment condition. Work of this type would apply to commercial buildings, warehouse facilities, materials handling equipment, and transport systems, with features such as airports, railways, dockyards, river and water management systems, motorways, bridges and roadway surfaces, traffic control equipment and fleets of vehicles. There is, of course, the military dimension, with all facilities and systems equipment on the agenda, and also commercial establishments with office blocks and warehouses, plus local authority services, retail centres and equipment, medical, educational and training establishments and many more besides.

When production is present, the effect on the maintenance schedule and vice versa is very strong and questions of downtime frequency refer to this as the obvious illustration. This does not mean that the maintenance is less necessary elsewhere. In every application, when the maintenance support is absent functional failure will subsequently occur, and where such a failure occurs additional costs will arise. When considering the cost and value of a maintenance facility, it may be convenient to think of the cost of providing a replacement service to deal with functional failure.

For true production its relationship with maintenance is very direct: both activities occur on the same site, share some facilities (including working space) and occupy the same time frame. Time reductions achieved in maintenance work lead to gains in production, raised outputs and reduced costs, which ironically, lead to increased maintenance needs.

An initial effect on one has a subsequent effect on the other.

They are truly interdependent.

9.1 Downtime

When considering the subject of professional maintenance, and particularly why it is carried out, 'downtime' is probably the most important concept of all. First impressions would agree that the downtime idea at its simplest is:

(1) A period of time when a normally productive plant is idle.

The results of such idleness, among many, are no production, no output and, of course, no production-generated revenue. First reactions would suggest that downtime should be rigorously avoided, and some companies, when their plant is new, adopt such a strategy. As plants age and equipment failure becomes more likely, however, downtime inevitably occurs, and the management attention changes from downtime by default to downtime by intent, emphasizing the positive reasons why the latter is adopted for so many plants.

BS3811 (see Glossary) defines 'downtime' as 'The time interval during which an *item* is in a *down state*', linking downtime to the individual item of plant, whereas our general definition is applied to the plant overall. At this item level the attention shifts to the 'down state', which is also defined as 'A state of an *item* characterised either by a *fault*, or by a possible inability to perform a required function during *preventative maintenance*'. Notice that in this latter definition the down state can be achieved either by a fault or through maintenance.

During a shutdown:

(2) Operational costs are generally lower[1].

The consumption of energy and materials, production labour and life support costs, are lower, while equipment wear and tear is reduced. However, they are not zero: plants use non-productive energy, are subject to wear, tear and degradation and incur labour charges even when the plant is idle.

(3) Downtime presents opportunities for work preferable or only possible when the plant is down.

Local refurbishment, repair or major maintenance work on production-critical items is frequently impossible when the plant is running. If the latter does occur it depends on the availability of built-in standbys or the possibilities of partial shutdowns. When major or critical maintenance is crucially necessary, full plant shutdowns are often the only answer.

We should note that Rule 2 above carefully refers to 'operational costs', while the opportunities referred to in Rule 3 will incur the job costs of maintenance, repair, and reconditioning. If the strategy of intended or planned shutdowns is adopted many tasks and projects will be specifically prepared for the shutdown period (see Section 5.8), and although reduced operational costs mitigate the effect, *in profitable operations a plant shutdown is a very costly event*.

If Rule 3 underlines the fact that downtime cannot be avoided, Rule 4 outlines the positive reasons for creating downtime on purpose.

(4) Carefully prepared downtime by intent will reduce or eliminate downtime by default, and reduce downtime overall.

---

1. Downtime could be regarded as a 'neutral' scenario when revenue is zero and operating costs are down. There is a worse, or negative situation in which plants are actively producing scrap. This is obviously a quality problem, but in this position revenue is zero and operational costs are normal.

By ensuring that shutdowns tackle many jobs simultaneously, which, if not done, would each cause unplanned downtime singly, overall downtime figures and economic penalties are reduced.

There is a danger here: if planned downtime proves effective, there will be inevitable pressure to increase it, changing the balance between planned and unplanned downtime. It is important to monitor these effects carefully, and the adoption of the planned shutdown approach is best accompanied by a regular downtime analysis system which examines all the reasons that have led to downtime and how they can be avoided in the future.

It may also be puzzling to see the strong reference to costs above in what is clearly a maintenance question, but there are two rules to emphasize:

(5) The avoidance of downtime is a principal reason for maintenance work.
(6) The manner in which downtime is treated is reflected in the maintenance method used.

These are fairly blunt statements. First, from Rule 5 the avoidance of downtime is not the *only* reason for maintenance, (there are many questions of operator safety, product quality, functional performance, equipment and plant asset value), but downtime is most important. It refers to the item and to the plant itself and all the reasons why work is done.

Rule 6 is more convoluted and we may go further.

(7) Decisions about downtime treatment can also affect the original plant design.

There are references elsewhere (see Section 9.7) to minimum facilities plant design, and apart from leaving many systems out altogether it includes no items of standby equipment and the acceptance of periods of downtime during working life.

(8) Minimum facility designs handle plant downtime by shutdown planning and rapid response.

There is a further consideration: some plants face mounting problems which increase as shutdown time progresses. The obvious example is ready-mixed cement. If the main output delivery pump fails, a standby must be used to restart the process at once or setting cement must be quickly removed from the original equipment, including pump valves and delivery pipework. Similarly, many furnaces are constantly lit, even when the plant is nominally shut down. Loss of the flame can mean a furnace refiring and the removal by hammer and chisel of the refractory lining.

(9) Plant downtime can continue to cause damage after the ensuing shutdown has occurred.

So far, plant downtime has been mainly considered, and in the operational language the unaccompanied word 'downtime' usually refers to the plant overall. Turning to 'equipment downtime', we can describe in a similar way:

(10) Equipment downtime is a period of time that an item of equipment is not usable.

So it may be down for maintenance, repair or failure; the important thing to note is that it cannot be used, whereas equipment awaiting changeout may be 'off' but usable.

Criticality

According to BS 5760, criticality is a combination of the severity and the probability of an occurrence and criticality analysis is applied as an extension of an FMEA to give failure modes, effects and criticality analysis (FMECA). See Section 3.4. In BS 3811 and the Glossary, criticality analysis is defined as: 'A quantitive analysis of events or *faults* and the ranking of these in order of the seriousness of their consequences'.

In this manual the terms 'critical' and 'criticality' are used more loosely as part of the maintenance language and they appear in two main forms, firstly as 'safety-critical', when the critical state is regarded as 'A state of an item assess as likely to result in injury to persons, significant material damage or other unacceptable consequences' (BS 3811), and secondly as 'production-critical', so two loose descriptions are used:

(11) If the item is production-critical equipment, downtime causes plant downtime.
(12) If an item is safety-critical its downtime has serious safety implications.

In this latter case an item can be classed as 'critical' when significant danger to personnel can be caused (i.e. action or evacuation) (see Chapter 12).

There are also other reasons why we may consider an item or state as critical, such as commercial imperatives, loss of product, short-term plant damage, and environmental consequences, such as escaping gas or toxic fumes, discharge of polluted fluids and damage to neighbouring public amenities. Criticality analysis is a formal method of analysis; here a more general interpretation is used.

We should distinguish between

(1) An item of critical equipment.
(2) An item of equipment in a critical condition.

The first is a question of the equipment *function* and the second refers to its *state*. Item (a) is fixed by the plant design and by its operational duty, while item (b) changes with use and deterioration.

In the overall plant design, only a few items are production-critical, and if any one of them fails then plant downtime results. If the planned shutdown approach is used, all critical equipment can be repaired or maintained fully at that time, and it is this concentration of critical equipment shutdowns into the same planned time slot which reduces their random occurrence and the overall plant downtime figure.

9.2 Access to equipment

Access is not merely a question of physical dimensions or 'accessibility', it is also a feature of authority, a component of the job which has to be specifically granted. If we think of it as a walled garden, access requires the gate to be unlocked (see Section 9.5).

Although maintenance teams are often based on the site, they are not the custodians of the plant, a duty which formally rests with the site manager. For an offshore oil platform this would be the OIM, the 'offshore installation manager', who would allow access to production or maintenance teams as the work programme requires. In some cases, non-production or service equipment, such as cranes or generators, with an indirect effect on production operations, may be permanently assigned to the services (which includes maintenance teams) group, who will make corresponding decisions about access to equipment, but even in such cases the formal assent of the OIM would be required.

The access decision tends to focus on the item of equipment itself and its immediate surrounds, which is not the first part of the story. Access has first to be permitted to the site overall or to the local zone. Access may be denied or retimed for a wide range of occurrences which have nothing directly to do with the job:

(1) Effect of industrial action.
(2) Site security restrictions.
(3) Loss of vital work support or safety services.
(4) Site or zone safety hazard, gas leak, current risk of fire or explosion and other safety restrictions.
(5) Preventive effect of separate work conducted adjacent to or overhead, such as overside diving operations nearby.
(6) Planned movement of hazardous wastes, corrosive or radioactive assemblies and contaminated materials in the area where access is required.
(7) Restrictions arising from the planned work schedule or the result of unplanned equipment failures.
(8) Access denied to work area for specific jobs by the deliberate non-issue of a permit to work.
(9) Arguably the most important of all, *restriction of access to site because of danger to personnel safety*. Usually this is the result of a safety officer's or supervisor's instruction (see Chapter 11).

Once general requirements are satisfied, the job-centred questions are usually addressed by a work permit system whereby access is denied until carefully defined preconditions have been met and additional safety hazards are shown to be absent. The steps are outlined in Section 11.9.

Final questions of access relate to the introduction of men and materials to the worksite and the provision of necessary equipment and services. It is sometimes forgotten that when space is limited, equipment may have to be disassembled and removed before sufficient space is available for replacement items to be moved into position. The following need to be considered:

(1) Do replacement parts or assemblies require disassembly prior to shipment because of transport or movement path restrictions?
(2) Can replacement equipment be placed in position prior to work start or will existing item removal be required before replacement transfer can take place?

(3)   Is separate repair or replacement component work to be done nearby, and if so is sufficient working space available for all work planned?

(4)   Can replacement and removed items pass each other moving in their respective directions to installation or transport from site?

(5)   Have modifications to site geometry occurred after installation which may restrict or prevent movement of materials?

(6)   Is there sufficient room for men to work in comfort and safety when wearing protective clothing, with space for replacement parts, working materials, tools and display surfaces for drawings and working instructions, preferably at comfortable working height?

(7)   Is access available for the safety officer, the supervisor and any personnel who need to pass, at all stages of the work programme?

(8)   Are routes of access and escape unrestricted and known to all members of the work crew, and in addition well lit and marked as required by Safety to cover cases of emergency evacuation?

(9)   Are materials handling, lifting and holding devices usable and available for the job, for the movement of materials required approaching and in the work location?

(10) Are special arrangements needed for the movement of tools, cleaning materials, safety equipment, or are special containers required?

(11) Is access restricted to one working group in the designated work space at the specified time, or are special arrangements necessary with other teams or departments?

(12) Are special 'hot-key' or similar arrangements necessary for the protection of workers during actual work? Is access to the internals of the machine required, restricting the ease of movement and escape? (Refer to the requirements of equipment isolation in Section 9.3, the permit to work in Section 11.9 and all pertinent safety requirements discussed in Chapter 11.)

(13) Consider the question of special safety equipment required, flameproof lighting, breathing apparatus and flameproof communications equipment. (Refer to the requirements of equipment isolation in Section 9.3, the permit to work in Section 11.9 and all pertinent safety requirements discussed in Chapter 11.)

9.3 Equipment isolation

Continuously running equipment at some point has to be shut down for any maintenance work which requires access to internals, exposure of moving parts and possible disassembly. For the item to be approachable it does not merely have to be stationary; it has to be free of inherent danger to the maintenance technician and unlikely to damage other equipment. This requires the dispersal of internal energy, such as working pressure or electrical energy, the removal of corrosive, combustible or toxic fluids, heat and other varied contaminants.

Neither the preparation nor the subsequent work can normally be undertaken until the item to be maintained has been removed from its operational duty, and in a 'live' process this is always difficult. The replacement of a ball valve seat, for instance, is uncomplicated work after the valve is out of circuit and all contaminants removed, but, if the valve is mounted in a main hydrocarbon line operating at even modest pressure, the process fluid will have to be diverted. Such diversion is not always possible: temporary flow rates and required capacity may be higher than the remaining piping or process vessels can provide.

Work of this kind requires the shutting down of at least some items of equipment concerned and their separation or 'isolation' from the remainder, perhaps active, parts of the plant.

As these notes should emphasize, the isolation of any item of equipment or part of a plant should only result from a formal procedure requiring the involvement of operational, maintenance and safety personnel. Procedural steps should also only be permitted through an isolation work permit, and items 1–12 below are abstracted from the HSE document 'Guidance on permit-to-work systems in the pertoleum industry', Appendix 3: 'System and Equipment Isolation Requirements'.

Isolation of hazards

(1) Once all the potential hazards associated with a particular job have been identified it will be necessary to consider how they can be separated or isolated from the equipment or plant to be worked on. Flammable, toxic, pressurized, high-temperature or low-temperature fluids will normally have to be removed from isolated plant before working on it.

(2) Equipment and plant that should be isolated:
    (a) **Machinery** – this should be isolated from its power supply (electric, pneumatic or hydraulic), or if engine-driven, the starting system or engine disconnected. Where necessary the machinery should be prevented from moving, e.g. from gravity fall or release of pressure, by positive physical means;
    (b) **Pressurized systems** – These may include:
        (i) hydrocarbon or other process systems containing fluids that may be flammable, toxic or corrosive.
        (ii) fluids under pressure that may range in temperature from very cold to hot.
    Pressurized systems of all kinds should be isolated, depressurized and, when necessary for safe working, drained, vented, decontaminated and purged as required. Checks should be carried out to verify that these requirements have been met before system entry;
    (c) **Non-pressurized systems** – which comprise hydrocarbon or other process systems containing fluids at equilibrium that may be flammable, toxic or corrosive, at temperatures ranging from very cold to hot, that are regarded as hazardous but are not subject to applied pressures (other than that resulting from static head to a maximum of 0.5 bar above atmospheric pressure).
    Hazardous non-pressurized systems of all kinds should be isolated,

and, where necesary for safe working, drained, vented, decontaminated and purged as required. Checks should be carried out to verify that these requirements have been met before system entry;

(d) **Electrical systems** – those capable of causing a hazard to personnel working on them or of igniting a flammable atmosphere should be isolated, proved dead and earthed and a notice relating to applied isolations should be placed at the isolation point;

(e) **Safety and emergency systems** – where these systems require isolation or inhibition for maintenance, careful consideration must be given to the continued protection of systems that may remain operational during the maintenance period.

Use of permits for isolation: isolation certificates

(3) The permit to work should clearly specify the necessary isolation procedures and requires a signature that they have been complied with, either on the permit itself or in a supporting permit or certificate (see Section 11.9), which may be required for work associated with a particular hazard.

(4) It is advisable that for any work requiring more isolation than can be detailed on the permit to work itself, a system of isolation certificates should be introduced. The purpose of an isolation certificate is to certify that the plant or equipment has been thoroughly isolated from sources of power or process fluids. These can be designed to cover the various types of isolation required, for example:

(a) mechanical/process systems.

(b) electrical systems.

(c) safety/emergency systems.

They may be used independently for operational purposes, but not as a means of work control on equipment and plant where hazards are present. Normally a permit to work would be initiated before raising isolation certificate(s) of any kind

(5) When the hazards have been identified, the responsible person should specify what additional permits or certificates are required, such as an isolation certificate. It is the duty of the responsible person to ensure that they are raised and the necessary isolations completed before a permit to work is issued to the person in charge of the work. The responsible person may delegate the actual isolation to another competent person.

(6) The permit to work should contain a section for completion by the responsible person which states the type and reference number of all additional permits or certificates raised.

(7) The permit to work should contain a section that allows the permit holder to state that they have completed the work for which they are responsible or have left it in a safe, suspended condition. (It should be noted, however, that suspension of permits may not be appropriate for some systems, e.g. complex power distribution systems.) It should also include a separate section for

signature, stating whether the isolations are continuing or have been removed. A further section should be signed by a member of the operating staff indicating that he or she has checked the condition of the plant and it is safe to resume operations.

Methods of physical isolation

Pressurized systems/process plant/vessel entry etc.

(8) The kinds of isolations used and the precautions taken will depend on the level of risk that the job is likely to involve. For details of how to carry this out see OIAC guidance: *The safe isolation of plant and equipment*[2].

Machinery isolation

(9) In general, hydraulic or pneumatic-powered machinery or valves etc. should be initially isolated at valves and then the supply and return pipes disconnected or otherwise made safe to prevent any possibility of movement of the machinery. When producing isolation procedures for hydraulic supplies, consideration must be given to the protection of plant from ingress of contaminants and the risks to personnel of breaking connections which are not depressurized.

(10) Engine-driven machinery should be isolated by shutting off the engine fuel supply and then isolating and disconnecting all starting systems at source.

(11) Where machinery power systems have been disconnected or engines prevented from starting, but there is still a foreseeable risk to people working on the machinery because it may move, then a device such as a properly engineered mechanical device should be fitted to lock the machinery in a safe static position.

(12) Isolation of electrical supplies should be maintained by means of a lock and key system or equivalent. Isolation should be maintained until all trades associated with the work have completed their tasks.

From sections 13–16 inclusive the document discusses some of the hazards and requirements of electrical isolation. This aspect of isolation is discussed in Section 11.10.

Because isolation work affects both process downtime and personnel safety, it is possible to outline the frequent elements included in the job itself, and in many cases the following steps can be included in a standard instruction or a standing plan. *Such plans will reflect the fact that each isolation routine is different*, even when the same type of units go through the same maintenance task. The identity and position of valves, the location of isolation points, the fixing junctions for pipeline blinds

---

2. Normally, active devices such as valves, activators, pressure relief valves, solenoids, limit or rotary switches and many others which can themselves fail must not be used as key barriers for isolation. Although item failures are infrequent, the time of isolation usually reflects maintenance or replacement need, when failure is much more likely. Also, duty changes apply, particularly at shutdown and restart, fluid differential pressures are higher, and relative movement between adjacent parts and lubrication pressures may drop to zero, making component jamming more likely.

and other factors will vary from one machine to another, and specific versions of the plan will be required. The following elements are likely to be included:

(1) The safety officer concerned with the work has been informed in advance and the job preparations carried out with his involvement. Established safety procedures are main ingredients of the work programme and are included.

(2) Relevant safety equipment, such as breathing apparatus, goggles, fire helmets, gloves, torches and other items are checked and provided (see Chapter 11).

(3) The applicable work schedule, work start time, duration, disciplines concerned, activity sequence and task delegation are all clearly identified. Parallel or associated priority jobs are known together with any inter-job or task effects and dependencies.

(4) Subject equipment and all affected items, some directly coupled to it, are subject to shutting down, decontamination, decoupling and removal specifications or procedures, where applicable (see Section 11.10, 'Electrical safety' and the HSE document 'Electricity at Work: Safe Working Practices', items 41 and 42).

(5) Shutdown and decoupling procedures applicable to the equipment to be isolated or merely shutdown available to the safety officer, maintenance and operational supervisors and technicians concerned.

(6) Equipment is isolated from the operational system using specific and attached isolating components, such as pipeline blanks, fixing bolts, coupling and anti-rotational clamps.

(7) After isolation has been completed but before maintenance work has started, fully competent personnel should attempt to start the equipment to confirm that isolation has been satisfactorily achieved.

(8) The plant control room has been informed and *out of use* warnings posted. The maintenance 'hot-key' is used to prevent inadvertent 'startup' or 'power on' occurrence while men are working on the machine[3,4].

(9) Requirements of the safety officer have been checked and met, safety apparatus and work site warnings or restrictions posted.

---

3. The permit is prominently displayed in the control room, at the points of isolation and the place of work. Other locations are also used as the job or safety officer requires.

4. The 'hot-key' is a well-established device, often used as part of the 'permit-to-work' discipline. In its most fundamental form a starter key or linking component is removed from shutdown equipment, making restart impossible until the hot-key is returned. This component then remains in the technician's charge during the maintenance work, sometimes while he is working inside the machine itself. Once the work has been completed, with the technician safely returned and the equipment boxed up, the hot-key is returned for restart. *Although it has been repeatedly effective, this method has its drawbacks, especially where several workers are engaged on the programme (sometimes out of contact with each other). Where only one key is used, the risk of premature start-up is clear. Also, some plants have wider and distant effects and the use of a single key may be inadequate. Other versions have been used employing several keys, technicians' tags or padlocked boxes, each employing the principle of restart impossibility while work is under way. It may be persuasively argued that hot-keys which act as part of the restart mechanism are more reliable than technicians tags which are a means of identification.*

Although such methods are initially attractive, their shortcomings must be recognized; many have difficulty catering with change and can induce a false sense of security. All should be implemented only as part of the full permit-to-work system (see Section 11.9).

(10) Shutdown and decoupling procedures applicable to the equipment to be isolated or merely shut down available to the safety officer, supervisor and technicians concerned.

(11) Intrinsically safe or flameproof equipment is required. Use when defined as necessary by the possible presence of combustible gases or materials for portable lights, indicators, power supplies for the electrical job and local equipment.

(12) Process tanks and vessels have been opened and vented with flameproof airblowers where internal access is required to gas or fume-containing atmospheres. Toxic/corrosive fluids, radioactive scale, contaminated sludge or sand have been removed.

(13) Equipment and working areas have been washed down and allowed to cool, with the degassing operation complete.

(14) The work site has been checked for additional but separate hazards; safety and communications equipment has been identified, checked for fully effective operation and provided subject to constraints listed, and technicians are familiar with its use.

(15) Residual energy has been removed or identified, high-pressure containment, electrical energy, mobile sliding or falling impact producing parts, displaced weights, potential energy and mechanical items in torsion or tension. Components that are sharp-pointed, hot, radioactive corrosive or abrasive to the touch.

(16) When welding or metal cutting is involved, ignition sources are created by sparks or hot surfaces. A designated fire watch presence is needed.

(17) Work site and work hazard warning signs have been erected, escape routes have been checked as clear and posted. Portable gas, fire, high-temperature and other warning instruments are checked within calibration dates, if applicable, and available for use.

(18) The bypassing of warning and failsafe devices for work access is recorded and included clearly in warning notices. All restoration steps similarly identified for startup.

(19) The interested certifying authority or third party inspectors have been advised in advance, with the implicit opportunity to attend and inspect work where statutory requirements are involved[5].

(20) Disengagement, safety work complete, alignment checks, circuit reconnection, power on and startup procedures have all been identified, checked and made available[6].

---

5. Following the Cullen Report, the duties and responsibilities of certifying authorities are in the process of change (see Chapter 12).

6. Because of ongoing construction project work (see Section 9.8) present on many extraction industry sites to cater for changes in the process itself, it is likely that plant modifications will occur between shutdowns. For this reason, it is imperative that *procedures followed are based on* **up to date** *piping and instrumentation diagrams and revised issues of relevant isolation procedures.*

9.4 Work instructions

All managers have faced a situation where useful work has apparently been done but it is not what was wanted. A simple example for maintenance is when the wrong machine has been shut down and stripped for maintenance immediately after precisely the same work was done by the preceding crew, and the alternate machine which needs the work has been left untouched. There are countless similar examples in all fields which demonstrate the value of defining precisely what is required.

In maintenance, more than in most fields of work, there are many sources of confusion. The performance and maintenance history of a machine can be confused with a unit of the same type, maintenance teams often work at times when conventional sources of information are not available, and major jobs, although known, occur infrequently, failing to generate knowledge and familiarity born of constant repetition. In addition, maintenance workers tend to be mobile, sometimes working alone and sometimes with a different team. They are moved from one facility to another, stand in for missing colleagues, are seconded to other work and are reassigned or promoted, all of which adds to the likelihood of new hands to the task and *the need for current, accurate and complete work instructions.*

Although such work instructions are intended to define what is wanted, they should also be regarded as sources of information. Major operating organizations do not pretend that these work instructions are perfect, but they are intended to provide the *best knowledge and information that is currently available.* When an oil business maintenance manager was recently asked why work instructions were being used his reply was direct:

To provide the best and most current information available, and to ensure that required work is correctly done in a known safe manner.

The inclusion of the word 'required' is interesting. One of the less obvious advantages of using a PPM (pre-planned maintenance) system is the known period of maintenance inaction between major jobs which can be used to correct work instructions and include the experience of using the existing edition on the last job. By this means, maintenance history, planning and work instructions are combined into a dynamic progress to ensure the currency and definition of maintenance tasks, together with the most recent lessons that have been learned.

In the following typical WIC (work instruction card) for Ninian Central Gas Turbines (see pp.202–225), there are 69 work steps for the mechanical discipline alone, and the complete document consists of 22 A4 typed pages of instruction. This figure is large but not unusual, and results from many years of operating and maintenance experience coupled with the manufacturer's recommendations. It becomes clear that WIC creation and upkeep is a less than obvious major component of maintenance management system costs. Remembering that workloads of 10 000 maintenance interventions per year are known, and assuming a conservative average of five instruction pages per document, we are considering

one and a half million pages of issue over a 30 year plant life span, without changes and additions. This means that the motive forces behind WIC creation and upkeep must be very powerful, especially as the current commercial environment has located and pruned every cost-saving opportunity, while the WIC remains in constant use. Perhaps surprisingly, we are forced to recognize that of all the rules and documents attaching to maintenance, for preplanned work the WIC, however apparently mundane, is the most important:

it defines in detail what has to be done and the action sequence to be followed.

A WIC may be described as a document issued by the maintenance planning group, describing the work required for regular specified maintenance jobs, applicable to defined items of plant.

When we look at the document, however, it is soon obvious that the WIC is much more. The appendix at the end of this section describes in detail what the typical WIC contains, while the text below describes why WICs are used.

(1)   The WIC specifies the work to be done and the materials, tools and technical disciplines needed.

   The WIC is a major aid to work preparation.

(2)   For the operator, the appraisal of scheduled maintenance work changes from 'what we did' to 'what we intend to do', together with plans and schedules.

   Use of the WIC permits organizational preparation.

(3)   At any time on site, a large proportion of work under way is known in specified detail to the operating organization, whereas unknown activity means danger and inefficiency.

   Use of WICs reduces the incidence of unknown and pirate work.

(4)   In the work environment many different and unrelated activities are conducted simultaneously and in neighbouring locations.

   Use of WICs and schedules allows expected events and their combinations to be examined before they occur.

(5)   Whatever maintenance is required and whichever contractual conditions are in force,

   use of a WIC eliminates any doubt about what work has to be done.

(6)   When properly kept up to date, WICs provide current details acting as a focus for information management.

   Use of a WIC helps to reduce the dangers of outdated and inaccurate information.

(7)   During work preparation, both time available and resource uncertainty are improved.

Use of a WIC reduces prework urgency and logistical inefficiencies.

(8) By defining the job content, sequences and resources in detail, work management can be improved.

Use of a WIC improves all actions of work management and supervision.

(9) Maintenance plans, WICs and schedules, combined with other site work programmes, enable the safety group to identify and incorporate specific safety-related work requirements.

Use of a WIC improves the timing and relevance of safety instructions.

(10) Any improvement in activity scheduling allows a corresponding improvement in assessing the future.

WICs can act as the basis of work estimates and financial forecasting.

(11) In most operations, work is initiated by the 'customer' organization and passed along the command chain to the point where it is carried out. Maintenance and repair work sometimes reverses this arrangment, when work is instigated on site by machine failure and passed up the command chain until sanction is received.

Use of WICs and work schedules helps to restore the operators' initiative and improve overall control.

(12) Because WICs provide the basis for resource requirements and hence job hours forecasting for working time, they can influence.

the planned to unplanned ratio
which is often used to judge the balance and effectiveness of a work schedule (see Sections 5.6 and 13.3).

(13) Operational and maintenance use of any item of plant, coupled with the manufacturer's technical developments and design improvements, lead to changes in the WIC, the work required and the method of carrying it out.

A properly supported WIC acts as the focus of information changes, work practices, safety regulations and logistical support arising from operational and technical changes.

(14) Changes in the supply world can lead to the loss of a key supplier, the introduction of a useful competitor, changes in quality of maintenance or information support for customers, the inclusion of a broadening range of irrelevant equipment types, changes in format, presentation and numbering systems, and many other things, some of which will apply to the maintenance organization and some will not.

The WIC interprets relevant information and presents work requirements in a familiar format.

(15) Work preparation sometimes includes the training of new or transferred technicians before the job starts. For these purposes, and the need for initial revision of the job contents for existing personnel,

the WIC acts as an ideal basis for prework training.

Appendix

The WIC referred to in the preceding pages is described below, starting at page 1 and is reproduced in its entirety on pages 202–225. Note that each page has an issue number followed by a date in the form 12–8/93. The complete WIC follows the description and is reproduced by permission of Chevron UK Ltd.

Page 1

(1) Where the WIC applies is identified, along with the functional purpose of the subject equipment group. This is done through the maintenance code, in this case, NC 3001 – Ninian Central Platform – Gas Turbine Generators.

(2) Tag numbers of each machine of the functional group to which the work instructions apply are specified, in this case the three Gas Turbine Generators on Central Platform, GE1401A, GE1401B, and GE1401S. (Notice the use of the affix 'S' to denote the machine originally regarded as 'Standby').

(3) The document carries a Safety Precaution statement applicable to all work detailed in the WIC or jobs specifically identified (in this case one of the stated precautions is referred to the 10 000 and 20 000 running hours maintenance jobs).

(4) Switching points, together with the relevant panels and boards, are tabulated for each tag numbered machine to facilitate switch-off and electrical isolation.

(5) In the table headed 'WIC Details' the disciplines required for each of the job frequencies contained are shown. For instance, in the frequency column 06.01, the technical disciplines M, E (Mechanical and Electrical) are shown as needed for the work to be conducted at that frequency and contained as such in the document.

(6) The right-hand side of the table shows which frequencies require a test record to be completed for the two disciplines, where it is specified. In this case (Instruments) and (F&G, Fire and Gas).

Note Test records are used when specified measurements are taken during the maintenance work and are retained for future reference. In some systems readings are recorded on computer as part of the maintenance history.

(7) The notes at the foot of the page can be used to present different information relevant to the WIC.

(8) The final entry at the foot of the page specified which pages are presented for the different disciplines included.

Page 2

(9) The whole of page 2 is used to define electrical supplies, drives and heaters, tabulating them against junction box tag numbers, switch panels and drawing numbers, where applicable. Also shown are the MCC feeder trip supplies for each turbine.

Pages 3 and 4

(10) These pages present the equipment and materials required for each of the major PPMs. Information for 06.01 (six monthly), 10 000 hours, 20 000 hours and 40 000 hours maintenance is given with details of manufacturers' maintenance books, relevant inspection reports and the quantity and vocab (stock) number of parts required for the detailed task.

Pages 5–11

(11) These pages define the steps M1–M69 to be carried out by the mechanical discipline. Crosses included in the right-hand tabulation show which work steps are included for each of the frequencies shown; notice that from page 6 the monthly PPM (01.01) and the three-monthly (03.01) do not appear as further steps and are not specified. Similarly, the six-monthly (06.01) does not appear after page 6.

Pages 12–17

(12) Steps for the Electrical discipline, E1–E15, are shown on these five pages. Note that the steps are written as though the disciplines are separate from each other, whereas some interdependence is certain. Interrelation between working groups and supporting services is arranged by the maintenance supervisor at the time of issue. In addition, all the elements of work management are present, including task delegation, coordination among groups, specialists and supporting technicians, and safety briefings, task training and special instructions.

Pages 18–20

(13) Instrumentation steps commence on this page. IN1 is a general requirement applicable to the subsequent actions and SP(E) refers to electrical standard procedures. Note that IN2 requires the examination of numerous instruments listed to the bottom of page 20, which are shown in groups of the same devices. The column headed SP(I) lists the numbers of the instrument standard procedures to be used when doing the work.

Pages 21–22

(14) This section details IN3–IN12 in the more usual presentation for the attention of the Fire and Gas Technician (F&G).

(15) The Heating and Ventilating Outline Drawing for the three generators is attached for guidance when working with the WIC.

(16) Two blank test records (three sheets) are included with the WIC at each issue on the schedule. Measurements are noted on the test records as the WIC is completed. (Refer to pages 224, 5 and 6).

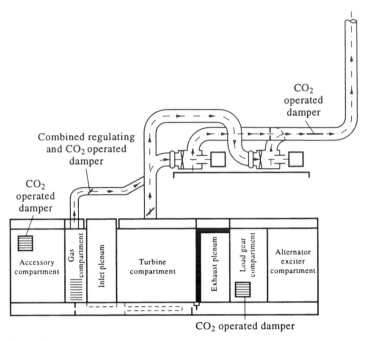

Figure 9.1
Outline drawing of gas turbine generator.

WORK INSTRUCTION CARD

Maint Code	NC 3001	**No. of Pages** 22
Tag No.	GE1401A	**Issue No.** 12-8/93
	GE1401B	
Equipment GAS TURBINE GENERATORS	GE1401S	

SAFETY PRECAUTIONS

"Ensure the equipment has been isolated and permits obtained as appropriate to the works being undertaken".

Ensure Gas Discharge System is locked off or on manual control while personnel are working in turbine/generator enclosure.

Isolate fuel systems, alternator from bus bars and diesel starter (10000 Hrs and 20000 Hrs only).

SWITCHING POINTS

SERVICE		SWITCHBOARD	SWITCH PANEL
GE1401A	Alternator O.C.B.	05-1	7
GE1401A	Motor Control Centre	05-3	LG1
125VDC	Supply	Isolate at Charger Feeder & Battery	
GE1401B	Alternator O.C.B.	05-1	13
GE1401B	Motor Control Centre	05-3	RF2
125VDC	Supply	Isolate at Charger Feeder & Battery	
GE1401S	Alternator O.C.B.	05-1	18
GE1401S	Motor Control Centre	05-3	LG2
125VDC	Supply	Isolate at Charger Feeder & Battery	

WIC DETAILS		TEST RECORD DETAILS	
Frequency	**Disciplines**	**Instruments**	**F&G**
01.01	M		
03.01	M, E		
06.01	M, E		
12.02 (10000 Hrs)	M, E, I & F&G	TR 1	TR 2
24.02 (20000 Hrs)	M, E, I & F&G	TR 1	TR 2
48.02 (40000 Hrs)	M, E, I & F&G	TR 1	TR 2

NOTES:

1. All 3 monthly and 6 monthly checks may be carried on an operating turbine.

2. Isolation for the 10000 20000 and 40000 hour maintenance is to be completed with due reference to the status of the system at that time and to the single line diagram E7498-00-51A. For isolation refer to the preceding safety section.

PAGE NUMBERS: M(2-11) E(12-17) I(18-20) F&G(21-22)

Equipment	GAS TURBINE GENERATORS			Page 2 of 22

Maint Code	NC 3001			Issue No. 12-8/93

SERVICE	JBE TAG NO.	SWITCH PANEL	DRAWING/SHT NO.
MISCELLANEOUS A.C. SUPPLIES			
A.C. Incomer	-	B7	7496-503-93C/10A
A.C. Voltmeters	-	A0	7496-503-93C/10B
10 Way Split 240 VAC Panel Board	-	AA1+	7495-503-93C/09A+
With No. 1 and No. 2 Transformer	-	AA2	7496-503-93C/09B
4 Way 120VAC Panel Board With			
No. 1 Transformer	-	AA2	7496-503-93C/09B
460V Panel Board & Transformer	-	C1-C6	7496-503-93C/17A
A.C. Under voltage Relay	27MC	AA0	7496-503-93C/15A
MISCELLANEOUS D.C. SUPPLIES			
D.C. Bus Bars & Incomer	-	BB5/BB6	7496-503-93C/14B
125VDC Panel Board	-	BB4	7496-503-93C/14A
Auxiliary Drives Master Alarm	49X	AA0	7496-503-93C/15A
Ventilation & Fire Detector			
Control Supply	-	AA0	7496-503-93C/15A
Ventilation & Purge Control Relays	-	AA0	7496-503-93C/05A
A.C. DRIVES			
Aux Lube Oil Pump Motor	88QA	A1	7496-503-93C/01A
Load Gear Vent Fans	88GV1/2	A2	7496-503-93C/02A
Turbine Ventilation Fan	88BA1	B1	7496-503-93C/06A
Turbine Ventilation Fan	88BA2	B2 & B3	7496-503-93C/06B
Aux. Hydraulic Pump Motor	88HQ	B5	7496-503-93C/08A
D.C. DRIVES			
Emergency Lub. Pump Motor	88QE	BB1	7496-503-93C/16A
Starting Diesel Starter Motor	88DS	BB2	7496-503-93C/12A
Hydraulic Ratchet Pump Motor	88HR	BB3	7496-503-93C/13A
HEATERS			
Lub Oil Heaters 15 Kw	23QT	A3	7496-503-93C/03A
Turbine Comp. Heaters 12 Kw	23HT	A4	7496-503-93C/03C
Accessory Comp Heaters 12 Kw	23HA	A4	7496-503-93C/03C
Alternator Anti-Con Heater 8.25 Kw	23HG	A4	7496-503-93C/03C
Control Cab Heaters 9 Kw	23HC	A5	7496-503-93C/03B

GE1401A MCC FEEDER TRIPPING SUPPLY　　　）
GE1401B MCC FEEDER TRIPPING SUPPLY　　　）　Isolate at Switchgear Tripping & Supervision
GE1401S MCC FEEDER TRIPPING SUPPLY　　　）　Distribution Cubicle as Appropriate

| **Equipment** | GAS TURBINE GENERATORS | | **Page 3 of 22** |
| **Maint Code** | NC 3001 | | **Issue No.** 12-8/93 |

EQUIPMENT/MATERIALS REQUIRED FOR 06.01 MAINTENANCE

DESCRIPTION	QTY	VOCAB NO.
GAS TURBINE		
HP Fuel Filter	3	08-2114-8203
Primary Fuel Filter	3	08-2114-6925

EQUIPMENT/MATERIALS REQUIRED FOR 10000 HRS MAINTENANCE

GE Inspection/Maintenance Guide GEK 28145.
John Brown Instruction Book No. 1052, Vol. 1.
Gas Detection Meter.

| Inspection Reports | ISE/GT-FF 6000, ISE/GT-FF 6002, ISE/GT-FF 6003, |
| | ISE/GT-FF 0004, ISE/GT-FF 6005 and ISE/GT-FF 6006. |

SPARES

DESCRIPTION	QTY	VOCAB NO.
TURBINE AIR INTAKE FILTERS		
Altair HV2 Bags	54	
GAS TURBINE		
Element - Control Oil Filters	4	08-2114-8203
Gasket Cover - Control Oil Filter	4	08-2114-8202
Element - Hydraulic Filter	2	08-2114-5203
Gasket 'O' Seal - Hydraulic Filter	2	08-2114-5204
Gasket, Cover - Hydraulic Filter	2	08-2114-5206
Diaphragm for Size 10-528 Actuator (False Start Drain Valve)		
- Fuel Nozzle Purge Valve	2	08-2114-4030
Main Lub Oil Filter	18	08-2114-6952
Coupling Oil Filters	6	08-2114-6502
Hydraulic Ratchet Pp Filter (Part No.225A9753P1)	1	08-2114-5025
Fuel Pump Hydraulic Filter (Part No. 225A9753P3)	1	08-2114-4251

EQUIPMENT/MATERIALS REQUIRED FOR 20000 AND 40000 HRS MAINTENANCE

As for 10,000 hrs maintenance plus the following:-

SPARES

DESCRIPTION	QTY	VOCAB NO.
Kit - Combustion Inspection	1	08-2118-1001
Hot Gas Path Inspection Kit	1	08-2118-2001
START DIESEL ENGINE		
Element - Air Intake Filter	2	08-2112-3270
Element - Fuel Oil Filter	1	08-2112-3257
Element - Fuel Oil Strainer	1	08-2112-3256
Element - Lube Oil Filter	2	08-2112-3501
Element - Crankcase Breather	2	08-2112-3282
Element - Air Filter, Boost Compressor	2	08-2114-3051
Belt-drive, Boost Compressor	3	08-2112-1969
Filter - Breather, Boost Compressor	1	08-2114-3032

Equipment	GAS TURBINE GENERATORS	**Page 4 of** 22
Maint Code	NC 3001	**Issue No.** 12-8/93

EQUIPMENT/MATERIALS REQUIRED FOR 20000 AND 40000 HRS MAINTENANCE

Cont...

SPARES

DESCRIPTION	QTY	VOCAB NO.
GAS TURBINE		
False Start Drain Valve	1	(08-2114-4801)
Part No. 226A1382P1		(15-0164-0101)
Fuel Nozzle Purge Valve	1	08-2114-4601
Part No. 235A5840P1		
Atomising Air By-pass Valve	1	08-2114-2202
Part No. 226A1006P2		
10th Stage Bleed Valve	2	08-2114-1701
20 PL Air By-pass valve	1	08-2114-4480
Part No. 917B495P3		(15-0651-3291)
VPR 2	1	08-2114-7051
Part No. DS 1360 P2		(15-0130-0101)
Maxi Torque Clutch	1	08-2112-3173
Spring - Gas Stop Speed Ratio Valve	2	08-2112-2373
Repair Kit - Gas Valve	1	08-2112-2501

Equipment   GAS TURBINE GENERATORS   **Page 5 of** 22

Maint Code   NC 3001   **Issue No.** 12–8/93

Item	Inspection Requirements *Statutory	01.01	03.01	06.01	Frequency 12.02 (10,000 Hrs)	24.02 (20,000 Hrs)	48.02 (40,000 Hrs)
	MECHANICAL/CONTRACTOR						
	NOTE: Maintenance packages 12.02, (10,000 hrs) 24.02 (20,000 hrs) & 48.02 (40,000 hrs) should be conducted with maintenance & Getsco personnel. Refer to John Brown Instruction Book No. 1052 installation and service standards for combustion inspection (12.02), Hot Gas Path Inspection (24.02) and major inspection (48.02) respectively.						
	MECHANICAL						
M1	Lubricate compressor bleed valve (20CB) solenoid plunger as per lubrication schedule.	-	-	-	X	-	-
	BOROSCOPE EXAMINATION						
M2	Carry out boroscope examination of combustion liners #2 and #8. Paying attention to crossfire tubes and collars. Examine transition pieces, 1st and 2nd stage nozzles and buckets. Prepare written report.	-	-	-	X	-	-
	Note: Ensure ratcheting gear is not usable during boroscope examination.						
	FUEL NOZZLES						
M3	Change out fuel nozzles as necessary.	-	-	-	X	X	X
	WASTE HEAT RECOVERY UNIT (NCP and NSP only)						
M4	Check security of waste heat recovery unit by-pass damper linkage. Lubricate link pins with high temperature grease as specified in lubrication specification.	-	-	X	-	-	-
	AIR DUCT VENT FANS						
M5	Grease air duct vent trunnion bearings. Grease Points.	X	X	X	-	-	-
	INLET GUIDE VANES						
M6	Grease (as specified in the Lubrication Schedule) the inlet guide vanes carrier at the five points available.	-	-	X	-	-	-
	CONTROL CAB AIR CONDITIONER						
M7	Inspect condenser and evaporator coils for accumulated dust deposits, clean as necessary. Check condensate drip pan drain clear. Wash air inlet filter in warm soapy water, rinse and dry before replacing. Check trouble free operation of unit.	-	-	X	X	X	X

Equipment	GAS TURBINE GENERATORS			Page 6 of 22
Maint Code	NC 3001			Issue No. 12-8/93

					Frequency		
Item	Inspection Requirements	*Statutory	06.01	12.02 (10,000 Hrs)	24.02 (20,000 Hrs)	48.02 (40,000 Hrs)

AUX GEARBOX COMPARTMENT

Item	Inspection Requirements	06.01	12.02	24.02	48.02
M8	Examine all auxiliary gearbox driven equipment for signs of fluid leaks. Look for signs of hot spots (paint discolouration) on atomising air compressor which usually indicates mechanical problems.	-	X	X	X
M9	Replace both main hydraulic supply filter elements. Replace Main Lub Oil Filters. Replace Coupling Oil Filters.	-	X	X	X
M10	Replace HP Fuel Filters. Replace Primary Fuel Filters.	X	-	-	-
M11	Start the auxiliary hydraulic pump (hand selection) and check it's associated bleed valve (AB-2) for signs of oil leaking from vent port. Replace valve if leaking.	-	X	X	X
M12	Visually examine the following motor driven pumps. In particular check coupling rubber bushes for deterioration. (a) Auxiliary Lube Oil Pump (AC). (b) Emergency Lube Oil Pump (DC).	-	-	X	X
M13	Examine accessory gear air ejector nozzle for erosion, prove air line clear of obstruction.	-	-	X	X
M14	Remove starting clutch coupling guard, examine for damage. Check minimum clearance between jaws in disengaged position - 0.38 inch. Line up clutch jaws for engagement, operate hydraulic system and engage clutch. Ensure smooth and rapid operation; check with jaws fully engaged that the clutch yoke pin is centred between lock-nuts with a clearance either side of 0.06 inch. CAUTION Lock out Hydraulic Ratchet Pump whilst guard is off. Refer to John Brown Instruction Book No. 1052, Volume 1, Tab 6, GEK.25823.	-	X	X	X

ATOMISING AIR SYSTEM

Item	Inspection Requirements	06.01	12.02	24.02	48.02
M15	Remove main atomising air compressor drive end gear inspection cover, examine gear teeth for damage, tooth fracture, excessive wear and uneven tooth contact.	-	-	-	X
M16	Prove atomising air separator orificed blowdown lines clear of obstruction, examine orifices for erosion and wear. Orifice diameter:- 0.328".	-	-	X	X
M17	Clean and examine atomising air line filters situated before solenoid operated purge valve (20PL) and air operated purge valve.	-	-	X	X

Equipment	GAS TURBINE GENERATORS	Page 7 of 22
Maint Code	NC 3001	Issue No. 12-8/93

Item	Inspection Requirements	*Statutory	Frequency 12.02 (10,000 Hrs)	24.02 (20,000 Hrs)	48.02 (40,000 Hrs)
M18	Remove gas supply line strainer element, examine for damage to mesh, wash thoroughly in solvent and brush clean, re-assemble using new gasket.		-	X	X

COOLING/SEALING AIR SYSTEMS

M19	Remove porous filter element (before solenoid valve 20CB), clean and replace. Flush through and clean all parts with solvent. Ensure constant blow-down orifice clear and manual blow-down line clear.	-	X	X

NOTE

Porous filter elements can be cleaned in any solvent, blow dry before re-instating.

GAS/STOP SPEED RATIO VALVE ASSEMBLY

M20	a)	Remove valve plugs and inspect for damage. Note - cover is under spring tension. N.D.T. plug springs by dye penetrant, check for cracks. Replace if required, Vocab 08-2112-2373. Renew seals etc.	-	X	X

ACCUMULATOR(S)

	b)	The accumulator(s) should require no maintenance. The gas pre-charge pressure should be checked and corrected as required at regular intervals. To do this, first shut the accumulator(s) shut-off valve, then open the accumulator(s) bleed valve to bleed all oil from the accumulator(s). Remove the pre-charge valve guard and cap and check the gas pressures with the pre-charge gauging device provided. Designated pressure will be found in the Turbine Control Specification found in the Equipment Details Section of the gas turbine instruction book. If additional pressure is required, obtain a supply of CLEAN, DRY NITROGEN and, using the charging hose assembly provided, bring the pressure back to the required psig. Replace the valve cap and guard and close the bleed valve, then crack open the shut-off valve for 5 minutes. Open the shut-off valve full at the end of 5 minutes.	X	X	X

This pre charge check may be performed at any time the turbine is not running. Do not attempt to perform this check during start-up or shut down or any time when the turbine load or speed are fluctuating.

DIESEL START ENGINE

M21	Check engine air box drain tubes are clear of obstruction, remove air box covers and clean air box of any water or accumulated oil.	X	X	X

M22	Examine engine throttle control linkage for security and possible "lost" motion; lightly lubricate pivot points with clean oil as per lubrication schedule.	X	X	X

M23	Replace engine air filter elements (qty 2). Check air inlet manifold connections for security.	-	X	X

M24	Replace engine crankcase breather element.	-	X	X

M25	Replace engine fuel strainer and filter elements. Vent filters of air and re-prime fuel system.	-	X	X

| Equipment | GAS TURBINE GENERATORS | Page 8 of 22 |
| Maint Code | NC 3001 | Issue No. 12-8/93 |

Item	Inspection Requirements *Statutory	Frequency 12.02 (10,000 Hrs)	24.02 (20,000 Hrs)	48.02 (40,000 Hrs)

STARTING ATOMISING AIR COMPRESSOR

| M26 | Examine compressor drive belts for wear, damage or deterioration. Check belts are correctly tensioned. | - | X | X |

| M27 | Replace booster compressor air inlet filter element, examine filter fabric for damage, ensure end seals are in good order. | - | X | X |

| M28 | Drain oil from booster atomising air compressor. Replenish to half level of sight glass; as per lubrication schedule. Examine bearing carrier vent holes for blockage. | - | X | X |

| M29 | Lubricate booster atomising air compressor inlet end bearings as per lubrication schedule using a low pressure gun, displacing "old" grease through grease relief fitting. | X | X | X |

HYDRAULIC IN-LINE FILTERS

| M30 | Replace the following in-line filters with new items. | X | X | X |

    (a)   Hydraulic Ratchet Pump Filter.
    (b)   Fuel Pump Hydraulic Filter.

DIAPHRAGM OPERATED VALVES

| M31 | Replace the following valves with new or re-conditioned items, or else renew diaphragms. | X | X | X |

    (a)   False Start Drain Valve.
    (b)   Fuel Nozzle Purge Valve.
    (c)   VPR 2.

AIR BY-PASS VALVE 20 PL

| M32 | Replace 20 PL with new or re-conditioned item. | - | X | X |

10TH STAGE BLEED VALVES

| M33 | Replace both 10th stage bleed valves with new or re-conditioned items. | - | X | X |

COOLERS

| M34 | Clean both lub oil coolers by water jetting tubes. Care must be taken when removing the end cover in order to avoid disturbing the tube bundle/sump joint. Clean atomising air pre-cooler by water jetting tubes. Clean Alternator Air Cooler. | - | X | X |

| M35 | Replace the control oil supply filter elements. (Duplex - 2 elements per side). | X | X | X |

ACCESSORY GEARS AND ASSOCIATED EQUIPMENT

| M36 | Run the auxiliary lub oil pump and check sprays unobstructed and correctly aligned. | - | X | X |

Refer John Brown Instruction Book No. 1052, Volume 1, Tab 6 GEK - 25766A.

Equipment	GAS TURBINE GENERATORS				

Page 9 of 22

Maint Code	NC 3001

Issue No. 12-8/93

Item	Inspection Requirements	*Statutory	Frequency 12.02 (10,000 Hrs)	24.02 (20,000 Hrs)	48.02 (40,000 Hrs)
M37	Remove accessory coupling cover. Clean coupling in situ and visually inspect teeth for wear and markings. Check coupling float by shuttling from end to end. Check radial clearances and record (0.004" max). Check coupling float and record. Check oil sprays with oil on. Refer inspection/maintenance guide GEK 28145 Pages 209-212.		-	X	X
M38	Remove accessory coupling, check alignments, clean coupling, inspect teeth for wear and markings, re-assemble and check coupling float by shuttling from end to end. Check radial clearance and record. Check coupling float and record. Check oil sprays with oil on. Refer inspection/maintenance guide GEK28145 Pages 209-212.		-	X	X
M39	Run auxiliary lub oil pump and check oil sprays for unobstruction and even spray formation. Renew PTFE Packings.		-	X	X
	LOAD COUPLING				
M40	Remove load coupling check alignments, clean coupling, inspect teeth for wear and markings, re-assemble and check coupling float by shuttling from end to end. Check radial clearances and record (0.004' Max). Check coupling float and record. Check oil sprays with oil on. Refer Inspection/maintenance guide GEK.28145 Pages 209-212.		-	-	X
	LOAD GEARBOX COMPARTMENT				
M41	Remove the particle trapping gauze from the core of the magnetic filters in lube oil supply line to the load gearbox and generator, and examine it for signs of metal pick-up: Clean the gauze and magnet in a suitable solvent, dry and refit.		X	X	X
	DUCTING				
M42	Examine the inlet, exhaust and silencer ducting arrangement externally paying particular attention to the following:- (a) Roller and guide supports. (NNP only). (b) General conditions of ducting. (c) Condition of expansion and section joints. (d) Condition of cladding and insulation.				
	AIR INLET HOUSING				
M43	Renew filter media and examine second stage coalescing panels for damage.		X	X	X
M44	Examine the seals between the modules and second stage filter coalescing panels for damage or deterioration, renew seals as necessary.		X	X	X

Equipment   GAS TURBINE GENERATORS   **Page** 10 **of** 22

Maint Code   NC 3001   **Issue No.** 12-8/93

Item	Inspection Requirements *Statutory	12.02 (10,000 Hrs)	24.02 (20,000 Hrs)	48.02 (40,000 Hrs)
M45	Examine the first and third-stage eliminator vanes for damage, thoroughly wash vanes with warm detergent/water mixture. CAUTION Ensure detergent used is suitable for use on aluminium alloys.	X	X	X
M46	Examine the seals between the filter frame and duct walls. Examine access door seals & repair/replace as necessary. Ensure that drains pipework is clear and functioning and that no water has collected downstream of the filter media.	X	X	X
M47	Wash down and clean housing side frames, examine paintwork for deterioration.	-	X	X
M48	Examine the seals between module flanges and mounting bulkheads and flexible duct connections for damage or deterioration.	X	X	X
	AIR INLET AND EXHAUST DUCTING			
M49	Inspect ducting and plenum chambers for structural damage or deterioration paying particular attention to silencers, turning vanes and diffusers in the exhaust plenum, and duct expansion joints.	X	X	X
M50	Inspect silencer panels for possible extrusion of infill media. Ensure all access doors are sitting squarely, and door seals are in good order.	X	X	X
M51	Examine anti-icing pipework distribution slots for salt encrustation, clean as necessary with warm detergent/water mixture.	X	X	X
M52	Examine module drains for debris, check insulation of drainage system is in good order. Check drain traps are fully primed. Check exhaust plenum drain primed.	X	X	X
M53	Replace pre-filter media.	X	X	X
M54	Examine blow-in doors for correct seating and check for ease of operation. Lubricate hinge pins with grease as per lubrication schedule. Examine door seals for damage/deterioration.	X	X	X
	EXHAUST SYSTEM			
M55	Examine exhaust duct bird guard for damage/deterioration of wire mesh.	X	X	X
M56	Check flame arrester gauze of vent from gas valve compartment is clean and in good condition.	X	X	X
M57	Check exhaust ducting support/expansion rollers are unobstructed. (NNP only).	X	X	X

Equipment	GAS TURBINE GENERATORS	Page 11 of 22
Maint Code	NC 3001	Issue No. 12-8/93

Item	Inspection Requirements *Statutory	Frequency		
		12.02 (10,000 Hrs)	24.02 (20,000 Hrs)	48.02 (40,000 Hrs)
	WASTE HEAT RECOVERY UNITS (NCP and NSP only)			
M58	Examine fins, tube and tube plates of heat exchanger for deposits. Clean with compressed air jet.	X	X	X
M59	Examine fins for evidence of burning.	X	X	X
M60	Examine edges of dampers for build-up of deposits, damage or pitting. Clean if necessary.	X	X	X
M61	Examine damper shaft in way of bearings for wear or evidence of binding.	X	X	X
M62	Check for security and wear of waste heat recovery unit by-pass damper linkage. Lubricate link pins with high temperature grease as per lubrication schedule.	X	X	X
M63	Witness correct operation of dampers before closing access door.	X	X	X
M64	Change diesel high pressure pump clutch. Check drive drum bearing is serviceable.	-	X	X
M65	Remove fuel injectors for servicing, replace with new or serviced set.	-	-	X
M66	Visually examine engine exhaust valve and injector rocker gear for general security of fastenings, springs for damage etc. Check valve tappet clearances and injector timing. (Exh Valves 0.016" cold). Refer - Detroit Operators Manual Pages 75-77.	-	-	X
M67	Drain engine lubricating oil sump (55 litres). Replace oil filter elements. Replenish engine sump with clean lubricating oil as per Lubrication Schedule.	-	-	X
	PRE-FIRED START			
M68	Carry out compressor wash cycle at crank speed.	X	X	X
	POST START			
M69	Fire unit to rated speed, bring to overspeed condition, check operation of electronic overspeed trip (5610 ± 30 rpm).	X	X	X
M69	Prove operation of the mechanical overspeed trip mechanism (5737 ± 50 RPM) by blocking the electronic overspeed trip, record trip setting. Do not exceed 5787 rpm. Refer speedtronic control specification 164B 2173.	X	X	X

Equipment	GAS TURBINE GENERATORS				Page 12 of 22		
Maint Code	NC 3001				Issue No. 12-8/93		

Item	Inspection Requirements	*Statutory	03.01	06.01	Frequency 12.02 (10,000 Hrs)	24.02 (20,000 Hrs)	48.02 (40,000 Hrs)
	ELECTRICAL						
E1	ELECTRICAL FUNCTION TESTS						
a)	Check operation of A.C. Auxiliary lube oil pump. Using test valve slowly bleed off pressure until A.C. pump starts. Check pump discharge pressure & cut-in point.		X	X	X	X	X
b)	Check operation of D.C. emergency lube oil pump. Using test valve slowly bleed off pressure until D.C. pump starts. Check pump discharge pressure & cut-in point. Check that A.C. auxiliary pump stops.		X	X	X	X	X
c)	Check correct operation of auxiliary hydraulic pump Select "Hand" and observe motor current.		X	X	X	X	X
d)	Check condition of alternator cable box desiccant and replace or reactivate it if more than half the desiccant has turned pink.		X	X	X	X	X
e)	Check the water cooled air cooler leakage detector pipes for signs of liquid coolant leakage. Tell tale signs are on the right hand side of the generator enclosure. Ref. J.B.E. Manual Vol 2 Tab 1 Page 3/6/A15.		X	X	X	X	X
E2	BATTERY AND CHARGER SYSTEM						
a)	Examine the battery and charger in accordance with SP(E) 8A & 8B.		-	-	X	X	-
b)	Discharge the battery at a current equivalent to 0.2C amps for five hours then boost charge at 0.2C amps until the terminal voltage remains constant for three hours.		-	-	X	-	-
c)	Test the battery in accordance with SP(E)8C.		-	-	-	X	X
E3	PRE-SHUTDOWN CHECKS						
a)	Check pilot exciter drive coupling and bearings for unusual noises etc.		-	-	X	X	X
b)	Clean the exciter pedestal bearing insulation of all dust and oil accumulations. Open alternator circuit breaker and ensure excitation is off. Check the pedestal insulation using a 12 volt supply and a 12 volt 3 watt lamp. Any current flow indicates a short circuit which must be cleared before exciting the alternator.		-	-	X	X	X

Note: Machine must be rotating during this test.

Equipment	GAS TURBINE GENERATORS	Page 13 of 22
Maint Code	NC 3001	Issue No. 12-8/93

Item	Inspection Requirements	*Statutory	Frequency 12.02 (10,000 Hrs)	24.02 (20,000 Hrs)	48.02 (40,000 Hrs)

E4   GENERAL EARTH BONDING

Check bonding at the following points in accordance with SP(E)1B.   X   X   X

a)   Control Cab
b)   Accessory Compartment
c)   Turbine Compartment
d)   Alternator Compartment
e)   Alternator Skid including H.V. Terminal Box
f)   Diesel Day Tank
g)   Neutral Earthing Resistor

For specific locations refer to Drawing 7496-503-76C.

E5   ALTERNATOR/EXCITER

a)   Check alternator stator/exciter anti-condensation heaters for security of fixing   X   X   X
and correct operation.

b)   Examine alternator drive end earthing brush box connections. Ensure brush is   X   X   X
free to move and renew it if it is less than 27mm long.

The brush grade must be correct (Morgan CMIS 25mm x 12.5mm x 40mm) and
the new brush must be bedded to the shaft profile.

c)   Check security and general condition of neutral CTs and their connections.   X   X   X

d)   Disconnect the alternator stator star point neutral connection at the neutral   X   X   X
earthing resistor. IR test stator in accordance with SP(E)1B.

Note: The "Brush" recommended minimum value is 5 megohms.

Caution: Before disconnecting the test instrument, short circuit all
three phases using the switch gear earth trolley to ensure
the windings are discharged.

e)   Disconnect the alternator rotor rotating diodes at the pigtail end and   X   X   X
disconnect the exciter stator. Using a multimeter check the diodes in
the forward and reverse bias directions and check the continuity of the
protection fuses.

f)   Short out the rotor diodes then using a 500 volt insulation tester check   X   X   X
isulation resistance of the alternator rotor, exciter rotor and exciter stator.
Record the readings obtained.

Note: "Brush" recommended minimum IR is 2 megohms. Reconnect exciter
stator and diodes and remove diode shorting links.

g)   Examine the undernoted items for ingress of dust oil or water and for signs of   -   X   X
tracking or localised overheating.

Stator End Windings
Rotor Rectifier Assembly
Main Exciter
Permanent Magnet Pilot Exciter

h)   Examine pilot exciter drive coupling for signs of wear.   X   X   X

Equipment	GAS TURBINE GENERATORS	Page 14 of 22
Maint Code	NC 3001	Issue No. 12-8/93

Item	Inspection Requirements	*Statutory	12.02 (10,000 Hrs)	Frequency 24.02 (20,000 Hrs)	48.02 (40,000 Hrs)

E6   <u>NEUTRAL EARTHING RESISTOR</u>

a)	With the alternator stator star point neutral connection disconnected at the N.E.R. examine the resistor grids for security of connections and damage then measure its ohmic value using a digital multimeter. Record the reading obtained which should be 15.87 ohms ± 0.1 ohm.		X	X	X
b)	Using a ductor type continuity tester measure the resistance between the earth end of the resistor and Platforms structure.		X	X	X

       <u>Note:</u>   Do not use earth lugs as test point. Maximum acceptance value is 0.1 ohms.

c)	Reconnect the alternator star point neutral connection at the N.E.R. Then using a ductor type instrument verify that the connection is good.		X	X	X

E7   <u>MOTOR CONTROL CENTRES</u> (For isolation and drives refer to Safety Section at front)

a)	Examine feeder switch panels in accordance with SP(E)1B.		X	X	X
b)	Isolate the switchboard supply to each MCC and isolate the D.C. supply at the relevant battery bank. Examine each of the MCC cubicles in accordance with SP(E)1B and SP(E)5B, and examine the associated drive motors in accordance with SP(E)7A where applicable.		X	X	X

       <u>Note:</u>   D.C. drives should be inspected generally in accordance with SP(E)7A but brushes, brushgear and commutator must be checked for free movement flats etc. and carbon dust should be removed using a vacuum cleaner.

E8   <u>GENERATOR CONTROL AND PROTECTION PANELS</u>

a)	Examine all components wiring and heaters in accordance with SP(E)1A.		X	X	X
b)	Visually examine all solid state assemblies and components for cleanliness tracking and security of components.		X	X	X

E9   <u>ANCILLARY EQUIPMENT</u>

a)	Examine lighting in the following areas in accordance with SP(E)9B.		X	X	X

       Turbine Compartment
       Accessory Compartment
       Alternator Compartment
       Load Gearbox Compartment
       Control Cab
       Filter House

       For isolation refer to Drawings 7496-503-93C, Shts 09A, 09B, 14A, 19A, 19B

b)	Inspect the remote control panel switches pushbuttons, meters lamps etc for security of fixing, correct operation etc.		X	X	X

cont'd

| Equipment | GAS TURBINE GENERATORS | Page 15 of 22 |
| Maint Code | NC 3001 | Issue No. 12-8/93 |

Item	Inspection Requirements	*Statutory	06.01	12.02 (10,000 Hrs)	24.02 (20,000 Hrs)	48.02 (40,000 Hrs)
					Frequency	

E9 Cont'd

c) Examine the electrical conduit and conduit fittings location in the following areas for security, damage and correct installation.

	06.01	12.02	24.02	48.02
c)	-	X	X	X

Turbine Compartment
Accessory Compartment
Alternator Compartment
Load Gearbox Compartment
Control Cab
Filter House

d) Examine JB5 and JB7 for ingress of water or dust condition of gaskets and security of connections etc. — — X X

E10 ALTERNATOR OIL CIRCUIT BREAKER

Carry out inspection and maintenance of the OCB in accordance with the undernoted Brush Operations Maintenance Manuals and Drawings:-   X — — —

Instruction manual No. 55/4027 for type R OCB.
Instruction Manual No. 55/4012 for types M14CJ1 and M14CK3 solenoid operated power closing mechanisms.
Instruction Manual No. 35/4029 for contact servicing on type VSI, VTD AND VB gear.
Instruction Manual No. 55/3011 Supplementary
Instructions for circuit breakers and mechanisms having frequent duty switching mechanisms.

Drawing No. C5323580 Sht 2 of 2 ARRG' of interlock (padlocking) (M14 Mechanism). This drawing shows the adjustments required for the circuit breaker trip defeat mechanism which is used during maintenance. It is important to check the clearances and to operate the trip defeat mechanism several times. If at any time the mechanism does not reset fully to the non-inhibit position, all linkages and return springs must be checked for freedom of movement etc. INCORRECT OPERATION OF THIS MECHANISM CAN CAUSE FAILURE OF THE BREAKER TRIP SYSTEM.

General Notes/Comments

(a) All OCBs installed in the Ninian Field have been modified for frequent duty switching.

(b) As a general rule, refrain from disturbing existing adjustments unless absolutely necessary.

Some nut fixings are marked with a red spot. These have been treated with Loctite preparation "Nutlock" and if they are disturbed they should be re-treated with this preparation for tightness/security on re-assembly.

Stiff nuts and locking plates should be renewed if they have been removed.

Equipment	GAS TURBINE GENERATORS	**Page 16 of** 22
Maint Code	NC 3001	**Issue No.** 12-8/93

Item	Inspection Requirements	*Statutory	06.01	Frequency 12.02 (10,000 Hrs)	24.02 (20,000 Hrs)	48.02 (40,000 Hrs)
E10 cont'd						
(c)	Particular attention should be paid to wear and travel components and over-travel mechanism, refer Brush O&M Manual 55/4012 Section 4.1; Solenoid Travel, refer Brush O&M Manual 55/4012 Section 4.2.		X	-	-	-
	Note: It is important that all dimensional checks as found and as left are recorded on NIMMS.					
(d)	OCB carriage; clean and lubricate the bearings, gears, jackscrew, guide pillars, carriage wheels and interlocks.					
(e)	OCB oil checks and test shall be carried out in accordance with SPE 2D.					
E11	**ELECTRICAL INSTRUMENTS**					
a)	In consultation with the Instrument Technicians examine and function test those instrument loops shown on the function test records.		-	X	X	X
E12	**PROTECTION RELAYS**					
	Caution: Never use files, knives or emery paper etc. to clean relay contacts. If cleaning is absolutely necessary then reference should be made to the vendors procedure for the use of the burnishing tool. Relay operation must be checked following any interference with contacts.					
a)	Examine all relays for ingress of dust and signs of contact burning.		-	-	-	X
b)	Remove dust from magnet pole faces, induction disc and armature using a fine brush which does not have a magnetic metal collar. Ensure no loose hairs are left inside the relay case.		-	-	-	X
c)	Check condition of relay case dust filter elements and clean or replace them as necessary.		-	-	-	X
d)	Carry out any further checks and finally secondary injection tests on the protection relays and record test results on test sheets - all as detailed in Protection Relay Maintenance and Test Manual.		-	-	-	X
	Note: The settings therein are taken from the relay coordination drawings and any deviation from them must be highlighted and investigated.					
E13	**SPEEDTRONIC CALIBRATION**					
	Carry out a full speedtronic calibration in accordance with SP(E) 19.		-	X	X	X

| Equipment | GAS TURBINE GENERATORS | Page 17 of 22 |
| Maint Code | NC 3001 | Issue No. 12-8/93 |

Item	Inspection Requirements	*Statutory	06.01	12.02 (10,000 Hrs)	24.02 (20,000 Hrs)	48.02 (40,000 Hrs)
				Frequency		

E14   PRE START-UP CHECKS

Item	Inspection Requirements		06.01	12.02	24.02	48.02
a)	Verify that all A.C. and D.C. power supplies are available and that the relevant battery has been reconnected.		-	X	X	X
b)	Check that all protection relay flags are reset.		-	X	X	X
c)	Lamp test all annunciators etc. and clear all annunciator flags and alarms.		-	X	X	X
d)	Ensure all selector switches are in the correct mode as follows;-		-	X	X	X

AC Lube Oil Pump	-	Auto
Load Gear Box Vent Fans	-	Auto
Lube Oil Heaters	-	On
Auxiliary Hydraulic Pump	-	Auto
Turbine Ventilation Fans	-	Auto
DC Lube Oil Pump	-	Auto
Diesel Engine Starter Motor	-	Auto
Hydraulic Ratchet Motor	-	Auto
Auxiliary Transformer No. 1	-	On
Auxiliary Transformer No. 2	-	On
Battery Charger	-	On

E15   GREASE TURBINE VENT FAN MOTORS

		06.01	12.02	24.02	48.02
GE 1403 AX - ASX		X	-	-	-
GE 1403 BX - BSX					
GE 1403 CX - CSX					

Equipment	GAS TURBINE GENERATORS		Page 18 of 22
Maint Code	NC 3001		Issue No. 12-8/93

Item	Inspection Requirements	*Statutory	Frequency 12.02 (10,000 Hrs)	24.02 (20,000 Hrs)	48.02 (40,000 Hrs)

INSTRUMENTATION

IN1 Note 1: Current instrument settings, ranges etc should be obtained from Platform Instrument Record/Data Cards.

Note 2: The explosion protected aspects of instrumentation and associated loop equipment should be maintained in accordance with BS 5345 and appropriate SP(E)'s & (I)'s, every 24 months or more frequently as conditions dictate.

IN2 Carry out an examination/test of the instrumentation listed below as and when indicated.

Tag No.	Description	SP(I)			
	PRESSURE GAUGES TAG NO. GE1401A/B/S				
PG01	DC Lube Oil Pump Discharge	05	X	X	X
PG02	Lube Oil Header	05	X	X	X
PG03	AC Lube Oil Pump Discharge	05	X	X	X
PG04	Main Lube Oil Pump Discharge	05	X	X	X
PG05	Compressor Discharge	05	X	X	X
PG06	Hydraulic Trip Circuit (Oil Fuel) Stop Valve	05	X	X	X
PG07	AA Compressor Manifold	05	X	X	X
PG08	Fuel Oil Before Stop Valve	05	X	X	X
PG09	Hydraulic Trip Circuit (Gas Fuel) Stop Valve	05	X	X	X
PG10	Lube Oil Filter Inlet	05	X	X	X
PG11	Lube Oil Filter Outlet	05	X	X	X
PG12	Lube Oil Filter Delta P	05	X	X	X
PG13	Coupling Lube Oil Filter Delta P	05	X	X	X
PG14	Fuel Oil Forwarding	05	X	X	X
PG15	Fuel Oil HP Filter Delta P	05	X	X	X
PG16	Fuel Oil Nozzle	05	X	X	X
PG17	Gas Fuel Upstream	05	X	X	X

Equipment	GAS TURBINE GENERATORS	Page 19 of 22
Maint Code	NC 3001	Issue No. 12-8/93

Item Inspection Requirements	*Statutory	Frequency 12.02 (10,000 Hrs)	24.02 (20,000 Hrs)	48.02 (40,000 Hrs)
Tag No. **Description**	**SP(I)**			

Tag No.	Description	SP(I)	12.02	24.02	48.02
Cont..					
	PRESSURE GAUGES TAG NO. GE1401A/B/S				
PG18	Gas Fuel Inter Valve	05	X	X	X
PG19	Gas Fuel Downstream	05	X	X	X
PG20	Hydraulic Manifold	05	X	X	X
PG21	Hydraulic Filter LP	05	X	X	X
	TEMPERATURE GAUGES				
TG01	Lube Oil Tank	12	X	X	X
TG02	Lube Oil Heater	12	X	X	X
TG03	Turbine No 1 Brg Oil	12	X	X	X
TG04	Turbine No 2 Brg Oil	12	X	X	X
TG05	Alternator No 1 Brg Oil	12	X	X	X
TG06	Alternator No 2 Brg Oil	12	X	X	X
TG07	AA Precooler Outlet	12	X	X	X
TG08	AA Compressor Discharge	12	X	X	X
	PRESSURE SWITCHES				
63QT	Lube Oil Low Pressure Trip	03	X	X	X
	DC Lube Oil Pump	03	X	X	X
63QN	DC Lube Oil Pump Stop	03	X	X	X
63QL	DC Lube Oil Pump Start	03	X	X	X
63QA	Lube Oil Low Pressure Alarm	03	X	X	X
63QD	Diesel Engine Lube Oil Pressure	03	X	X	X
63DM	Diesel Engine Fuel Oil Pressure	03	X	X	X
	TEMPERATURE GAUGES				
63FG	Gas Fuel Supply Pressure + No. Contact in Annunc. CCT	03	X	X	X
63FL (63FD)	Oil Fuel Supply Pressure	03	X	X	X
63HG	Hydraulic Trip Circuit (Gas Fuel)	03	X	X	X

Equipment	GAS TURBINE GENERATORS		Page 20 of 22
Maint Code	NC 3001		Issue No. 12-8/93

			Frequency		
Item	Inspection Requirements	*Statutory	12.02 (10,000 Hrs)	24.02 (20,000 Hrs)	48.02 (40,000 Hrs)

Tag No.	Description	SP(I)			
Cont..	TEMPERATURE GAUGES				
63HL	Hydraulic Trip Circuit (Oil Fuel)	03	X	X	X
63HQ	Hydraulic Low Pressure	03	X	X	X
63QQ-1	Lube Oil Filter Delta P	03	X	X	X
63QQ-2	Coupling Oil Filter Delta P	03	X	X	X
63AD	AA Compressor Delta P	03	X	X	X
63HR	Hydraulic Ratchet	03	X	X	X
63AF	Inlet Air Filter Delta P	03	X	X	X
63CS-2X	Inlet Air Filter House Delta P	03	X	X	X
	TEMPERATURE SWITCHES				
26QT	Lube Oil Header (Trip)	13	X	X	X
26QA	Lube Oil Header (Alarm)	13	X	X	X
26QN	Lube Oil Tank (Normal)	13	X	X	X
26QL	Lube Oil Tank (Low)	13	X	X	X
26QM	Lube Oil Tank (Moderate)	13	X	X	X
26AA	AA Compressor Discharge	13	X	X	X
26HA	Accessory Compartment	13	X	X	X
26HC	Control Compartment	13	X	X	X
26HT	Turbine Compartment	13	X	X	X
26BA	Turbine Compartment (High)	13	X	X	X
26TP-1	Inlet Plenum High Temperature (Trip)	13	X	X	X
26TP-2	Inlet Plenum High Temp (Alarm)	13	X	X	X
71QL	Lub Oil Tank Level Low (Alarm)	07	X	X	X
71QH	Lub Oil Tank Level High (Alarm)	07	X	X	X
71WG	Gen. Cooling Water Leakage (Alarm)	-	X	X	X
96CD	Comp Disch. Pressure Transmitter	04A	X	X	X
94FG	Fuel Gas Pressure Transmitter	04A	X	X	X

Equipment	GAS TURBINE GENERATORS		Page 21 of 22
Maint Code	NC 3001		Issue No. 12-8/93

			Frequency		
Item	Inspection Requirements	*Statutory	12.02 (10,000 Hrs)	24.02 (20,000 Hrs)	48.02 (40,000 Hrs)

FIRE AND GAS

IN3   Inspect the following in accordance with SP(E) 1B.   X   X   X

   (a)   Gas Zone Modules and Gas Detector Heads (4 in number of each) and
         connections between.

   (b)   High Temperature Thermostats (7 in number turbine 2 in number generator) and
         connections to/from Pilot Solenoid valves (2 in number) and pressure switch on
         each bank of CO_2 cylinders (turbine and generator) and connections to/from.

IN4   Carry out inspection/test of the following high temperature thermostats in accordance   X   X   X
      with Test Record.

      Isolate 125 VDC supplied by removing fuses in turbine MCC, after having informed
      Main Control Room.

      Relevant Standard Procedures are to be followed. Report if as found operating
      temperatures differ by more than 10% from required.
      Set Point

*	Acc. Compt Thermostat	45.FA-1	162° C/325° F
*	Acc. Compt Thermostat	45.FA-2	107° C/225° F
	Turbine Thermostat	45.FT-1	232° C/450° F
	Turbine Thermostat	45.FT-2	232° C/450° F
	Turbine Thermostat	45.FT-3	316° C/600° F
	Turbine Thermostat	45.FT-4	316° C/600° F
	Turbine Thermostat	45.FT-5	232° C/450° F
	Generator Thermostat	45.FG-1	162° C/325° F
	Generator Thermostat	45.FG-2	162° C/325° F

      *   Acc. = Accessory

IN5   Test the circuit from each of the thermostats in IN7 above as follows and complete   X   X   X
      Test Record. Relevant Standard Procedures are to be followed.

   (a)   Ensure that the pilot cylinder discharge lines (4 in no.) are disconnected
         from each of the CO_2 banks. Re-instate 125 VDC supplies by replacing
         fuses removed in 4 above, after informing Main Control Room.

   (b)   Apply an electrical short circuit to the thermostat connections.

   (c)   Check CO_2 discharges from relevant pilots.

   (d)   Check that the fire alarm bell sounds in the accessory compartment, the
         annunciator indicator FIRE activates, the alarm bell sounds in the control
         compartment.

   (e)   Reset by:-
         Removing the short circuit.
         Resetting the annunciator.

   (f)   On completion of these tests and those in 4 above replace each thermostat
         in its own housing.

Equipment	GAS TURBINE GENERATORS		Page 22 of 22
Maint Code	NC 3001		Issue No. 12-8/93

Item	Inspection Requirements *Statutory	Frequency 12.02 (10,000 Hrs)	24.02 (20,000 Hrs)	48.02 (40,000 Hrs)
IN6	Test each manual release unit and complete Test Record.	X	X	X
	43 cm-1 (Gen.) 43 cm-2 (Turbine) 43 cm-3 (Turbine)			
	(a) Operate each CO_2 manual release trigger in turn (turbine, generator).			
	(b) Check as in 5 (d) (accessory compartment bell, annunciator FIRE flag, control compt bell, and that both appropriate solenoid operated discharge heads have operated.			
IN7	Carry out inspection/test of the following (CO_2 released) pressure switches in accordance with Test Record and relevant Standard Procedures.	X	X	X
	CO_2 pressure switch Model 41644 Part No. 11765 fitted to:-			
	(a) Turbine CO_2 cylinder battery. (45 cp-2).			
	(b) Generator CO_2 cylinder battery. (45 cp-1).			
IN8	After completion of item 7 above and replacement/reconnection of each pressure switch into its system, test the control circuit of each as follow. (One to the turbine CO_2 cylinder battery, one for the generator CO_2 cylinder battery), and complete Test Record.	X	X	X
	(a) Pull the pressure switch button.			
	(b) Check that the system responds as in 5(d) above. The appropriate CO_2 discharged LED will be illuminated on the main control room F and G panel.			
	(c) Reset by pressing the pressure switch button and reset the F and G panel.			
IN9	On completion, replace pilot cylinders. Return partially discharged items to beach for refurbishment.	X	X	X
	Check weigh each CO_2 cylinder. Replace any whose content is less than 90% of indicated quantity. Check inspection date stamped on each bottle, if last inspection is more than 5 years ago, then replace bottle.			
IN10	Check the pressure on each CO_2 pilot cylinder gauge. Report if indicated pressure is lower by more than 50 psi (3.45 bar) than the following:-	X	X	X
	650 psi/45 bars at 50°F/10°C. 840 psi/58 bars at 70°F/21°C. 1250 psi/86 bars at 105°F/40°C.			
IN11	Visually examine the CO_2 cylinder batteries and the turbine and generator enclosures for damage and security of attachment.	X	X	X
IN12	Carry out inspection/test of the following pressure gauges in accordance with Test Record, relevant Standard Procedures are to be followed.	X	X	X
	Gauge fitted to each of 2 pilot operated solenoid valve assemblies on Turbine CO_2 cylinder battery and Generator CO_2 cylinder battery.			

INSTRUMENT TEST RECORD

Equipment	GAS TURBINE GENERATORS	TR No.	2
Platform	CENTRAL	Tag No.	
Maint Code	NC 3001 - 12.02 (10000 Hrs), 24.02 (20000 Hrs) 48.02 (40000 Hrs)	Sheet 1 of 1	
Discipline	INSTRUMENTATION (F&G)	Issue No.	2-10/8

TAG NO.	INITIALS/ DEFECT NO.	AS FOUND/ AS LEFT	TAG NO.	INITIALS/ DEFECT NO.	AS FOUND/ AS LEFT
FIRE DETECTION THERMOSTATS					
45-FA-1					
45-FA-2					
45-FT-1					
45-FT-2					
45-FT-3					
45-FT-4					
45-FT-5					
45-FG-1					
45-FG-2					
MANUAL RELEASE UNITS					
43-CM-1					
43-CM-2					
43-CM-3					
PRESSURE SWITCHES					
45-CP-1 (Generators					
45-CP-2 (Turbine)					

INSTRUMENT TEST RECORD

Equipment : GAS TURBINE GENERATORS	TR No. 1
Platform CENTRAL	Tag No.
Maint Code NC 3001 - 12.02 (10000 Hrs), 24.02 (20000 Hrs) 48.02 (40000 Hrs)	Sheet 1 of 2
Discipline INSTRUMENTATION	Issue No. 2-10/87

TAG NO.	INITIALS/ DEFECT NO.	AS FOUND/ AS LEFT	TAG NO.	INITIALS/ DEFECT NO.	AS FOUND/ AS LEFT
PRESSURE GAUGES					
PG 01			PG 18		
PG 02			PG 19		
PG 03			PG 20		
PG 04			PG 21		
PG 05					
			TEMPERATURE GAUGES		
PG 06			TG 01		
PG 07			TG 02		
PG 08			TG 03		
PG 09			TG 04		
PG 10			TG 05		
PG 11			TG 06		
PG 12			TG 07		
PG 13			TG 08		
PG 14					
PG 15					
PG 16					
PG 17					

INSTRUMENT TEST RECORD

Equipment GAS TURBINE GENERATORS				**TR No.** 1	
Platform CENTRAL				**Tag No.**	
Maint Code NC 3001 - 12.02 (10000 Hrs), 24.02 (20000 Hrs) 48.02 (40000 Hrs)				**Sheet** 2 **of** 2	
Discipline INSTRUMENTATION				**Issue No.** 2-10/87	

TAG NO.	INITIALS/ DEFECT NO.	AS FOUND/ AS LEFT	TAG NO.	INITIALS/ DEFECT NO.	AS FOUND/ AS LEFT
PRESSURE SWITCHES			TEMPERATURE SWITCHES		
63 QT			26 QT		
DC LUBE OIL PUMP					
63 QN			26 QA		
63 QL			26 QN		
63 QA			26 QL		
63 QD			26 QM		
63 DM			26 AA		
63 FG			26 HA		
63 FL			26 HC		
63 HG			26 HT		
63 HL			26 BA		
63 HQ			26 TP-1		
63 QQ-1			26 TP-2		
63 QQ-2			71 QL		
63 AD			71 QH		
63 HR			71 WG		
63 AF			96 CD		
63 CS-2X			96 FG		

9.5 Authority and responsibility

With good reason, classical management theory insists that authority and responsibility are balanced. In effect, no person can be held responsible for an action that he does not have the authority to carry out. There are countless examples in all areas of human organization which illustrate this principle, and the rule is just as applicable to maintenance activity as any other, but it is more easily violated. In the cockpit of pressure and broken equipment, such disagreements can easily result in bitterness and blame.

Maintenance work is regrettably notorious for disputes of this kind, and all maintenance managers have seen the effects on staff of pressure to cut corners, to restart before completion or to run up while the covers are being put back on. Apart from the damage to sound work management, there is an enhanced safety risk and the lingering unease of the maintenance technician himself. The latter can erupt into work rejection, with tools flung down in anger and despair. In spite of appearances to the contrary, it is usually the technician who is in the right.

For most major operations, equipment is released from production for attention by the maintenance team, and this release is accompanied by a transfer of responsibility. While work is carried out, the team are responsible for the technical integrity of the equipment, for the safety of personnel on board endangered by maintenance actions or equipment failure, and for the safety of their own personnel. This general responsibility is sometimes overlooked, especially in smaller organizations where the payroll does not extend to safety officers who regularly watch the wider safety implications of maintenance work[7].

Careful presentation is required before the work is started. Even for the smallest of jobs, danger is not proportional to job size[8]. All the principles of good work management, including clear definition, job preparation and task delegation, should apply, coupled with definite isolation, shutdown and handover procedures.

The 'isolation' process described in Section 9.3, apart from the technical step-by-step details, helps define the areas of maintenance authority and responsibility because of the technical division arising from the act of isolation itself. For maintenance work, delegated authority to act usually has a time limit specifying when corrected equipment is expected to be returned to operation (see note 8 and item 3 below).

Following the handover from operations to maintenance, responsibility for the equipment, the work required and the authority and opportunity to carry it out lies with the maintenance manager and his team. Up to this point the transfer to maintenance could be refused, although given the production group's famous reluctance to release equipment for maintenance it is hard to see when this would

---

7. Even this approach gives no guarantees. Technical people tend to think in detail, concentrating on the equipment concerned and the job itself. If the safety officer does overlook the wider effects in bigger plants, it can have massive implications.

8. According to the rules of risk, as assemblies increase in complexity the likelihood of failure increases. Danger, however, is another matter, and bears on human behaviour. It is sometimes the simple tasks or machines which receive inadequate attention and this becomes dangerous by default. (Refer to Table 2.2 item 3 priority modifiers page 32).

occur. It is more likely to happen in reverse; that is, after the maintenance work has been completed and the transfer handover back to production is refused. There are many reasons why this could happen, such as the job not done, the machine perceived as unsafe or a failed restart requiring further maintenance effort.

The important point to underline here is that, in spite of the usual way things are done, *the handover from production to maintenance and vice versa is not automatic. It is a formal step whereby both responsibility and authority are transferred from one part of the organization to another.*

In the outside world such transfers are vividly illustrated by colourful routines like the 'changing of the guard' or 'the ceremony of the keys', where the original reason is often lost in the traditional routine. Unfortunately, having carefully rehearsed the way things ought to be we must recognize that maintenance is different.

(1) In the 'ceremony of the keys', responsibility passes with the handover when the keys pass from hand to hand, but for maintenance work, when, after completion, equipment is handed back, *responsibility for technical integrity initially remains with the maintenance team.*

(2) Maintenance work is often carried out by a single individual: either a specialist or the only person needed. In addition, the work may have been conducted with minimum supervision, remote from other workers in the team. For these reasons, maintenance technicians are independent and *responsibility effectively passes further down the command chain to the individual maintenance technician*, which makes the work more interesting but requires more responsible behaviour from the individual.

(3) Although the authority to act is nominally passed to the maintenance team at handover, it is effectively limited. It is commonly found after initial equipment strip-down that the job is more extensive than expected, and estimated work durations will be greater. If the difference is important, this news is passed to production, who may demand return to service with limited or no work done and a longer intervention planned later. This is an effective veto on the maintenance work programme, and can be regarded as a limiting of the authority to act.

So, Maintenance work is often conducted

where the authority to act is limited but the responsibility is not.

(4) Maintenance or repair work is frequently applied to equipment that is about to expire. The work can involve worn-out mounting frames and castings, rusted moving parts, nuts and bolts fused together, changes in the molecular structure and properties of materials, dimensional changes through creep, temperature distortion or vibration, wall thinning through corrosion, and massive disassembly forces required for the removal of worn out components. It is not surprising therefore that estimated work time sometimes proves inadequate and failure to restart or unacceptable performance can follow work completion.

This latter effect is the so called *maintenance-induced failure*, where the implied assertion is that if the maintenance had not been done the failure would not have occurred. This is a difficult argument to disprove, and it does underline the importance of the correct frequency. Maintenance, however, is largely a consequence of operation. The problems referred to above will follow incorrect maintenance or incorrect use. Because of the vigour of maintenance work, approaching failure increases defects to be tackled after maintenance work.

(5) Maintenance work is sometimes conducted informally, without clearly defined objectives or work preparation. Although the operator suffers through inefficient work management, the maintenance team will suffer low achievement, poor motivation and the blame when things go wrong. The result is a self-fulfilling prophecy: because of low expected standards, inefficient service and poor expectations, that is what inevitably follows, and both maintenance and production teams suffer as a result.

9.6 Relations

It is sobering to realize that all efforts and improvements can be negated or overwhelmed by poor human implementation, and that gains available through operational change can be diminished by lukewarm attitudes or poor intergroup relations. This latter influence affects all groups with tasks in the workplace and some of these have been examined in Section 9.8. This section, however, refers to the interaction betwen the project and operational means of work management, where projects are temporary and of known duration and where some work difficulties are to be expected. There are no such excuses for poor relations between the three main operational groups (production, maintenance and safety), and somewhat unsurprisingly, maintenance, with strong links to them both, often suffers most when such difficulties are at their worst.

In some situations poor intergroup relations are created by the operating companies themselves, and the following are some of the classical ways of doing it:

(1) By denying sufficient resources for the required tasks.
(2) By assigning key resources simultaneously to more than one group in the work place.
(3) By setting destructive group targets, where one group's target can only be met at the expense of another.
(4) By assigning targets which are not realistically achievable.
(5) By failing to adjust targets and procedures to take account of other groups' activities in the workplace.
(6) For maintenance alone, the failure to adjust targets and resources assigned to work programmes when plants grow old and the maintenance work load sharply increases.

Any of these features will produce hostility or indifference, especially if management disapproval is certain from the outset.

Poor performance is certain if teams are not permitted to be successful.

Maintenance systems are most successful when shortcomings are used as opportunities for cooperation rather than reasons for dissent.

Highly effective systems are a great asset when good intergroup relations exist, but are of far less value when relations are poor.

The most obvious relationship to consider here is that between maintenance and production as the demands on the production department rise, spare production capacity decreases and the risk of failure multiplies. The temptation to retain equipment in production instead of releasing it for maintenance work is very strong. When machines are subsequently overrun, the difficulty and amount of maintenance work increases, extending the equipment downtime and ensuring even greater reluctance to release the item for maintenance work in the future.

These dilemmas are clearer when considering the ways that group performances are assessed. Production is measured in terms of 'output', while maintenance is measured in terms of 'availability', and it is not always realized that they are contradictory.

As production increases so does the maintenance need, Time required goes up and opportunity goes down.

It is a recurrent fallacy that increased production requirements can be met, at least in part, by squeezing the time assigned to maintenance (as opposed to improving maintenance efficiency)[9]. This approach can produce short-term results, and at low output requirements there is usually time for abundant maintenance work, but as machine loads and running hours increase, maintenance is constrained or neglected, failures are more frequent and availability is inevitably reduced, so the production problem gets worse. It is at this point that relations between groups can suffer, and maintenance will often find it worth repeating that

abiding by maintenance schedules will increase rather than decrease output when considering the longer term.

Here is the root of disagreements between maintenance and production: that as production strives to meet difficult output targets, reluctance to abide by maintenance schedules makes the 'availability' problem worse.

Safety first

Relations between safety and production are also vital, the common interest being the many ways that faulty machinery or its incorrect operation can injure human beings. One of the unnoticed assets of routine inspections, safety or maintenance, is the regular observations of machinery in operation. When machines are running smoothly it is easy to overlook the destructive power of stored energy and massive rotational forces, but safety officers or inspectors have the power of ultimate veto, whereby unsafe production operations may be shut down.

---

9. Increasing the 'availability' of equipment is a main purpose of maintenance activity and improvements are constantly being sought. The comment here refers to the temptation to reduce time made available for maintenance work without the corresponding maintenance improvements which would allow it to happen.

The aftermath of many safety demands is maintenance action, and the latter has two distinct interests:

(1) A direct regard of the technical status of the subject equipment.
(2) The incidence of additional maintenance opportunities while dealing with safety-initiated work.

Maintenance and safety requests for equipment attention often coincide. Poorly maintained equipment presents an operating hazard, endangering the primary function while putting personnel at risk. Unhappily, production perceptions are different, and in some cases safety and maintenance are the casualties of increased production demands.

Where maintenance and safety often agree on a job needed, the production perception of the situation is different.

Maintenance itself

There is a further example of intergroup difficulty, which ironically refers to different maintenance groups, between maintenance planning/engineering and the site maintenance team. Although broad agreement is usual, there are several subjects of contention. When disagreements erupt it is likely to feature one of the following:

(1) *Maintenance job frequency*
   When using calendar-based maintenance schedules, the frequency of regular tasks should be adjusted according to the results of the routine inspections. When equipment is performing well, in good condition and with few failures the work may be relaxed in one of three ways:
   (a) Make the whole task occur less often by reducing the frequency, e.g. making a 12-monthly PPM task occur every 15 months.
   (b) By transferring groups of the more predictable operations to another less frequent PPM routine.
   (c) By creating a new less frequent routine specifically for those tasks referred to in (b) above.
   If such adjustments are not made regularly, both production and maintenance groups will complain that the routine is done too often and there is no reward for sound performance.

   It is particularly difficult to persuade production to release a machine when the site maintenance team do not believe that it is necessary.

The difference between 'planned' and 'actual' routine frequencies also causes disagreements. In the formal sense, a PPM routine occurring every year in the same month has a planned frequency of 12 months. However, if it is performed in January the first year, April the second year, in July the third year, being three months late each time, while the 'planned' frequency is 12 months the 'actual' frequency is 15 months.

It is best to either adjust the frequency or avoid the delays. The option above is a source of false information.

(2) *Inaccurate records*

Incomplete or inaccurate information will result from poor data input disciplines when computer-based maintenance systems are used. It is not possible to detect trends or examine causes or equipment performance without a complete equipment history produced from accurate work records. Unfortunately, those gifted in technical work often find data disclplines most tiresome, and disagreement can result.

Data disciplines are more welcome to maintenance teams if direct benefits are seen to result from this work.

Incomplete records can also result in arguments about how recently the work scheduled has been done, with the assertion that the work record is obviously missing from the equipment history or incorrectly dated.

Once the maintenance history is known to be inaccurate or incomplete, challenges to the current work schedule will become more common.

(3) *Over-maintenance*

Adjustments to a PPM routine frequency can take many years to have a beneficial effect, especially if the change applies to a major work. If, for example, the two-yearly PPM produced two exemplary work records which resulted in reducing the frequency to every three years, the work would have been completed three times in seven years. If however, the three-year frequency had been applied at the start, the same work would have been completed three times in nine years. This is the effect of an active frequency adjustment on an originally high figure. If, however, the frequency adjustment does not occur, the resulting over-maintenance becomes permanent.

Selection of the correct maintenance frequency at the outset, together with regular adjustments, makes ideal frequencies happen more quickly.

Because all operational situations are different and maintenance history is not available at the outset, selection of the correct figure is difficult. Manufacturers' figures are generally over-cautious because of the penalties arising from poor equipment performance. The experiences of other users of the same equipment in similar processes can be most beneficial and are independent of manufacturers' concerns.

(4) *Equipment condition*

In addition to arranging production release, it is always difficult to turn off apparently healthy equipment for maintenance because of the requirements of a schedule. Such decisions are far easier for non-critical equipment, where production output is not affected, producing the unfortunate tendency for better maintenance of the less important equipment.

Arguments about equipment condition are often eliminated when condition-monitoring systems are applied to critical equipment. These give early warning of impending failures, initiating preventive work and lowering the risk of failure, without relying on a calendar-based schedule. Such systems are most beneficial when applied to a *modest number* of machines while retaining routine maintenance planning for the higher numbers of non-critical items. In addition, it is also true that, in many situations, tasks are overdue and the need for maintenance work cannot be seriously questioned.

(5) *Responsibility*

One of the benefits of a schedule is the removal of chance when work is required of different teams. A careful planned maintenance schedule will smooth work and resource requirements during the year and distribute the workload evenly among working teams. This technique associates specific tasks (such as the three-monthly PPM on the North Crane) with one team, say, Red Crew. If, by a quirk in the schedule, the task falls to Blue Crew, not only are they less familiar with the task, they may also reject it as not belonging to them: 'this is a Red Crew job and not one of ours'.

There are of course a thousand other reasons for disagreement, and perhaps a surprising condition is that arguments are not more frequent. When they do arise, they should be regarded as symptoms of deeper problems. Good working relations not only make working life more pleasant, they are a symptom of effective operations[10,11].

9.7 Treatment of standbys

Built-in standbys are not included in plant designs to make maintenance management easier. Although this is an important effect, they are intended to be a constant alternative to an operationally 'critical' item of equipment.

The principal feature of a standby is to be available.

The purpose behind high availability is of course to keep productive plant running

to ensure that parent equipment failure does not induce major downtime.

Notice that the motto does not say *no downtime*, as in some cases the switch to a standby machine may require at least a partial shutdown or limited downtime for the changeover itself but compared to the time required for a unit changeout or local repair these times are trivial.

There is some equipment, such as water injection pumps, which will reduce production output in the event of failure rather than cause downtime, and the text does not always make such distinctions clear, pointing mainly to the most severe cases.

---

10. Human reliability, a key influence in overall reliability and very relevant to maintenance affairs, is examined in 'Human aspects of reliability', in BS 4778, section 3.1, 1995.
11. Refer also to 'Human factors in industrial safety', HSE publication HS(G)48.

Whether standbys or surplus capacity are used to counter this possibility depends on the plant design.

Because of the need for instant functional availability, idle standbys must be frequently checked and maintained to avoid invisible failure. The nature of such failure, and a constant worry of the maintenance team, is that

invisible failures affect inactive machines and become evident at startup.

If a parent machine, during its normal operational service, breaks down or has to be taken out of service it is sometimes a case of switching *processing connections* to the standby machine:

(1) Partial production shutdown procedure giving throughput reduction.
(2) Parent machine 'switched off'.
(3) Production throughput changed over to standby machine.
(4) Production restart procedure, including 'switch on' of the standby machine.

Both the sequences and content of these steps will vary according to the plant design, the state of the parent machine and the processes at work. Shutdown and startup procedures are usually intricate and have to be followed meticulously; it is not simply the case that one machine is switched off and another is switched on. In production flow, the throughput quantities may have to be reduced overall before a switch to standby can be made, and the provision of the standby opportunity should not be jeopardized by uncertain techniques for bringing it into use. For the mechanics of transfer it is sound practice to

ensure that the switch from parent to standby is carefully defined in an operational procedure which has been previously tested and is known to be effective.

More importantly, there are safety implications (see 'For the maintenance manager', (item 1, below) which are applicable to all such changes, and such influences should be formally considered, *but men and machines are especially vulnerable if the fabric of process equipment is affected. Full 'isolation' and 'permit-to-work' procedures may be essential* (see Sections 9.3 and 11.9). It is also necessary to ensure that the operational portion of the process circuit is clearly identified, especially after changeover to standby and especially if there is a crew changeover at this time.

*The attempted disconnection of an operating item which is **believed to be transferred to standby can be fatal.***

Such procedures are likely to involve other machines interconnected through the process system and require adjustments to control set points, overall rates of throughput, diversion of selected feedlines and changes of valve positions. Other steps specific to the plant will frequently be included. If, immediately following the changeover, the troubled parent is removed for repair or reconditioning off site (note the safety implication referred to in maintenance managers' notes below), then action must follow at once, because at this point there is no standby and *any failure will induce downtime.*

Apart from the absence of the standby machine, two other problems arise:

(1) The existence of a standby can encourage unwarranted confidence of maintenance teams concerning the subject equipment, and

such attitudes can persist even after parent equipment has failed.

(2) When site crews change over it has been known for incoming personnel to be unaware of the status change of the standby concerned. Most supervisors are most dogmatic on this point:

the operational status of site standby equipment must be specifically reported during the end of tour handover.

Figure 9.2 shows three feedstock pumps, A, B and S, of which A and B are regarded as usual production units and switched on, while S is the built-in standby and usually switched off. In this situation, valves 1, 2, 3 and 4 are all open and valves 5 and 6 are closed. Fluid flows from the input header through the A and B flowlines to the output header, pressure increases and hence fluid transfer is affected by the operation of the pumps A and B.

The outcome of a failure of pump A for example is pump A shut down, valves 1 and 2 closed, valves 5 and 6 opened and pump S run up; similarly, should pump B fail the result is pump B shut down, valves 3 and 4 closed, valves 5 and 6 opened and pump S run up.

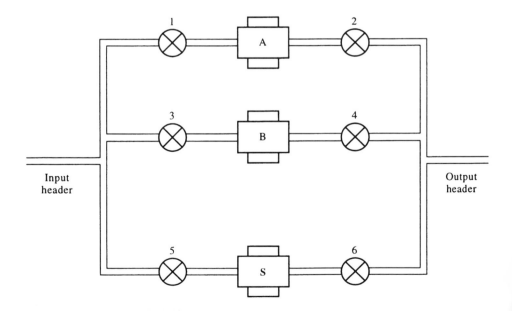

Figure 9.2
Feedstock pumps.

In this case, one standby supports two parent machines and some systems employ four or five pumps connected in parallel, employing one standby in the circuit. In such a situation, as the number of parent machines increases the case for a built-in standby diminishes. If a standby is considered as alternative capacity:

similar flexibility can be achieved by raising the number of working machines and running constantly below their rated capability.

Under-running usually extends equipment life, and the production loss of a machine failure can be made up by raising the output of the remainder. The risk of course is the greater chance of multiple failures, and standby machines tend to be more reliable than standby capacity. Also, the change in unit output needed can be severe, especially if the equipment group is small, leading to subsequent failures of other units in the group. As an example, the failure of one pump in a group of five identical units operating in parallel will cause a 20% loss in throughput, which can be met by increasing the output of the remaining four by 25%.

We have assumed here that the built-in standby machine is idle, and this may not be the best policy.

The under-running of all units can lower the load punishment on the group and reduce the risk of failure overall.

Such an arrangement is particularly attractive if loads are severe and fluctuating, causing frequent failure of parent machines.

In addition, there are many designs where the duty of the standby is partly to provide alternative capacity and partly to increase output during periods of peak demand. The operation of the standby machine does improve its reliability and it reduces the risk of invisible failure. Regrettably, the risk of failure is highest at the peak of output demand, when all machines are running at maximum and there is no further standby available. Furthermore, the resultant increase, far from being occasional, is often compounded by frequent demands for the same output, especially if demand growth can be met without further investment in production capacity.

Rising output demands increase the risk of equipment failure, the pressure on standby capacity and cost of downtime should it occur.

For the maintenance manager

(1) There are strong safety implications in preparing for and carrying out changeover, and these must be formally considered. The act of transfer from one operational arrangement to another is a change between two (hopefully) stable states, and it is the transitional phase which is the most uncertain. *In cases where process disconnections and reconnections are required, complete 'isolation' and 'permit-to-work' routines will be necessary.* In other cases, the operation of changeover valves may be all that is needed, but the safety implications then centre on subsequent changes required for repair or reconditioning. In all cases, check the requirements of isolation and permits to work in

conjunction with safety and operational personnel before deciding whether or not these procedures are required (see Sections 9.3 and 11.9). Refer also to 'Guidance on permit-to-work systems in the petroleum industry', HSE Books.

(2) Make sure that it is abundantly clear, through appropriate safety signs or operational indicators, *which items are in operation and which are on standby.*

(3) Ensure the constant availability of standby machines by frequent maintenance and off-duty running. *Check for invisible failures.*

(4) Make sure that maintenance routines are currently developed for such critical machines and that required maintenance schedules are followed.

(5) Ensure that 'switch to standby' operational procedures are fully documented and kept abreast of any platform changes. Ensure that they are known and understood by maintenance supervisors and part of regular training and new staff induction.

(6) Make sure that *'standby machine status'* reporting is part of handover procedures when site crews change out.

(7) Develop and document the removal, handling, shipment repair and return to installation of these relevant items of equipment to apply in the event of failure.

(8) Keep watch for signs of overconfidence with regard to built-in standby installations. The extra costs of such arrangements must not be negated by human forces.

(9) Consider alternative maintenance procedures if the active duty of standby machines changes.

(10) Be prepared to recommend investment in additional capacity when output demands steadily increase, eroding standby availability and reliability through increased running.

Required increases in output capacity are sometimes met by raising the number of identical units installed (for example changing the number of water injection pumps from six to seven), and maintenance personnel often think first in these terms, especially given their detailed knowledge of the existing equipment.

Such modifications, however, require careful evaluation, including fresh forecasts of production downtime, purchase of additional items and the installation project costs. It is possible that higher performance items could replace existing units, lowering the downtime and installation costs, but raising the purchase costs of the replacement machines, provided that requirements of power, piping, floor support, couplings, and cooling water or air do not introduce fresh limitations.

9.8 Construction projects

There are times when the most vigorous of actions to get things done end in anger and exasperation when the overall organization, after months of aiming for the near impossible, seems to turn objectives upside down, either making success impossible or trivializing key targets by suddenly replacing them with something else.

Such situations occur when several key work programmes compete for the same work space, the same support and some of the same resources. The assessment of overall priorities can result in a sudden change of emphasis and, if handled badly, can lead to the serious demotivation of key working teams and the permanent reduction of work output.

There are many major influences resulting from construction work, whether it is work from scratch or that applied to an existing installation, some of which are examined below. However vital the questions of project progress, technical achievement and cost performance are, it is the matter of the safety of all forms of construction which dominates management concerns and will help define the content and sequence of work carried out. (Refer to the HSE booklet 'Health and Safety in Construction', HS(G)150.)

The introduction of construction projects into plant continuing in operation are sometimes subject to such difficulties, where unrelenting pressure to meet tight targets can magnify difficulties. Although great care is taken to define the operational and production requirements of a major plant during the design stage, operating company requirements change and construction projects are introduced to modify the physical structure of the plant. This may arise because of geological changes, the introduction of a new product line, an increase or decrease in production capacity or a change in environmental aspects of the process. Some of this latter category have the force of statutory obligation, for instance the fitting of subsea platform isolation valves to oil platform gas input lines following the Piper Alpha tragedy.

(1) Construction projects affect normal operations. The requirements of working space, access to equipment, use of materials handling equipment, temporary storage, safety and life support are often necessarily provided by complementary reductions in the availability of these features to existing operational teams.

(2) Construction teams are project-organized, mainly independent of existing site command chains and employ different or transitory personnel. Non-project supervisors in particular often feel bypassed, especially if their usual workgroups are temporarily reassigned.

(3) Regular work programmes are often adversely affected. Production personnel strongly motivated by operational targets feel such effects most strongly finding it difficult to regain former enthusiasm once the project is over.

(4) Construction teams need different skills, employing the black trades in cutting, welding, local fabrication, erection, assembly and installation, working with many new items and materials plus some existing equipment. After installation work is complete, commissioning engineers are sometimes contracted from the original manufacturer.

(5) Construction projects work to short-term timetables. A single project duration of six months would be long if production operations were adversely affected.

(6) Because of the interruption of 'normal' work routines, both existing plans and working standards can be disrupted. The latter has most force with

safety standards and procedures; regrettably, it is not confined to the duration of the construction project, but if not addressed it will persist long after the project team have departed, and one of the recognized dangers is that the drop in standards proves difficult to redress and is potentially permanent.

With such features it is hardly surprising that working relations are often cool, and it is the most active and successful construction teams that produce the most impact. It is unfortunately the case that maintenance and construction group relationships are often the most affected.

It is easy to forget *that existing maintenance programmes will be affected*; there is an uneasy accommodation of the construction project's needs for space and resources and existing site-centred work programmes. Clearly, some maintenance tasks scheduled to be actioned during the construction project period will need to be rescheduled to before project start. This is particularly important if the construction project work involves the modification of critical equipment, requiring a partial production shutdown and work level increases for those production facilities which remain outside the construction work programme. Formal safety requirements, including 'isolation' and 'permit-to-work' routines, will be applied as needed to the construction project team by the safety officer, but he will also address the effects on other personnel, particularly those working nearby.

Such changes will probably include:

(1)   The diversion of existing flowlines or production routes.
(2)   Changes in duty for equipment remaining in production.
(3)   The change of some equipment from standby to full production work.
(4)   The potential overloading of all equipment engaged in production during the construction project.

These influences occur when site maintenance teams are demanned or reassigned and external maintenance contractors have limited access.

The failure of remaining operational equipment during the programme is more likely and the opportunities to put it right are diminished.

If maintenance teams remain on site it is sensible to plan their work in regions away from the construction work and be sure

that maintenance project work is reduced locally during the construction project duration.

The observations may seem obvious, but it is stunning to discover that construction projects and ongoing maintenance plans are often treated as totally separate, any mutual effect only becoming apparent when one or other work programme is affected.

During the run-up to the project there are two further factors for the maintenance manager to consider:

some equipment may be removed from site as part of the project, but must remain in full operation up to the time the project starts.

This is a fine judgement, especially if maintenance of the item concerned is already overdue and it is likely to fail. Nobody wants to conduct site maintenance work when the item is about to be refurbished or removed altogether. However, this thought should be underlined: *don't gamble with project costs*. Penalties can be much greater than the additional cost to the maintenance programme if there is a real concern do the work first.

> *Failure just before the project, having less time to recover, can ruin manning and shipping arrangements and must be avoided.*

The second matter refers to maintenance work after the construction project has been completed:

> *on many occasions, maintenance requirements will have changed.*

While the project is under way it is a useful time to revise future work schedules, and the following points need to be considered.

(1) Equipment due for maintenance work during the project, but delayed, should feature early in the restarted programme.

(2) Work will be needed for newly installed equipment to be added to the maintenance plan.

(3) 'Running hours' clocks and corresponding records must be returned to zero for equipment replaced by new or reconditioned equivalents.

(4) Similarly maintained aspects of such machines should have their datum points reset[12].

(5) Status inspection or condition monitoring routines should be changed to add extra applicable machines and omit reference to those that have been removed.

(6) Amend serial number/tag number records for affected machines which have serial number-assigned maintenance histories.

(7) Update maintenance histories of all equipment worked on during the project or affected by it using special work orders for the records of work with a maintenance or operational implication.

---

12. For example, a machine replaced with a reconditioned equivalent by the project should not be the subject of a major maintenance work order one month after the end of the project. Typically, if a PPM two-yearly is applicable to the machine, it should be timed for two years after the installation of the reconditioned item, not two years from the previous event, which applied to a different machine.

Chapter 10

Maintenance and quality

Quality management is a topical feature of the present business environment, with rising recognition of the importance attaching to the way that work is done. All activities conducted by the company, whether rebuilding broken equipment or typing a letter, are regarded as key elements of a cohesive whole, each significant in its own way. Should the quality of any one be diminished, sequential steps are adversely affected and overall organizational quality is reduced. Quality management recognizes all necessary activity in the workplace as contributing to a harmonious balance where the effective operation of one element encourages and depends upon satisfactory performance of the others. Maintenance is a prime beneficiary of such thinking, with specific improvements in its own activities and gains of departmental coordination and support.

Most of the topics discussed here are addressed more fully in other sections of the manual and the cross-references are noted. There is, however, a different emphasis: by looking at quality the spotlight falls in those areas of maintenance work where early improvements are most likely. It can, of course, be legitimately argued that it is the heart of maintenance, the actual spanner work (Section 10.6) where quality improvements are most beneficial, and that this is where the true ambitions should be. While there is truth in this, we should immediately note that quality applies to all steps and all activity in a company's operation. Secondly, the main constraints on maintenance quality are opportunity and work pressure. Improvements in job preparation and work management raise the effective use of maintenance team skill and improvements in work quality will naturally follow. Also, by focusing the attention on quality issues, early evidence of improvement becomes self-reinforcing, with the team demonstrating that it can be done.

Many companies have followed the operational and organizational changes required to achieve certification to BS EN ISO 9001/2/3, an effective method of quality management into a working system of standards, assessment and continuous improvement. After some initial adjustments, the application can be shown to work, and it is worth considering some of the more obvious aspects where quality improvement improves the maintenance work. (Refer to BS EN ISO 9000–1, Quality management and quality assurances standards. Part 1: Guidelines for selections and use, formerly BS 5750: Section 01.)

(1) *Quality of reporting*

Maintenance tasks are often triggered by a non-maintenance input, perhaps from performance statistics, condition monitoring or safety incident reporting. Improvement in detail and accuracy of such reports helps define the maintenance work required.

(2) *Supporting information*

The more pertinent and useful the job-supporting information, the quicker and more effective maintenance work becomes. All types of information benefit when the quality is improved, such as work instruction, layout drawings, system diagrams, operational specifications and performance standards. Engineering drawings and circuit diagrams merit special care: done well they act as the focus of effective work; done badly they introduce danger, uncertainty and inefficiency.

(3) *Computer records*

High-quality maintenance histories based on accurate data records created at the time ensure that perspective evaluation is applied to current failures and more effective forecasts warn of future events. Such histories are based on accurate data input; conversely, when data errors or omissions are present, fault interpretation is often wrong and the resulting repair or maintenance work is increased unnecessarily.

(4) *Effective preparation*

Maintenance technicians are required for their knowledge of plant and equipment, not for locating, shipping and gathering key work materials, where different skills are required. Improvements in the quality of preparation shorten the maintenance task, reduce downtime and make maintenance work more efficient.

(5) *The training of personnel*

Knowledge and skill lead to confident action and more effective work, which in turn leads to job time contraction and improved quality. The latter provides the bonus of less frequent breakdowns and fewer maintenance-induced failures.

(6) *Quality of work done*

The question of work done deserves some expansion. Although there are different ways of measuring quality, like the man hours required for a standard task, total job cost, or overall response time, one feature stands out from the rest – *the quality of the work itself*. Maintenance work is frequently subject to time pressure. The temptation to regard the work as temporary, something needed immediately to get production moving again, is very strong and the result is *reduced work quality*, increasing the number of failures and leading to higher costs and a further increase in pressure next time around.

(7) *Quality of supply*

It makes sense to ensure fully effective quality of goods and services supplied. Any failing causes extra unnecessary effort and time loss by the customer, and where maintenance supplies are involved problems are magnified as work is delayed, extended or needs to be done again. Further influences on production usually follow. Control of customer-supplied product is a comprehensive requirement of ISO 9001, 2 and 3, and helps to reduce a crucial element of uncertainty in maintenance work. (Refer to Annex D, BS EN ISO 9000–1: Part 1, 1994, and to Sections 10.1 and 10.5 below.)[1]

There are an almost countless number of working aspects where quality improvement aids the maintenance required and the maintenance that is done. Although neither the conduct nor the need for maintenance work is relished in many quarters, it fully deserves attention and respect. Quality management regards maintenance as a key component of operational success: it needs to be done well. There are other features which affect the quality of the host organization and, indirectly, the maintenance activity. These are tabulated in Annex D of BS EN ISO 9000–1 and their appearance in other parts of the ISO 9000 family is shown. The table is reproduced as Table 10.1.

The following notes describe situations where improvement in the operational quality will have an immediate discernable effect and where its absence can introduce losses and inefficiency.

10.1 Replacement goods inspection

One of the early company actions which directly refers to work quality is the inspection of incoming spare parts, and six aspects underline the importance of this work (see Sections 8.1 and 8.6.)

(1) *The right part has been ordered and delivered*
If the delivered part does not fit the broken machine and cannot be used elsewhere, however high its intrinsic quality its effective quality is zero.

(2) *The delivered part meets the technical requirements*
It is not unusual for replacement parts to be of a lower quality than the original, especially if it comes from a different source.

(3) *Time of order and time of use are sometimes years apart*
As discussed in options (2) and (3) in the next table, parts are often purchased in anticipation of their use. As equipment failure increases with operation and age, it can be years before the replacement part is needed.

---

1. Because of the infrequency of major maintenance tasks, preparation and training take on a special significance. Not only are the gains proportionally more dramatic, but they both offer further insights and familiarity which would otherwise not exist.

Table 10.1
Annex D of BS EN ISO 9000–1

External quality assurance				Clause title in ISO 9001	QM guidance	Road map
Requirements			Application Guide			
ISO 9001	ISO 9002	ISO 9003	ISO 9000–2		ISO 9004–1	ISO 9000–1
4.1 ●	●	0	4.1	Management responsibility	4	4.1; 4.2; 4.3
4.2 ●	●	0	4.2	Quality system	5	4.4; 4.5; 4.8
4.3 ●	●	●	4.3	Contract review	X	8
4.4 ●	X	X	4.4	Design control	8	
4.5 ●	●	●	4.5	Document and data control	5.3; 11.5	
4.6 ●	●	X	4.6	Purchasing	9	
4.7 ●	●	●	4.7	Control of customer-supplied product	X	
4.8 ●	●	0	4.8	Product identification and traceability	11.2	5
4.9 ●	●	X	4.9	Process control	10; 11	4.6; 4.7
4.10 ●	●	0	4.10	Inspection and testing	12	
4.11 ●	●	●	4.11	Control of inspection, measuring and test equipment	13	
4.12 ●	●	●	4.12	Inspection and test status	11.7	
4.13 ●	●	0	4.13	Control of nonconforming product	14	
4.14 ●	●	0	4.14	Corrective and preventive action	15	
4.15 ●	●	●	4.15	Handling, storage, packaging, preservation and delivery	10.4; 16.1; 16.2	
4.16 ●	●	0	4.16	Control of quality records	5.3; 17.2; 17.3	
4.17 ●	●	0	4.17	Internal quality audits	5.4	4.9
4.18 ●	●	0	4.18	Training	18.1	5.4
4.19 ●	●	X	4.19	Servicing	16.4	
4.20 ●	●	0	4.20	Statistical techniques	20	
				Quality economics	6	
				Product safety	19	
				Marketing	7	

Key:
● = Comprehensive requirement
0 = Less-comprehensive requirement than ISO 9001 and ISO 9002
X = Element not present

(4) *Deterioration while in stock is a real possibility*
When long storage periods apply, shelf life is an important consideration. Environmental factors can cause deterioration of apparently enduring items, so correct packaging and regular stock examinations are essential (see Section 10.2).

(5) *At the time of use the item may be out of production*
In many cases the pace of new machine development is more rapid than the demise of the items they replace. By the time a spare part is used, several new machine generations may have been in production, and the installed design has been discontinued.

(6)   *The right part was chosen in the first place*
We cannot seriously expect to stock everything, but it is most galling if a comprehensive and well-protected item of stock was missing the vital part which had actually broken down. This problem can be countered by the regular review of stock held with respect to more recent maintenance and operating experience.

When incoming replacement parts are delivered to the warehouse, they will probably be in response to one of the following situations:

(1)   The host equipment has failed and is either awaiting parts before stripping down or is already stripped and parts are needed to complete a repair.
(2)   A routine maintenance PPM is approaching and expected work arising will require the parts ordered.
(3)   A shutdown is being prepared and the spares are required for refurbishment.
(4)   Host equipment is approaching a critical age in its operational life when failures are expected, and pre-emptive work is planned to prevent failure.
(5)   Following the acquisition of new plant, replacement spares have been purchased to safeguard future operation.
(6)   It is also possible that parts have been ordered to top up stocks to existing levels which have dropped below a specified reorder level in the stock control system.

The main effect on maintenance here is the extra support and confidence given by a system well managed. However, the purchase of new parts is in response to previous issues and will probably refer to a job already completed. Stock control systems using this technique monitor historical stock movements to estimate future requirements and raise purchase orders accordingly. The method is very effective where stock levels and movements are statistically high and quantitative judgements can be made. The use of stocks for maintenance and repair is usually the opposite, and as the statistics of small numbers apply this approach is seldom used for maintenance stocks.

Although in each case the goods inwards inspector will apply similar technical criteria, his emphasis on more general aspects will vary. When incoming parts are required to complete a repair, as in (1) above, it is likely that the host unit was purchased several years previously and the replacement part is nominally brand new. Many manufacturers produce spare parts only during the production life of the parent machine, using the same production facilities for both and producing an estimated quantity of spares to be held in stock.

There are some incoming materials which will require a more thorough technical inspection, such as items manufactured to the customer's drawing, fragile parts where damage in transit is more likely, and items where a spread of performance is expected and sampling measurements are in use. When such techniques are applied to spares, extra processing time will be needed and maintenance engineers may be called upon to raise or approve technical concessions.[2, 3]

It is likely that the part is several years old.

The following steps are either to be dealt with by general goods inwards procedures or specifically applied to the handling of replacement parts. The notes refer to maintenance and repair materials. From a quality viewpoint the use of incorrect or damaged parts merely exchanges one problem for another. There is another point: when dealing with spare parts, stock turnover is low and storage times are extended, often for years (see above), so the stock control operation relies heavily on information from maintenance engineering.

Do not expect stock or warehouse management to hold your hand. If the wrong parts are held in stock it is largely a maintenance failure.

(1)   Before handling or lifting the item, review the method and equipment to be used. Pallet mounted items are usually moved by fork-lift truck, while larger assemblies will require a crane. Any lifting cables attached to the crate or container should be removed and replaced. Replacement cables must be in good condition, free of rusted or broken strands and within the currently defined period of six-monthly examination.

(2)   Check that the inspection code or inspection instructions applied to the order by the purchasing department conform. Always be prepared to query with maintenance engineering in the case of uncertainty.[4]

When dealing with spare parts maintenance engineering instructions are to be sought

(3)   Make sure that the incoming package is marked with the purchase order number, with the delivery note and relevant certification attached.

(4)   Locate the goods inwards copy of the purchase order and confirm the general contents of the delivery. Remove the certification, attach it to the purchase order copy and retain safely in the file.

---

2. The question of age introduces another problem, that of plant and equipment changes by the operator and the attendant effects on stock. Parts already in stock may no longer be usable, and different replacements are not ordered because existing stock levels look satisfactory. Alternatively, wrong parts are urgently ordered at the time of failure and the error is not discovered until the refit is attempted.

3. Replacement parts may be stored for long periods. Do not remove antivibration packing or break into sealed plastic covers without the inspector's sanction. Such decisions should be made after delivery, and in some cases a maintenance engineer's judgement will be needed. When parts are required for immediate use long storage considerations will not arise, but shipment to site still needs to be considered and in most cases units will have to be resealed and packed after the inspection has been made, which means perhaps that either the replacement parts have been held in stock for several years, or an alternative has been supplied.

The production of spare parts after the manufacture of the main unit has been discontinued is expensive for the manufacturer. It lacks both high value and high volume sales incentives, and when continued is done for customer support. Some companies specialize in the manufacture of discontinued spare parts originally produced by others and may buy design and manufacturing details from the original equipment manufacturer (OEM). Unfortunately for the customer, such items are expensive and often inferior. In the case of ppm the purchase is based on anticipated requirements with the frequent need for long-term storage, where their original supply condition must be retained.

4. Not all companies follow this practice, preferring to rely on goods inwards' experience and engineering instructions.

(5)   Where the removal of unit packing and protective material is required, detach with as little damage as possible, marking loose items and retaining carefully for reuse.

(6)   When unit repacking is needed before shelving, parts will need to be transferred to a repacking area. Consider whether temporary storage is required, and if original packing is not reusable ensure repacking meets the full protective standards required.

(7)   Confirm that purchase order conditions have been met and that the part numbers and type numbers are correct. Ensure that serial numbers have been recorded and apply to certification supplied.

(8)   Check for special handling and storage instructions. Confirm that transfer and stacking methods in the warehouse will not cause damage. Take note of fragile 'wineglass' or 'upside/downside' notation on the packet.

(9)   Conduct a thorough technical examination, identifying faults quickly, to avoid delay and allow correction while facilities remain available. Faults discovered years later are often impossible to correct.

(10) Check carefully if an alternative product or manufacturer has been supplied. If this has occurred, the maintenance engineer should be told. Although the part may look the same, several questions have to be answered:

(a)   Will the alternative meet the functions of the original, i.e. is it *fit for purpose*? Remember that there are probably secondary functions to be considered. Replacement pumps recently offered for offshore use, capable of the required delivery and correct mounting, were too short to provide support for adjacent structures.

(b)   Can it be installed? Are mounting and coupling details correct? If the replacement is bigger, is access to the installation possible? If the unit is heavier, is the available lifting equipment rated sufficiently and currently certified?

(c)   Are the materials and finishes used the same? An increase in physical properties does not necessarily make the item suitable. Hydrocarbon fluids containing hydrogen sulphide (H_2S) have to be contained in vessels of specific hardness value material when steel is used. Higher or lower values outside the correct range lead to material corrosion and eventual failure[5].

---

5. A major motor car manufacturer supplied cars with engines containing square-topped pistons. When the compression ratio of the engine was changed in a later model, new pistons were chamfered around the top circumference with corresponding changes to the cylinder head. The bore and stroke of both engine versions were the same. It had been carefully arranged that as the stocks of replacement old pistons ran out, new items could be used with only a marginal loss of power in old engines. What had not been expected was the use of old pistons in new engines. This happened in several cases and caused severe damage. There are of course four possibilities:
Old pistons in old engines.
New pistons in new engines.
New pistons in old engines.
Old pistons in new engines.
Three give no problems, but the last can destroy the engine.

These and similar questions described above will normally be answered by the Maintenance Engineer rather than the Inspector but the latter will require details of what to look for and in some cases a written specification.

(d) Can it be used without damaging the process or the host machine?
There are occasions when the restored performance of a replacement part transmits higher than customary loads to another part of the system and creates another failure.

10.2 Shelf life, handling and storage

It is frequently the case that the warehouse where maintenance parts and materials are stored and the site where they will be used are at different locations and require the transport of goods from storage to workplace when issued for a job. Such a step occurs after initial delivery from supplier to warehouse, often after a storage period of several years and after the material has become company property. Although this situation introduces several problems, it is inevitable, even when the warehouse is on site that the final leg of the journey is often more severe, and 'just in time delivery' arrangements currently in vogue are not suitable for maintenance materials.

Section 8.4, 'Packing, marking and inspection', describes the dismal consequences of incorrect purchasing of maintenance parts and the effect of them arriving on site. Here we assume that items have been correctly ordered and see that other factors have to be taken into account.

(1) Because of long storage periods, maintenance parts are particulary vulnerable to deterioration in the warehouse.

(2) Following delivery, parts are inspected for damage in transit to the warehouse, often requiring the removal of protective packing and increasing the likelihood of subsequent transit damage when shipped to site.

(3) This effect is compounded by the need for identity confirmation at the time of initial delivery, when sealed packets may be opened.

The problem starts with crates and packages arriving at the warehouse goods received section sent by the OEM or supplier in response to a company purchase order. Some items are delivered in small packets and some have to be lifted from the delivery lorry by crane or fork-lift truck. Major engineering assemblies usually supplied individually are often required on site in a hurry, while others may be part of a larger number, such as oil-processing Christmas trees, which will spend time in storage as replacement stocks for issue in rotation and subsequent return to the warehouse.

Items supplied in specially designed crates and packing assemblies cannot be easily checked by goods inwards inspectors without destroying the outer protection. Those manufactured to a detailed customer design specification will at some stage qualify for dimensional measurements, physical checks and quality assessments before acceptance, and the presence of the maintenance engineer is often required, in addition to the goods inwards inspector (see Section 10.1).

Items delivered in packets or envelopes do not usually suffer these problems. Goods are lighter and less susceptible to transit damage and can be more easily repacked if they are opened for identification. Goods supplied in clear plastic bags within the antivibration and outer packing may often be checked without breaking the moisture seals and be readily returned to the incoming protective box.

Very large units are treated differently again. Items like production pressure vessels are often the size of a small lorry, made of steel inches thick and weigh several tons. When lifted from the delivery truck to the warehouse yard, if one is dropped it is the building rather than the vessel which is likely to be damaged.

Such items are often delivered in open-sided crates where they are clearly visible and there is no doubt about their identity. Such transport and handling situations are rare and usually managed by the construction team working to preplanned steps through the whole project process.

Action for the maintenance manager

(1) In keeping with the outline in Section 8.4, make sure that all maintenance items on order are technically identified in detail and the parent plant item is also defined.

(2) Where items are manufactured to the customer's purchase order, make sure that detailed technical drawings and specifications are provided.

(3) For major engineering items, especially those referred to in point 2, provide for in-process inspection and release both during final inspection and after packing by your own company inspectors or third party inspectors appointed by you.

(4) Make sure that items in all groups are carefully and correctly identified, with company purchase order numbers painted and documents attached to the outside of the case. Second copies should be sent under a separate cover.

(5) Make sure that handling, lifting and storage instructions are marked on the outside of the packing, with overall weight, upside/downside, centre of gravity, standing surface and lifting points all shown.

(6) Do not overlook the thorough examination requirements which apply to company (used lifting appliances (ropes, beams, cranes) for materials handling at the warehouse[6].

(7) Draw the attention of the warehouse/goods inwards inspection supervisor to marking, packing and inspection requirements already on the purchase order and ensure that goods already inspected and released on site by your own inspectors are not unnecessarily opened.

(8) Check maintenance items held in storage, review their packing and shelving,

---

6. This is often a maintenance rather than warehouse responsibility. Lifting items in use must have been examined within the current six-month period and re-tested if modified or remounted.

When goods are delivered by lorry to the warehouse they often bring their own lifting ropes with them. Some operators refuse to allow such ropes to be used and treat any other lifting or moving appliance which arrives with the delivery with great circumspection. They are also very wary of boxes, pallets and containers checking for signs of deterioration before allowing the lift from the lorry to company property. Also be careful if the local handling and movement of material is contracted to a separate company which operates on your site. Responsibility for the regular examination of lifting appliances may reside with them.

and check for damage from adjacent items, water leakage/ingress, local vibration, local contamination, condition of working surfaces, poor packing, lack of protective grease and all requirements described in Sections 2.5 10.1 and 10.3.

(9) Prepare a specification covering the range of different maintenance items from the very small to the very large, and describe the handling and shipment techniques to be employed during transport of equipment from the OEM to the warehouse and for onward transfer from warehouse to site.

10.3 Packing, shipment and transport

For computers, electrical goods, electronic instrumentation and other fragile assemblies, careless shipment can be more destructive that its normal operational service. Although the dangers have become more appreciated in recent years, a fully crated major control panel dropped from a lorry trailer to a concrete quayside will cause severe damage to the unit, requiring its return for repair, an extension of the installation programme and furious arguments about money and responsibility. Apart from the more obvious effects of severe impact, other environmental variables will damage items in shipment, and not all effects are evident when the packing is removed[7]. (Refer to notes on shock and the shock logger in Section 3.6 and below.)

Temperature variations

Movement of goods from cold to hot parts of the world include a possible temperature rise of 100°C and vice versa. All component dimensions of close-fitting parts will be changed. Thermal expansion will increase in direct proportion to the relevant dimension and the temperature change, so large closely toleranced machine parts will witness the largest effect. When mating parts are constructed of the same material, relative thermal changes are small; however, when different materials are in use differential expansions apply and voids and working clearances are reduced, either increasing tensile loading or introducing compressive stresses, higher inter-component friction and increased wear. Similarly, the viscosity of lubricating oils changes if the temperature variation is wide or non-viscostatic oils are in use. Some oils poured easily in the tropics can be rock solid in Siberia.

The following typical applications and different effects will contribute to problems in all types of material and equipment. Almost nothing is immune and we forget them at our peril.

(1) All close-fitted machine parts in relative motion when used, journal bearing clearances, outer bearing engagement in mounting castings, sliding of pistons in cylinder blocks, cam on cam-follower clearances, metal to metal seals, and gate valve sliding clearances.

(2) All parts fitted under compression, such as interference fitting of components, the sealing of interfacial surfaces by compression of 'O'-ring seals or gland packing, compression fittings for copper piping, electrical cable glanding, the

---

7. Requirements for handling, storage, packing, preservation and delivery are specifically listed in BS EN ISO 9001: 1994, Clause 4.15.

tight fitting of screwed or push-fit studs into castings such as cylinder blocks and differential housings.

(3)   Apart from thermal growth, many materials' properties will change. Some will extrude more readily under pressure, while others will fuse with constraining parts. Thermoplastic materials soften at modest temperatures, savagely reducing their mechanical properties and permitting return to distorted shapes once temperatures are reduced.

(4)   The density and state of materials, particularly liquids, change as their temperature changes. The most obvious being water at 0°C or 100°C. Water trapped within a confined volume and expanding with temperature reduction below 4°C can destroy containing metal parts, split pipes, rupture vessels and burst through restraining seals.

(5)   Some seals perfectly effective against liquids at modest temperatures will fail to prevent the leakage of steam, petroleum vapour or gas. Some heavy oil products benign at normal temperatures can degas and ignite in the presence of air when heated by a welding torch or cutter.

(6)   All components including residual tension, such as metal springs, lifting hawsers, control cables, electrical wiring, pull-down fastenings such as cylinder head bolts etc.

(7)   All vessels and systems containing fluid during shipment, particularly lubricating oils, hydraulic fluid and coolants, will be affected. Static fluid levels will change, as will the likelihood of leaks and the affect on any wetted item. Shipment of toxic or corrosive fluids merits special containers and packing, and all recognize the increasing activity of cnemical processes with underlying increases in temperature. Increases in the pressure of closed vessels containing liquid are particularly dangerous, and glass vessels containing acids or strong alkilis packed in straw to combat vibration may explode at elevated temperature, destroying the vessel and allowing active chemicals to escape and cause more damage, while gas steam or liquid droplets can be borne upon the wind to attack exposed surfaces or seriously contaminate the air we breathe.

Most manufacturers are familiar with these problems, and have made design adjustments accordingly. However, a company used to distributing only in the temperate zones may face such effects for the first time, or perhaps you have included their unit in your assembly.

Temperature cycling

Repeated temperature reversals arising from perhaps daytime/nighttime variations will similarly reverse and reintroduce the effects described above. Temperature extremes are likely to be lower because of temperature gradients established through component packing evident in the timing effects of external temperature change, and the frequency of the reversals plus the rate of change will have an influence. Such effects are worthy of consideration when they are sudden, such as airborne transport from a cold to a hot climate, the journey finishing with a descent from the low temperature of high altitude.

Generally it is the journey start and journey end temperature which merit the most attention, but temperature cycling should not be overlooked. If temperature reversals are numerous, and over a wide temperature band, changes in the molecular structure of materials are inevitable. Plastics, rubbers, some metals or industrial porcelain can be taken beyond the normal plastic or elastic temperature range, making items more brittle and open to damage, especially during final shipment and installation when, with no internal heating, temperatures remain low. Finished surfaces, such as paint and plating, can be affected by leaking fluids or temperature cracking.

Pressure variations

Atmospheric pressure changes on different parts of the Earth's surface are small compared with the pressure-bearing capabilities of most industrial equipment. However, problems can arise with *differential pressure effects* during transport. Atmospheric pressures in non-pressurized cargo holds of aircraft at high altitude are in the region of 35 to 48 kN/m^2 (5–7 lbf/sq in), around half the figure at sea level. If containers are firmly clamped shut before shipment this pressure differential will apply across the containing skin if pressure release is not available, and will be in addition to any initial pressures when the vessel was filled. Changes in pressure during transit accompany temperature changes, and effects will be combined. For example, a 1 metre square hinged cabinet door at 48 kN/m^2 (7 lbf/sq inch) pressure differential experiences a tensile force on a single retaining catch of nearly 48 kN. When containment seals are at low temperature and the whole is subject to vibration, explosive decompression is quite possible.

Vibration

Vibration is a major influence on equipment in general and its careful measurement is used in condition-based maintenance (Section 3.6), where it is treated as an important topic. During shipment, equipment will encounter high-frequency vibration in aircraft and lower frequency more damaging vibration from rail, ship or road transport. Careful packing can form effective isolation of equipment from the source using soft compressible materials, trapped air bubbles, elastic supports etc., usually sealed in a waterproof plastic container contained in a rigid wooden or steel container. The latter can often accept impact damage to the rigid outside of the container without allowing its transmission to fragile internals.

Damage can be caused by surface fretting between adjacent components, change in the crystalline molecular structure leading to vibration cracking of brittle materials, and damage during lifting and installation. Impact arising from shifting heavy loads can shear appendages such as mounting feet, coupling flanges, inlet and outlet piping, attached instruments and lifting lugs.

Shock

Shock is the very sharp acceleration or deceleration of an object caused by the abrupt end to high-speed travel, the dropping of the item onto a hard surface, impact caused by explosion or the collision of moving objects. Tapping a coin on a desk can produce shocks of 400–500 ms^{-2} (approximately 40 or 50 g) and a car smash at over 160 kph (approximately 100 mph) could induce 100 000 ms^{-2} acceleration (approximately 10 000 g) in the vehicle. The surprising thing about shock is that although it is a frequent cause of damage during transit, high shock values do not necessarily correlate to high damage. It is the combination of acceleration and duration which characterizes destructive inputs. The duration of the shock is the 'width' of the pulse when the imparted shock is presented as a spike on a curve of acceleration on the y-axis[8] against time in seconds on the x-axis. The area under the curve is a measure of the total energy of the shock pulse and the greater the energy the greater the damage that can be sustained by an unprotected item.

This explains why high amplitudes of shock and vibration generally cause less damage at high frequency, because as the frequency increases the duration of the pulse is reduced.

Shock is the most frequent culprit when damage during transit occurs. Such damage may be the result of poor handling, sharp inputs from transportation or the collision of heavy and fragile items. Three important responses to transport-caused damage are being widely applied to packers and shippers of fragile products and equipment, encouraged by insurance bodies, transport specialists and manufacturing companies.

(1) A growing use of packing and shipping specialists whose aim is to ship fragile and expensive items to the customer without damage[9]. There is a change of emphasis here. The service sold is *transport without damage*, effectively competing with speed or price as the main business criterion. (Refer to the description of antivibration packing in the section on vibration above.)

(2) A much wider understanding of the damaging potential of poor packing and heavy-handed shipment following wider training, experience and customer demands for better service.

(3) Recently introduced devices can log the shock experienced by a transported item during handling storage and shipment, identifying precisely the quantitive values sustained and the time that they occurred. Such 'shock loggers', the size of a matchbox, can be downloaded to a PC at the end of their journey to display the full history of the shocks that they have sustained. There is much excitement, and the writing is on the wall. The old chestnut that it was 'like this before it arrived here' can be clearly supported or dramatically disproved (see Figure 10.1).

Moisture ingress

One of the most familiar causes of transit damage, moisture can affect most types of industrial products, but can be especially destructive when wood, paper, printed documents, books, powders and almost all foodstuffs are transported. It is particularly

8. '*g*' is the measure of acceleration equal to that caused by the gravitional pull of the Earth. It is equal to 9.82 ms^{-2} (32 ft s^{-2})

9. One of the attractions of shipment by canal in Britain, and an original support for their construction, was the movement of glass, porcelain and pottery by barge without breakage.

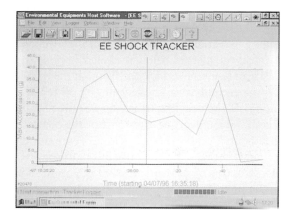

Figure 10.1
A shock logger and its output displayed on a PC (reproduced by courtesy of Environmental Equipments Ltd, Newbury, Berkshire).

effective because many packing materials, by soaking up surface water, will transmit moisture to all parts of the shipment.

Damage can include surface staining, rupture caused by swelling, surface corrosion, wood rot and cracking due to the formation of ice.

In general

There are other features which apply less often during transit, such as high-frequency sound, salt-water spray, smoke, dust, bacteria, termites and the more usual one of careless handling. It is the combination of these different influences which contaminates goods in transit and frequently appears finally as maintenance problems. One item which must not be forgotten is deterioration arising from prolonged storage, usually in a customs warehouse at the centre of a disagreement, when it is not safe to assume that equipment will not be damaged.

10.4 Site handling

The question of site handling of materials and equipment refers to the transfer between the place of installation and the site boundary. When considering an offshore oil platform, the site boundary can, of course, be 200 feet (60 metres) below the cellar deck of the platform to the deck of a supply ship. We also tend to overlook the fact that the transition happens in two directions, with new or reconditioned items being imported and used equipment for scrap or reconditioning off-site being exported. As we have already noted, these actions are usually linked with the removal and transport of the old unit, and require the transfer in of its replacement. This balance of action occurs during the management of normal operations and is perhaps the most usual way to consider site handling. However, during initial construction or subsequent plant extension such a balance does not occur, and although the singular movement of materials is simpler, the quantity in motion and its storage becomes a problem.

Site handling during construction

Although we indicate elsewhere (Section 2.5) that maintenance work begins prior to or during installation, the major portion of maintenance work follows installation and commissioning. For maintenance operatives the main interest in handling during construction is the possibility of damage causing subsequent maintenance problems, or a request for maintenance assistance to correct a problem which would otherwise prevent such installation work proceeding.

The construction project comes to an end at handover to operations, and because the time pressures are immense, demonstrated plant performance after commissioning is the short-term target, and minor or apparently trivial damage is not seen as a reason for delay. Construction work is often chaotic, with men and materials competing for limited workspace and access to overworked site resources. The sequencing of work to meet the various activities of the project plan depends on the availability of equipment and materials, often gathered at the workplace when space permits, thereby introducing exposure to damage or contamination. Waiting units may be adjacent to heavier more robust items themselves likely to be moved and with minor perturbation to inflict major damage on delicate items of instrumentation or control equipment while showing no sign of such an incident themselves.

Working space is further limited by temporary structures, short-term passageways, local material storage and mechanical handling equipment in their areas of operation. The transfer of equipment during these activities can be conducted in ways not tolerated or possible during subsequent maintenance operations. Use of chains or steel lifting cables attached to, or in contact with machined surfaces or easily fractured parts can cause damage (often invisible) after the assembly operation is complete. The crushing of seals, scoring of mating surfaces, fracturing of cast bodies or mounting feet and many other problems may only become apparent after months of normal operation.

Equipment and material quality is more difficult to maintain during construction projects, and some of the points listed below apply more strongly to such situations.

There is a further point to remember: the transition from construction site to normal operational site does not happen overnight. Indeed, on some sites construction crews remain as a near-permanent presence as continued plant modifications require. Facilities and practices needed during construction may remain as part of working methods, even though their use could not be justified by operational need. Putting efficiency questions to one side, it is the quality effects of such practices which will question the way in which things are done.

(1) Treat the transfer of equipment in either direction as a specific activity to be carefully prepared and carried out. Refer to the mechanical handling, the access route to be used, the timing of the exercise, its duration and the proximity of other structures, particularly replacement or removed assemblies.

(2) Taking the replacement unit from its protective packing may often be usefully delayed until installation is required and after the transfer to the assembly point has been made. If this is to happen quickly, the removed packing may sometimes be reapplied to the disconnected item to be exported for reconditioning.

(3) Directions for lifting and handling the package should be observed. These should be noted separately if the packaging, and hence packet outer symbols, cannot be preserved. The needless destruction of this material during unpacking is a regrettable common practice, mistakenly employed in pursuit of speed. Apart from the loss of the material itself, the packet often uniquely displays handling and storage instructions from the manufacturer.

For example, a water-pumping unit is designed to be mounted on a bed plate with compressive loads taken on eight cast 'plummer block' feet, four at each end of the unit.

The lifting of the unit by attaching a steel chain around one of these mounting feet applies the full weight in tension at one point, as opposed to one eighth in compression that the design is intended to take.

It is usual to lift such items by attaching steel cables to four corners of a bed plate or pallet, which in turn carries the pump lifted from underneath. Overhead load spreader beams are sometimes also used to ensure that applied loads and lifting cables are vertical and strength reduction of the cables through angular loading is eliminated.

(4) The quality and location of site storage areas need to be examined, especially when short-term storage is required. Removal of outer packing should be specifically avoided if extended storage is likely, as exposed machined surfaces and machine internals will be damaged by environmental effects.

Storage in surface water or soft ground, exposure to wind-blown sand, stacking on fragile or unstable structures and in conjunction with heavy or coarse material should all be avoided.

(5) Stack temporarily stored items of machinery correctly, oriented individually on their mounting surfaces. They should not be stacked on top of one another unless specifically designed to do so.

(6) Locate locally stored equipment close to supplies of air, water and electrical

and hydraulic power. Follow a pre-use maintenance routine, as described in Section 2.5, and use the correct tools for any preparatory work required.

(7) Check site stock, either located in temporary stores or positioned adjacent to installation points, to ensure freedom from local damage, proper stacking and preservation. Look for 'forgotten parts' (items delayed or overlooked), and consider the materials rather than the project activities which will call for them.

(8) Look out for replacement units that have been robbed of working subassemblies before they were ever shipped for repair. The broken units that have been replaced will also be on site, but may be elsewhere and may be discarded as scrap[10,11]

10.5 The performance of outside repairers

In Section 8.1 we referred to a maintenance man's golden rule as the 'life of operating machinery depends on the availability of spare parts'. A similar but more powerful dictum applies to the plant as a whole, but although it has more significance it suffers at the hands of the departmental organization of large users, where the presence of strong commercial imperatives mean that it is repeatedly regarded as somebody else's responsibility. The rule is simple and powerful.

The operating life of a plant depends on effective repair facilities.

Remember that we are not considering a young process, when equipment is new and in reliable condition; we are thinking of the length of operating life of a mature plant when equipment fails and things go wrong

When new plants are commissioned, huge capital investments are based on predicted technical performance and on estimated long-term commercial advantage. The subsequent loss of key operating machinery can destroy these assumptions and turn a wise investment into catastrophic ruin[12].

Under the present influence of intense competition, many plants require several years of successful operation before they move beyond capital breakeven, at the very time that the plant has matured and repair facilities take on their formidable role.

Once the rule is spoken it seems like a statement of the obvious, but it is surprising how often adversity is needed to make the lesson plain. In 1986, when Saudi Arabia became tired of being OPEC's swing producer, she opened the valves and forced the international price of crude oil down to around $9 a barrel (the present price varies between $17 and $22). Apparently with little choice, several international

---

10. The car business calls such assemblies 'workshop queens'. Vehicles awaiting repair lose key components if incoming replacements are delayed. By the time the new replacement is fitted, something else has been removed. The result is a permanent casualty in the workshop. Operating sites can be like this: lost or broken stock can be discarded into dark corners and overlooked as the pressure to act moves elsewhere.

11. In the manner of a chain being as strong as its weakest link, the use of robbed 'second-hand' subassemblies can impair the performance of an apparently new unit.

12. With this in mind it may make sense to justifiably revise the maintenance purpose from keeping equipment running to *keeping the plant alive* (see Section 3.7).

oil producers cut the placing of work to outside companies, retaining activity in-house wherever possible and taking the view that survival of the major customers was essential if the sector was to remain in long-term operation. Although this tactic was probably the only one that made overall sense, some supply companies never recovered, and the service sector had to be specifically rebuilt several years later when the crisis had passed. The actions of rebuilding specifically recognized the importance of the service sector, and when thinking mainly about maintenance work this includes repairs.

Notice that the rule refers to effective repair facilities in general, without specifying whether they are based internally or externally or whether they are separate commercial organizations. As the story above would underline, the role of commercially separate contractors is very strong. In fact, most major operators require a balance of such facilities, selecting by specialist skill, resources and availability according to the need.

Having said all this, why is this topic, if it is so important, in a chapter devoted to quality? Firstly, it is *because* quality is so important that it is in this section (the logical result of this thinking is to put quality first, with everything else subordinate to it). Secondly, let us take a more practical maintenance man's view of it with perhaps another simple dictum.

Poor quality means doing the job again.

Even this is a bit presumptous. Some repair jobs cannot be done again; you get one chance and it must be correctly done. Under these circumstances a bad repair is the worst result of all: a critical production item can go down and stay down.

What emerges is that when engaging in repair work quality is the dominant feature, when contemplating a conventional purchase, choice resides with the buyer, if the product or service is of poor quality, he can seek an alternative, with a repair this is far more difficult. The item to be repaired already belongs to the user, and because of operating pressures the repair cannot be ignored. The choice is far more restricted and time is pressing. Furthermore, the job uncertainty and intrinsic safety conditions are exported to the repairer, but although he is primarily responsible for the quality and safety of the work, the user's responsibility does not end entirely. The state of the equipment handed over must be made clear to the repairer, since many repairs will be conducted on the repairer's premises away from the site and knowledge of experienced users. When the repairer receives the equipment he may be seeing it for the first time.

Do not assume the repairer's knowledge of equipment condition or experience of danger.

Requirements of the repairer

Whether the repairer is a separate commercial organization or not, a strong relationship between customer and supplier is essential, and more necessary than conventional contractual requirements. Quality refers to all the features of the service and none of the following should be overlooked.

(1) *Essential skill*
Because repair work is so directly linked to plant operations, it is essential to avoid risk. In spite of the enthusiasm of the repairer or his previous success in other areas, it is best not to undertake repair work where the specific skill is not present.

(2) *Resource availability*
In many cases repair work is required in a hurry, especially when it relates to critical equipment. Tools, factory space and replacement parts must be available if short turnarounds are to be met.

(3) *Service experience*
Customer pressure applied to service organizations is specific and is usually couched in the user's rather than a manufacturer's language. The commercial outlook is different from most others, and experience of this type of activity is important.

(4) *Knowledge of customer needs*
Beyond the information required in normal contractual dispositions, repairers benefit from an extensive knowledge of their clients' operations. Although the repair skills offered are very similar, a wide variety of operational circumstances can apply to the various orders.

Once a repairer has been selected and his services found to be satisfactory, it helps his readiness and commercial viability if he is provided with a regular flow of work, even when production-critical items are involved. Use of replacement machines and a regular rolling reconditioning programme is an effective way of maintaining an essential repair facility and preventing equipment failures, with all the attendant risks. When arranging for repair work to be carried out, refer to the following steps:

(1) When engaged in a rollover programme, be sure that a reconditioned (or new) and tested replacement is available, together with the means of disassembly and immediate reinstallation plus the means of transport of the faulty item to the repairer (see Section 8.9).

(2) Complete commercial arrangements with the repairer prior to the arrival of the subject equipment on his premises, including all normal purchase order details, even where an overall contract exists. Define:
    (a) Normal details of price, time, availability and transport arrangements.
    (b) Repair services required, outlining general disassembly steps and follow-up actions.
    (c) Report observed symptoms, restrictions in unit performance and reasons for the repair.
    (d) Give current status and knowledge of details of any wash-down or preparatory steps that have been taken.
    (e) Repeat known hazard warnings, handling difficulties and potential dangers, making certain that the repairer is specifically informed of relevant safety knowledge gained during operations.

(f) Specify level of technical reporting required at different stages of disassembly and repair, together with customer in-process inspection and approval steps included.

(g) Methods of final inspection and test to be applied after the repair work has been completed, particularly requirements for end customer or third party inspection and certification.

(h) Specify return delivery schedule or local storage needs.

(i) Define requirements for delivery/storage packing crate marking plus method of handling and transport.

(3) Ensure that the repairer is aware of the order's significance to your operation. When emergency steps are needed it is vital that he knows the position. Be careful not to demand an emergency response which is not needed; using the facility only when needed will preserve its potency.

Apart from the requirements of individual orders outlined above, we have to address the original question of performance, and the following steps will aid performance and service repairer selection. Consider the value of company partnering arrangements, and in respect of each company maintain records which:

(1) Record normal company details plus full details of the repair service, including normal work location, response times and resource availability.

(2) Identify the key skills and experience on offer, adjust the entries as they change and indicate those areas where future offerings would be evaluated.

(3) Specify service contact names, company regular communication, out of hours fax and telephone numbers, and geographical areas of operation.

(4) Detail support workshops, their sizes, location and facilities, including access, storage and lifting.

(5) Describe the repairer's familiarity with specified manufacturers. Do they specialize in certain types of equipment or manufacturer, or do they engage in certain processes?

(6) Keep a record of all work placed with them, classifying and recording the results of such work. Link these records to the maintenance history, and pay particular attention to work rejection or repeated repairs on the same unit.

(7) Keep the repair company under regular review. Conduct visits and audits, recording and completing all audit requirements.

(8) Develop the relationship between the two companies. Make sure that good service is recognized and that the company regards itself as a contributor to the overall exercise. Make sure that the service company is kept informed of changes which affect his business and that he is fully advised of emergent opportunities.

10.6 Quality of maintenance work done

Most people associated with technical organizations will be familiar with methods of quality control, with the rigours of detailed inspection and the precise definition and application of technical processes. This concept is production-focused, aimed at producing a high-quality product or service and minimizing loss or waste. Such methods are an effective component of a quality management framework and always feature as part of engineers' concepts of quality systems. Unfortunately, however, they favour manufacturing processes where new parts can be measured and studied in ideal conditions prior to assembly into a working assembly. Judgements about the latter are then based on performance testing of the unit to a clear test specification in the secure knowledge that only correct and inspected parts have been used in the production process.

Maintenance work is far from this ideal, and traditionally such inspection has not been regarded as helpful. Companies like to use maintenance crews of their own for a variety of reasons, and even when OEM maintenance personnel are used original manufacturing rigours are not always known, nor are they particularly relevant. Measurements and corrections applied at the site are aimed at restoring the unit to efficient operating health, not returning it to its condition when it left the factory.

There is little doubt that some means of maintenance work inspection would help the crews monitor their own standards and reduce subsequent equipment failures when maintenance work improved. This is not to say that maintenance work needs correction; it is a reflection on any human activity. When it can be fairly measured then it can be monitored and improvements will naturally follow.

This is far easier said than done. Many maintenance practitioners could with justice claim that this is great in theory but impossible in practice, and before contemplating the quality of maintenance work done it is only reasonable to consider what is to be expected. Maintenance work, for example, could not be faulted if it failed to restore a 25-year-old machine to its original performance. Such thinking makes sense, and this needs to be understood at the beginning to annihilate the view that the maintenance done was somehow incorrect. When the aim of maintenance work is clear there are steps that will help.

(1) *Limit perplexity*

Maintenance technicians often receive incomplete and inaccurate reports, with the production group's request to 'fix it'. The machine to be examined is usually part of an operating system and the first point to establish is:

Is the subject equipment the cause of the problem?

There is nothing more infuriating than stripping a machine in sound working order to locate a problem caused by something else.

More usually, the technicians know what has to be done and how to go about it, but for major pieces of work it is beneficial to review intentions at the beginning, and in particular,

Consult the maintenance work history.

With infrequent work it is likely to be done by a different crew, and experiences of the previous team can be most helpful.

Subsequent disassembly can sometimes lead to surprises. Expected failures are not apparent and other damage is present. Assistance for the maintenance team is best given with:

(a) Full OEM performance information, including areas of likely failure, expected symptoms and realistic measurements.

(b) Supporting supervisory attention. Do not abandon the team in the face of difficulty. It may be that no one knows the answer, but if the team can call on support and the operator's resources, the solution will be found.

(2) *Define targets*

Before starting the work, be certain that the team is clear about the purpose and targets for the work. Such targets should be set at the outset by the team and its own supervision.

(3) *Avoid unsafe practices*

Work may be conducted without the presence of a safety officer and the temptation to minimize the time required by reducing safety procedures must not be permitted or encouraged.

(4) *Look for secondary damage*

The main subject of the exercise may be clear, but less dramatic secondary problems should not be ignored. If corrected at this point, they will save hours of hardship later.

(5) *Report as the work proceeds*

There may be other similar units on the plant and the job may be repeated in the future anyway. It is best considered as what will be necessary to interpret your own report in the future.

(6) *Avoid reusing second-hand parts*

This step is sometimes unavoidable when production is down, and replacements are not available. The fitting of new parts is also an opportunity to reduce future failures.

(7) *Reinstall with care*

The effects of long-term use and the environmental conditions will affect the condition of the equipment. Parts will become worn and material structures may change, older machines can be easily damaged. When new or repaired components are reinstated, excessive force may be applied.

Do not introduce fresh new failures while repairing the old one.

(8) *Box up systematically*

When the repair is done it is tempting to close up quickly, especially if production are waiting.

(a) Check for safety. Make sure all hot-keys are returned and all personnel accounted for.

(b) Check for all repair tools and equipment. Do not restart if items are missing.

(9) *Test before handback*

While maintenance work is being conducted the machine is held under the authority of the maintenance group. Once work has been completed, with all checks and safeguards respected, the unit should be tested in the presence of operating supervision before the unit is handed back and the responsibility is transferred. In some cases, a specific test routine will need to be actioned to demonstrate that the maintained item can safely be returned to its operational use[13].

(10) *Complete job records*

Paragraph 1 above refers to the need to consult the maintenance work history, which simply cannot be done if the preceding job records have not been prepared. Maintenance teams are usually technically biased, with a strong practical flavour. The gathering and preserving of job records, while appreciated when it is done, is not a popular activity and is tempting to omit. What makes it worse is the long delays before its use and the chance that the record maker will be engaged in other things. If the machine is due to be replaced, it may not be used at all.

We should dismiss such temptations. There are many company tasks, including equipment replacement, failure analysis and maintenance planning, which lean heavily on an effective maintenance history. In addition, the company needs time records, a basis for future estimates and a knowledge of how its resources are used.

More important than all of these is the improvement in work efficiency when we know what we are doing. Knowledge of pitfalls to be avoided, hazards to beware of and savings to take advantage of will all contribute to efficient work, and, remembering the long time between major maintenance jobs, there is a motto:

Absent maintenance history means learning the job again.

10.7 Temporary work

Quality is aimed at the system and mechanisms used when work is done. Quality improvements are achieved by ongoing regular improvements to the component parts of the enterprise and thereby to the whole. It is made easier by frequency and repeatability, and for these reasons quality improvements to maintenance work and its organization are more difficult to achieve and more dramatic when they can be done.

In spite of these difficulties, there is one area which is noticeably worse, which is of course *temporary work*, where the prospects of quality improvement are undermined by certain characteristics,

(1) Temporary work is usually done once. It is not repeated and the lessons learned during job completion cannot be put to subsequent beneficial use.

---

13. This step is just as important when the item is returned to a standby status. A clear advantage of a return to operational activity is that any failure or maintenance-induced problem is quickly apparent. A key purpose of standby, however, is immediate availability (see Section 9.7), and an additional reason for equipment testing after maintenance is to ensure that there are no invisible failures.

(2) Job records are scanty or simply not done. There is no previous job history by which preparations can be made, pitfalls avoided, performance measured and achievements gained.

(3) Estimating work times, setting targets and judging duration are all more difficult than for repeatable work.

(4) Temporary work is not part of the mainstream. It is by its definition different from regular work, and comparison and similarities have little meaning.

(5) Because the work needed is not widely understood, the job is not recognized nor its achievement applauded. Successful completion is at best met with relief that it is over.

(6) Work teams assigned to such work are themselves often temporary. Knowledge of team strengths, skills and weaknesses are missing and group motivation is poor.

(7) Once completed, the job is usually forgotten. There is little scope for analysis or praise, few stories of repeated satisfaction and no opportunity to offer assistance or support to other teams facing similar tasks.

(8) Work tools, procedures and skills are improperly known, standards do not exist and the process of work itself includes strong learning elements.

(9) The work is more dangerous, the outcome of actions is uncertainly known and dangers are unforeseen. Safety awareness, so vital in regular work, is diminished by lack of relevant experience.

(10) From the point of view of work management, temporary work does not fit well into the normal prescription of regular work. Jobs are often selected quickly in response to unexpected developments, their duration is expected to be short, work content is uncertain, and both work and safety standards are imperfectly applied. There are also more serious aspects to work of this kind.

(11) Temptation applies to temporary work, with a slogan of haste and a lack of care. When the job is short and never to be repeated, temptation says that quicker means less danger and careful work is not needed for so short a time.

How reasonable and how plausible this last point seems. The best of us could be persuaded when the right course takes so much longer. Danger, it seems, has the best dialogue, but the 'short' route is more sinister because it comes as the trap is closed and precedes the inevitable disaster.

When the accident happens in temporary work men do not know what to do.

It is for these negative reasons that quality is so important. Far from seeing this job as transitory and unimportant it is vital to get it right.

Careful quality of temporary work is designed to protect us all. When the work is done well, quality gains in other fields are secure.

Temporary work often requires the bypassing of safety circuits, warning instruments, automatic devices and cut-outs. Not only does the loss of such protection increase workplace danger, it can also affect other areas not involved in the temporary work itself, and may need to be specifically restored when the work is complete.

For example, the fluid level in a tank is controlled by a level switch driving a drain valve. When the tank output was being connected to a new manifold, the main outlet valve was temporarily closed and the level in the tank allowed to rise by inhibiting the level control. After the new manifold connection had been made the outlet valve was opened and the level control specifically restored. Failure to complete the latter step would have left the level uncontrolled, with flow fluctuation in the system and risks of spillage or downstream pump cavitation.

Keep in step with the following:

(1) Always supervise temporary work closely. Even though work is unfamiliar its importance should be abundantly clear.

(2) Be as formal as preparation time and work management will permit. Never succumb to casual attitudes or ill preparation.

(3) Prepare your team personally. Be sure that purposes and objectives are clear and underline the need for cool experience and unremitting judgement.

(4) Define as much of the job and the steps required as possible. Underline any areas of uncertainty where on-the-job decisions are required.

(5) Ensure that the safety department and all safety procedures are included. Do not allow work to proceed without their knowledge and involvement. Temporary work is known to be the focus of safety problems and safety officers act accordingly.

(6) Do not allow demands for immediate action push you into work against your better judgement. If things go wrong, the responsibility is suddenly all yours. Check with your team; make sure everyone knows what is required and that preparations are as complete as possible.

If they are uneasy, listen; if you are uneasy, seek more information and wait.

(7) Check other site supervisors and carefully consider the effects of temporary work on their activities and yours. When you have thought about this,

think about it again – accidents often come from the unforeseen.

(8) Consider the requirements of statutory regulations. Check general safe working of those regions most affected.

(9) Do not use broken tools or shoddy materials because the correct items are delayed. If you have overlooked something, hold your hand up and get it. Do not fudge or whitewash – it is better to be bawled out than locked up.

(10) Be wary of using temporary teams. Most of the difficulties with temporary work arise from lack of information and uncertainty. If temporary teams are necessary this is simply more of the same. If it is unavoidable, take time to learn about the personnel assigned and train them in the working system and the working environment[14]

---

14. If labour limitations are a problem, consider using temporary personnel on routine work, which can be more easily monitored, and assigning regular personnel to the temporary work so that team ability and knowledge of operations make the job more certain.

(11) Be a supervisor or a manager. Do the job right; popularity can wait. Remember these steps:

> Make sure the world knows.
> Inform your staff properly.
> Make team choices and listen.
> Clarify and clarify again.
> Look into the future.

There are, however, dangers that exceed all of these steps. There are features of temporary work which put it into a class of its own, and it is for this fateful reason that temporary work has featured so often in accident and tragedy.

10.8 Local repairs

The first thing to recognize is that repairs conducted locally are done in non-ideal circumstances, usually by teams that have general experience rather than specific skill. The size of the item needing repair, together with the difficulties of handling and shipment, often makes local repairs the only feasible option. Job durations are longer than figures quoted by OEMs and man-hours required are higher, with the work likely to be affected and interrupted by other activities simultaneously occurring on-site. Like all such work, the job preparation, work management and safety requirements all apply, as discussed elsewhere in this manual. There are, however, several specific questions ideally to be settled before the job begins.

(1) *Safety before everything*
Although safety matters are explored in more detail in Chapter 11, they need to be considered first whenever the question of work is considered. There is another point for the maintenance manager or supervisor:

Local repairs are directly under your jurisdiction.

The manager is firmly in the safety chain of responsibility, and his attitude to the subject

will have a profound effect on team behaviour.

(2) *Consider materials movement*
How are replacement components or assemblies to be moved and temporarily stored, imported and exported? Is movement space available? (see Section 9.2.)

(3) *Unit disassembly*
How will the unit be isolated (see Section 9.3) and prepared for disconnection? Will partial disassembly be required before materials movement can take place? Does the incoming replacement require the same treatment?

(4) *Working space*
What are the dimensions of the unit with covers up and doors open? What working space is needed to remove key items like a crankshaft as long as the parent machine?

(5) *Human access*

Is there sufficient room for access to the machine to store tools and materials and conduct work? Consider door swing, head height, clear line of sight, avoidance of cramped working, lack of floor space, all to be reckoned when safety equipment is being worn when necessary (see Section 11.2).

(6) *Load carrying*

Is special lifting or moving equipment required? What is the weight of removable covers, relief valves or other subassemblies? Are overhead lifting beams positioned over the centre of gravity of items to be lifted? Are lift spreader beams needed? What is the weight of units to be moved? What are the rated safe working loads for floor panels, stair treads and lifting equipment?

(7) *Interposing equipment*

Will other units get in the way of disassembly or movement? What about instrument cables, cable trays, process piping and temporary equipment?[15]

(8) *Other site work*

When the job is under way, will other site work be prevented or compromised? Resources are of course a fundamental question: what about common tooling, other job access, power required, temporary lights, safety equipment and sanction for both by the permit-to-work system?

(9) *The permit to work*

Repair work is a regular feature of maintenance work and its correct handling is a recurrent theme of this manual. Apart from the recognition of the important roles of preparation, work information, materials, tools, scheduling and so many other aspects, two criteria stand out. The first is the permit to work. When repair work is to be done, some jobs are trivial in a safety sense and others are highly dangerous, and the latter usually need a permit. One of the supervisor's tasks is to decide not which jobs need a permit but rather which jobs can do without (see Section 11.9).

(10) *Isolation*

The second vital criterion is isolation. It frequently surfaces as a need when repair work is contemplated, and again the importance and need for sound supervisory judgement is vital. The first question about any repair is whether the isolation is required and how it is to be done (see Section 9.3).

(11) *Work management*

The definition, planning and preparation of work required plus the identification and allocation of resources required, ideally tackled in a regular work management process, are covered elsewhere in detail (see Chapters 2 and 8).

---

15. Instrument cabling and small pipes are easily overlooked. They become evident when lifting assemblies snag on interfacing components during disassembly. Wires are sometimes cut to remove an immediate impedi ment, disabling detectors or control devices. While the main equipment is shut down for maintenance, the absence of such instrumentation may not be apparent. Provided the cabling is restored prior to retest little is lost, but the absence of control or protection would prevent the run-up of the machine after work is complete.

Chapter 11

Maintenance and safety

This chapter should be regarded as a commentary or pointer to a major work ingredient, not as a comprehensive treatment. 'Safety' refers to probably the most important aspect of all work, to the processes, plant and equipment and to the junction of human beings with the technology that they create. As we have observed, the human beings referred to include operational teams, maintenance technicians, site-located staff, visitors, delivery to site personnel, residential neighbours and the environmental population at large.

Chapter 14 of this manual is a partial list of HSE publications of particular interest to maintenance work. Some of these have been referred to in the preparation of this chapter and are repeated in the bibliography. However, this subject is too important to treat in the abstract; the HSE documents should be consulted directly and regarded as a selection of material on a much wider subject (see also Chapter 12).

Methods used for the evaluation of reliability are also used for the assessment of safety, though it is not always true that improvements in one automatically lead to improvements in the other. Additional complexity for safety reasons can lead to lower reliability. (Refer to Section 4, Fundamentals of reliability assessment, BS 5760, Part 2, 1994.)

Until the early 1990s, the regulatory stance required employers to provide a safe working environment and engage in safe working practices. While this approach has been repeatedly and recently underlined, a newer emphasis points to the necessary involvement and motivation of employees themselves towards safety targets. In addition, the changes described in Chapter 12 arise from the operating companies and the HSE working to create a new positive or proactive safety regime, which is leading to the introduction of new regulations of a less prescriptive nature. Unfortunately some suppliers are attempting to convince industry that only the purchase of new equipment will enable customers to meet these new regulations, particularly the 'Provision and Use of Work Equipment Regulations' (1992) (PUWER). The Health & Safety Commission has issued blunt warnings and commented, 'Employers who already meet existing legal requirements are unlikely to need to do any more to meet the requirements of these regulations' (*Plant and Works Engineering*, September 1996).

Major companies employ professional safety officers whose presence contributes to the desired safety culture and whose duties constantly require the participation of all in safety matters. With safety being recognized as the province of us all,

safety is no longer somebody else's problem.

Maintenance work is often conducted by non-company personnel who, in varying degrees, are unfamiliar with the site. The jobs themselves are often infrequent and these features combined make regular safety training and site awareness more difficult. Most oil and gas producers in the North Sea impose a safety briefing on all visiting personnel as soon as they arrive on an offshore installation, which necessarily concentrates on safety essentials such as muster points, escape routes, survival routines and safety drills. Work-related safety varies according to different groups and entails specialized arrangements, and for maintenance work is the most difficult.

When a contract company has been engaged to carry out maintenance work it is reasonable to expect that their staff have been properly trained. In general it is unwise to make such an assumption, and current regulations are emphatic where an employer's own staff are concerned[1,2].

Most sectors of industry and business have workers who are peripatetic. Because they do not have a fixed workplace and often work away from direct supervision, they may face or expose others to risks arising either from substances on the premises where they work or from substances that they take with them or draw from stock on the premises.

Although this quotation refers to the control of harmful substances, it captures the HSE attitude to maintenance workers clearly, i.e. the nature and manner of their work puts them at risk. The same document describes the responsibilities of an employer:

Under the Control of Substances Hazardous to Health (COSHH) Regulations, employers are required to protect the health of their own employees wherever they may work[3]. They also have a similar duty (so far as is reasonably practicable) towards employees of other employers and members of the public both on and off the premises. . .

An unending aim of effective safety processes is to reduce accidents which often follow the combination of unusual factors. Some responses to an accident would imply that it was totally unexpected and its occurrence could not reasonably be predicted nor the consequences avoided. How many times have we heard the phrase 'to ensure that it does not happen again'? While this is a noble sentiment, it is equally certain that many accidents happen which are a surprise to no one. We could usefully say if an accident occurs without surprise then

---

1. HSE Booklet HS(G)77, 'COSHH and Peripatetic Workers', refers to maintenance engineers among others.
2. The HSE produces an employers' guide: 'Managing Contractors' HS(G) 159.
3. Refer to HSE publication HS(G)97, 'Step by Step Guide to COSHH Assessment'.

the safety management system has failed.

All the event investigation and analysis, while absolutely vital, does not change the damage that has been done and the injuries that have been caused; the accident is in the past. As we have already observed (Section 5.3), anticipation is a key quality of good supervision, and in this area it is sensible to look for trouble before it happens.

Put safety first and be wise before the event.

Let us dispose of a myth. It is commonly believed that safe practices raise the job cost, that cut-throat competition necessarily requires us to cut corners to cut costs. This is seductive but untrue, and accidents often follow a misconception. An effective manager will not agree to impossible time scales or succumb to dangerous restart pressures. He will insist on safe conditions of work and safe practices by his team. Consider accident-created legal costs, damages arising, cost of equipment repairs, downtime loss of revenue and the effects on the working team.

Accidents have serious commercial consequences. It only takes one to destroy a business.

At the more detailed level, for any job, safety requires thinking in advance, preparation and the assignment of correct skills and resources, which are all recognizable as cost-efficient requirements.

Safe work does not cause extra cost.

Two aspects of maintenance work which constantly feature in safety matters are the questions of hazard and risk. Repeated below are the relevant descriptions from BS 4778, Section 3.1.

Hazard – concept (item 6.1)
A general concept of a hazard as applied, for example, to an industrial process, or a commercial organization, is the potential for adverse consequences of some primary event, sequence of events or combination of such circumstances.

Hazards can be classified according to the severity of their potential effects, either in terms of safety, economics or other consequences. Different industries use different classifications. Such classifications alone are purely subjective and usually require qualification, by definition of the precise form of the hazard and a quantified evaluation of the consequences.

Risk – concept (item 7.1)
A general concept of risk is the chance, in quantitative terms, of a defined hazard occurring. It therefore combines a probabilistic measure of the occurrence of the primary event(s) with a measure of the consequences of that/those event(s). Criteria for acceptability of some predicted risk or measured risk can be set voluntarily by the organization responsible and/or subjected to the hazard, or be set mandatorily by some regulatory organization.

These two concepts, as we would expect, feature in the concept of safety itself.

Safety management (item 11.2.2)

The application of organizational and management principles in order to achieve optimum safety with high confidence. This encompasses planning, organizing, controlling, coordinating all contributory development and operational activities.

Safety concept (item 11.1)

Safety relates to the freedom from risks that are harmful to a person, or group of persons, either local to the hazard, nationally or even worldwide. It is implied that for the consequences of an event to be defined as a hazard, i.e. a potential for causing harm, there is some risk to the human population and therefore safety could not be guaranteed, even if the risk is accepted when judged against some criterion of acceptability.

Major companies assign key personnel to formally identify hazards and risks present on company sites, particularly in operations. Maintenance hazards can be the most difficult to spot, partly of course because they may not arise in regular activities. So maintenance workers have to contend with hazards from the work environment plus those inherent in the work itself. Maintenance work is less well known. As we have seen, major jobs are less frequent and may be carried out by contractor personnel. In addition, extra tools, repair and replacement materials, greater demands on access and floor space (see Sections 9.2 and 10.4), mechanical handling equipment, disassembled components, exposed internals, materials and stored energy are all present when major maintenance work is being conducted.

This is only the beginning of the list of likely dangers, such as electrical hazards of shock, heat, arcing, ultraviolet radiation and others and problems of fumes, gases, very hot objects, escaping fluids, toxic substances, corrosive liquids, falling bodies and others of growing number and complexity. Any commentary which attempts to identify the full range and complexity of such hazards is bound to fail: new dangers and varieties are occurring all the time. There are two things that can be done here. First, point out some key examples to underline the subject and help us all think. Secondly, repeat the warnings of subject scope and complexity, that safety requires constant reading, concern and attention. *Whenever men are required to work they must not be despatched by their employer into an unsafe work situation.*

There is a temptation to believe that maintenance safety concerns major hazards and that normal safety inspections will take care of the regular 'environmental' questions. This is an understandable but dangerous assumption. Look at a simple example. Maintenance which requires the disassembly and reassembly of a rotating machine will almost certainly release lubricating oil in the process, which will gather in pools on the work floor to contaminate the boots of passing workers, changing frictional characteristics of the tread and introducing simple danger when the affected workers climb ladders, stairways and exposed scaffolds. (Refer to slips and trips in Chapter 14 page 333.)

Maintenance work requires special and additional safety attention.

11.1 Safety and maintenance work

While the maintenance team have the same general safety considerations as everybody else, there are additional factors which influence the safety dimensions of maintenance tasks.

(1) Many maintenance jobs are in direct response to the requirements of working safety.

   Safety requirements add to maintenance tasks.

(2) Many maintenance tasks are themselves hazardous and introduce hazardous solutions.

   Maintenance work is a cause of safety problems.

Item 1 can be interpreted as one result of the safety management system working successfully, while item 2 requires further examination. What are the aspects of maintenance work which give it this dubious safety reputation?

(1) A lot of maintenance work occurs infrequently, particularly major PPMs and machinery failures. There are therefore fewer opportunities to discern safety problems and to introduce remedies. The reduction of risk following a developed skill and job familiarity is also missing, as is the reduced cost benefit of using well-practised personnel.

(2) Maintenance work is frequently conducted in remote locations where teams work in small numbers at unsociable hours late at night. They work in danger and isolation, and while the latter may be tempting it is also recognized as extremely dangerous. The onset of fatigue, new dangers and the temptation to take risks can pass unnoticed, and accidents arise unknown to distant colleagues who could help.

(3) Work in unfamiliar surroundings means that hazards at the location, such as missing gratings, rusted handrails, sheared door hinges and broken light fittings may pass unnoticed until they are needed or an accident has already happened. (Refer to 'Working Alone in Safety, Controlling the Risks of Solitary Work', IND(G)73(L), HSE, 1989.)

(4) The job is sometimes caused by the failure or misbehaviour of a machine. Reported fault symptoms are often vague and unreliable, resulting in dangers unknown until arrival at the scene and more difficulty in predicting the behaviour of faulty equipment.

(5) Maintenance work requires the movement of tools and materials to the workplace plus access travel by the technician, sometimes via an infrequently used route. When work is applicable to a pipeline or railway maintenance crew, movements of several miles are quite normal and are themselves a source of hazard.

(6) Heavy and bulky materials are transferred from a maintenance warehouse to the workplace using lifting and transport equipment, sometimes outside a strict maintenance regime. A trite example of familiarity breeding contempt, failure of such equipment multiplies both dangers and work required.

(courtesy of Chevron Oil UK Ltd).

(7) Although regular communication with maintenance technicians is more necessary, it is also more difficult. The job may require work out of sight, behind or beneath other structures and machines, perhaps underwater, on top of factory chimneys, flare booms, electric pylons or tail buildings.

(8) Maintenance work often requires the disassembly of corroded parts, the manhandling of cumbersome heavy components and work in dark, confined spaces. In addition, the presence of combustible and toxic fumes, process by-products or hazardous residues requires the use of breathing apparatus, protective clothing and safety equipment.

(9) Whenever previously working machinery is disassembled there is a real risk of releasing stored energy (see Section 2.1).

(10) Maintenance work is sometimes conducted inside or underneath machines such as pressure vessels, storage tanks, large rotating machines, air ducts, large diameter pipes, supporting structures and furnaces. Apart from the difficulties of communication and special safety apparatus mentioned below, there are additional problems of biological cramp, claustrophobia, muscular strain, limited vision, isolation and fatigue. There is also the danger of lingering process contaminants and sudden mistaken startup while the work is still in progress.

(11) Maintenance work sometimes results from a previous change itself, either incomplete or improperly reported. In any situation where equipment isolation and subsequent disassembly is required, the availability of current and accurate information is paramount and its absence increases the maintenance task and jeopardizes the technician.

(12) Maintenance work is often suddenly required, presenting the need for immediate action with limited time to prepare. The resultant work is more difficult and, of course, more hazardous. When this is coupled with downtime,

the result is inevitable pressure for early completion and restart – all features which contribute to increased risk.

11.2 Safe access

The following list is intended to stimulate more detailed investigation of the site and the particular job. There is an important point here:

maintenance work content varies with the location.

This means that the requirements of the 'same job' will be different at different sites and for different locations on the same site. Work instructions, although generally the same for a group of the same type units, must cater for the variations caused by changes in:

(1) Location.
(2) Manpower access to the workplace.
(3) Tools, equipment and material access.
(4) Access to the equipment itself (see Section 9.2).
(5) Access constraint caused by local structures.
(6) Sources of power.
(7) Environmental effects.

Many supervisors and workers who are well trained to deal with safety in the workplace find it difficult to arrange safe access to the site or during necessary absences from it. This includes the movement of workers from base to site, from the warehouse to the workplace and many other locations. During movement and lifting of materials, the navigation of narrow passageways and stairways plus the temporary passage of unrelated workers through different workplaces are required[4].

The following apply:

(1) *Site to workplace access*
     Human approach to any part of the site should not require climbing over work material, such as drill pipes or scaffold poles. Passage will be required over different surfaces, such as steel flooring, packed earth, gravel and metal roadways. All should be clearly marked, firmly based, free of obstructions and well lit. Trackways and roadways must be suitable for proper traffic in adverse environmental circumstances, such as heavy rain, ice and snow, high winds and darkness. Surface contamination by process leaks, spillages and oil should be hosed away or clearly marked if they cannot be removed. Slippery walking or climbing surfaces should be mopped up and made safe at once.

(2) *Passageways*
     Gangways and passageways must be well lit, well marked, free of obstructions, have a clean non-slippery surface and be well maintained. Step treads, stairs, walkways and guard rails should be fully functional, rust-free and in sturdy working condition.

---

4. The 1961 Factories Act specifically charges employers with ensuring a safe means of access to the workplace.

(3) *Doors and doorways*

Also well lit, doors should be properly functional, with locks, latches and pull handles securely attached and hinges or sliding rollers correctly lubricated and effective. Display boards denoting access restrictions, danger signs, need for ear defenders or raised door sills must all be clearly marked. Similarly, door closure restrainers must be engaged and functional, especially when wide hinged industrial doors, usually made of metal, can close suddenly in high winds.

(4) *Access/manhole covers*

Covers must be properly located by firm supports, rock free and flush with the gangway surface when in place. Covers should be free of rot, rust or corrosion and be of known load-bearing ability. Access covers should not be removed unless proper hazard guards and warning signs are in place.

(5) *Ladders*

A frequently used means of access to the maintenance task, ladders are best regarded as tools required for the job to be specified in work instructions. Ideally, they should be numbered, added to the tools register and regularly inspected. Load-bearing figures, usually available at purchase, should be retained for reference, while overloading and other misuse must be avoided. Like all load-bearing devices, ladders deflect under load, reducing the 'reach' and introducing a danger if the ladder was initially too short. The ideal length allows an upward climber to step off the ladder at the proper ascent stage whilst retaining a steady grip on the ladder itself. The upper point of contact with the structure should preferably be wired or clamped firmly, and with the lower feet secured, slipping sideways, tipping backwards or simply falling down should all be prevented.

The vertical and side load capabilities of conventional designs are rarely the same. When ladders are used at low angles or horizontally, like bridges, working loads should be reduced. Rungs and rung attachments should be checked for rot, corrosion, damaged root welding or sheared rungs. Faulty material or components should be replaced at once (see BS 1129, Timber Ladders; BS 2037, Aluminium Ladders).

(6) *Scaffolding*

Maintenance teams frequently carry out work using scaffolds erected by a contractor engaged for that purpose, and the following should be confirmed:
(a) That the correct scaffold is in place. There are different types of scaffold for light, heavy or general duties, and scaffolds erected for previous work may be unsuitable.
(b) Correctly selected and installed scaffolds still need regular inspection by a competent person, often the supervisor, at least once a week and before a new job begins.
(c) Attachments to supporting structures, component clamps, floor supports and floor planking should all be examined during the regular inspection.

(d) Because scaffolds are usually seen as temporary they often fail to receive the attention needed. Results of inspections should be recorded, with signs of deterioration checked and noted every time. It is the temporary nature of the structure which can allow rapid degradation. Scaffolds are unusual, as they are used for access and as places of work. In addition to load support and structural integrity, space is needed for the temporary storage of tools and materials plus working space and room for personal access or manoeuvre (See Construction Regulations, 1966).

(7) *Roofs*

The overloading of fragile roofs, usually by a man's weight, causes accidents which could be readily avoided. For infrequent access by a maintenance technician such problems often arise because roofs supported from underneath look safer than they are. A further consideration is what lies underneath; high roofs over construction or factory spaces are often intended to keep off rain and snow only, and access by maintenance personnel requires climbing boards, fixed ladders and safety lines.

(8) *Man-lifting devices*

In recent years the erection of some inspection scaffolding has been replaced by the use of ropes, harnesses and other climbing equipment more often designed for mountaineering. Such techniques are particularly useful applied to tall structures, both internally and externally, for routine inspections rather than major repairs or maintenance work. Above the waterline inspections of steel support jackets of offshore oil production platforms are a good example. For the maintenance manager there are the additional advantages that this method can be more quickly put to use rather than facing the inevitable wait for scaffolding to be erected, is often less costly and has less impact on other work (see BS 2830, Construction Lifting Regulations, 1961).

For all sections and additional safety information refer to the HSE publication 'Health and Safety in Construction', HS(G)150 and Section 9.8.

11.3 Removing dangerous substances

Maintenance work often involves equipment or machinery that has been exposed to environmental contamination, hazardous process materials or other substances required for process or maintenance activities. Each type of hazard and the methods of its treatment must be identified for the site and the process before men are committed to maintenance tasks (see IND(G)67(L), HSE 1988, 'Hazard and Risk Explained'). Before opening any vessel or pipeline, the contents, its pressure and temperature must be checked. Accidents have followed the unbolting of sealed junctions and the release of hot, high-pressure, sometimes toxic fluids[5].

---

5. The total force on a circular ⅔ m (2 ft) diameter access hatch holding back 21000 kN/m² (3000 psi) is approximately 6000 kN (600 tons). Partially released catches or fixing bolts in sudden tensile fracture can go off like bullets.

A frequently encountered environmental hazard in extraction operations is radioactive scale, which forms deposits on the internal faces of fluid processing equipment. In oil production operations such scale is removed by water jetting prior to subsequent shipment or repair.

In other cases flammable vapours, if escaping from a leaking joint and denser than the surrounding air, can 'roll' over a work surface down into subfloor spaces and remain undetected until ignited by sparks from a welding torch. The reverse also applies: combustible lighter than air gases can rise into a closed roof or ceiling space and remain there in near contact with the atmosphere. This explosive mixture can be removed by pumping to the outer air; gas and vapour should be contained in the vessel before it is opened and the problem prevented by venting closed roofs where such material can accumulate. Also, when fluid-containing vessels are emptied for inspection there is a weight change which can flex floors or support structures and permanently distort piping connections, valve seats and the smooth operation of access hatches, pressure relief doors or valves.

Process liquids

Equipment such as tanks, vessels and pipelines will usually have contents until internal work is called for. Process fluid can be diverted into the upstream point of an alternative production train or, when safety and environmental regulations permit, to a suitable closed drain.

Diversion to other trains is usually chosen and, when followed by water flushing, some processes can also tolerate transfer of the contaminated water. In other cases solvents, detergents, hot water, steam or a neutralizing fluid will be necessary before all traces of the process fluid can be removed.

Some cleaning fluids are toxic, and manufacturers' warnings must be observed with care. Refer to the regulatory requirements of the Control of Substances Hazardous to Health (COSHH) Regulations (1988) and HSE publication HS(G)97, 'Step by Step Guide to COSHH Assessment'.

In addition, flammable liquids are dangerous when they come into contact with the air, especially if sparks can be created during cleaning. Inert gas is often used as a blanket to prevent mixture with oxygen.

Gas and vapour

Having removed all process liquids from vessels and associated pipework, toxic or flammable gas or vapour may still remain, and extensive flushing with steam or inert gas should be completed before human entry is permitted. The decomposition of fluid residues during subsequent maintenance work, especially if cutting, welding or grinding are involved, can cause fire, explosion or toxic contamination of the breathing air. Workers entering a vessel should wear breathing apparatus and safety clothing (see Section 11.5). Only electrically safe lighting should be used and to prevent sparking on bridging insulating flanges or contact with earthed equipment, the power supply to a cathodically protected pipeline must be disconnected for at least 24 hours in order to allow time for depolarization. The pipeline must be bonded to earth before starting work (See Townsend, A. *Maintenance of Process Plant*, Institution of Chemical Engineers.)

It should also be remembered that the concept of

Flame (ignition source)
Fuel
Oxygen

present before an explosion or combustion can start applies at low temperatures, and some liquids, gases or vapours will combust without flame if they come into contact with hot surfaces.

Solids

Some sludge left after cleaning may produce flammable or toxic gas or vapour if it is disturbed by subsequent work. The region should be treated as Zone 1 for safety purposes and residue removed with wooden non-spark-producing tools. Combustible fine powder or airborne dust can also cause an explosion; the vessel should be purged and washed completely using water or inert gas.

11.4 Safety officer's perception

It sometimes pays to view a workplace or project in progress as it would be seen by a safety officer, even if the company does not employ one, to anticipate a response before it happens and to improve the management or supervision of a maintenance job. There is a further dimension to this thinking: it reduces the likelihood of disagreement, encouraging the maintenance team to act more safely and to balance the inevitable pressures to skimp on their work. It also underlines the time-scales of urgency: to the safety officer, the problem once identified must be corrected before someone gets hurt; to the maintenance technician the same fault will need attention in the future; and to the production operator, as long as output can be maintained there is no cause for alarm.

Almost all jobs are made easier if we can prepare for them in advance to collect tools and materials, organize personnel work-time, gather information and contemplate the sequence of actions. For this reason, most of us are reluctant to start complex jobs quickly unless they are already part of our regular work routine. For maintenance work this is rarely the case, and reluctance to start at once is a familiar reaction.

Companies that employ safety officers significantly improve their chance of avoiding accidents or incidents, and constantly underline:

(1)   The value of current knowledge especially when applied to ongoing work.
(2)   The need to constantly monitor normal operations, not to assume that safety is automatically repeated and to check for correct regular implementation.
(3)   To target special and unusual work, particularly maintenance and construction, where safeguards may be removed and unusual hazards introduced.
(4)   To emphasize the immediate need to act: not 'soon', as the maintenance engineer wishes, or 'later', as the production team need, but 'now', as safety and personal danger require.

(5) To attend to the training and induction of new staff, emphasize safe practice, detail reactions in an emergency, and instruction about methods of survival and escape.

(6) To watch for change, look for new dangers and devise or revise procedures.

(7) In such a working environment, the safety officer will implement the following routine steps.

Work monitoring

To routine checking of work in progress at all locations, the correspondence with safety requirements and the known location of all personnel.

Work site temporary marking

In preparation for a job, during the work programme and in accordance with safety procedures.

Site safety patrolling

To identify emergent safety problems quickly.

Safety equipment checking

To ensure all personal safety and installed equipment is ready for immediate use.

Information checking

Out of date, irrelevant or incorrect information is a known source of accident, quite apart from the extra difficulties attaching to the job itself. Poor site layout drawings can mean maintenance guesswork when data is missing. Unrecorded changes to structure, unknown revision to process conditions, unmarked energy sources and many others all spell unnecessary danger.

Safety system operations

Checking of correct operation of inbuilt safety systems.

Noting of equipment or structural modifications.

In anticipation of reference to this information during subsequent related work.

Marking of isolation points

Usually conducted in conjunction with both process and maintenance supervisors. (Refer in detail to the description in Section 9.3.)

Checking of permits to work

Both during actual jobs at the workplace and of the system itself, its requirements and implementation (see Section 11.9)[6].

Hazard evaluation

Both as a formal contribution to the operator's hazard evaluation procedure and as a routine check for unexpected influences and persistent dangers (see Townsend, A., *Maintenance of Process Plant*, Chapter 6, Institution of Chemical Engineers).

---

6. The latter two items must be considered for maintenance tasks specified by the maintenance supervisor or the safety officer. Refer to HSE publication 'Guidance on permit to work systems in the petroleum industry'.

11.5 Safety of maintenance workers

It is general practice that processes, machines and equipment are designed to be safe, that safeguards are applied to the machine rather than to the operator, and the living and working environment of the human being is not adversely affected by machinery. The fact that this is not always achieved, as traffic congestion, fume-enhanced asthma incidence in children and oil contaminated shellfish or seabirds testify, is clear evidence that it can break down, and dramatic consequences follow. So a constant action of the safety officer in the working environment is to monitor the well-being of personnel and ensure that protection remains in place.

For the maintenance worker, however, there are times, such as the entry of confined spaces and for all workers on some occasions, where the protection is applied to the person rather than to the process. The fact that personal protection requirements are unusual makes them all the more important when they are used, especially for maintenance personnel, where dangerous and unusual tasks require that precautions should be thorough.

Personal safety is needed when environmental safety is absent.

All of the specific dangers mentioned below are countered by protection and careful monitoring. However, no precaution is adequate if the worker decides to dispense with it. For this reason, monitoring by a separate person is essential, especially for maintenance personnel nominally working alone. Safety officers are constantly on their guard for those factors which affect human behaviour; for example.

fatigue and cold can lead to fatal decisions.

Protected breathing

It is sometimes forgotten that when breathing normally the lungs are in direct contact with the environment, while the 'outside' human skin may be protected by clothing, helmets or goggles. For this reason explorers in very low temperatures face the danger of freezing from the inside out as cold air affects the lungs. Respiratory protection will be needed to combat the following:

(1) *Airborne contaminants, dust, toxic gas or vapours*
Some survivors of the Bhopal incident suffered damage to their lungs caused by airborne acid droplets.

(2) *Oxygen-deficient air*
Many processes, particularly fire, consume oxygen. If the air volume is limited, oxygen deficiency can cause asphyxiation.

(3) *Air at extreme temperatures*
Outside air temperature at night in Scandinavian winters can drop below −40 °C. Air temperatures at the scene of major forest fires can burn the skin.

(4) *Flames or radiation*
Direct contact will lead to immediate damage. Some maintenance and repair workers at Chernobyl died within weeks of direct exposure.

Separately supplied air lines should be used for major contamination or oxygen deficiency. Fully inspected breathing apparatus can be used by trained personnel. For maintenance work, extended periods in the location are often necessary and a properly installed and effective ventilation system should be provided.

Respirators relying on air filtration are only suitable for light contamination where sufficient oxygen is present. There is a further danger here, as the filter needed depends on the contaminant, which is not always known or may appear unexpectedly.

The inspection and certification of breathing apparatus is a legal requirement (see HSE Certification of Breathing Apparatus Testing Memorandum TM3).

Protective clothing

As well as protecting the wearer from the airborne hazards above, protective clothing will guard against abrasion, chemical damage or cracking of the skin, dehydration of the body at higher temperatures and loss of body heat in cold air.

Body heat can also be lost by conduction when working with low temperatures, metal parts and moisture condensation on cold surfaces. Idle machinery generates no internal heat and as warmth is soon lost by conduction through large mass systems, machine temperatures soon drop to that of the surrounding air.

Conversely, internal air temperatures will rise under the effect of high daytime temperatures, especially direct sunlight on glass-confined volumes.

Ventilation systems in use will replace higher temperature air when work is confined. Other parts of the body recognize special protection, which is also needed for head, hands, feet and ears[7].

(1) *Helmets and hard hats*
Most site operators insist that hard hats are worn by all personnel on site. The risk of head injury, especially when working in cramped conditions, is very real. Loss of consciousness can lead to even more serious dangers apart from the risks of concussion or brain damage. In some cases special helmets will be required, such as work underwater or when ear covers or faceplate attachments are included.

(2) *Ear defenders*
Sometimes attached to the maintenance technician's helmet, they are required in some environments where machine or process noise can damage the ears.

(3) *Gloves*
Maintenance work requires the handling of tools, old and new equipment,

---

7. Under revisions to the regulations presently being introduced, PFEER (Prevention of Fire and Explosion, and Emergency Response) identifies a requirement for written schemes of examination under Regulation 18:
Regulation 18 – Suitability of PPE for use in an emergency.
A WSE is required for Personal Protective Equipment provided to personnel under the Personal Protective Equipment at Work Regulations (1992), but only for specific items of equipment as identified under Regulation 18. This relates to equipment for protecting persons against effects of fire and smoke, and in the event of immersion in the sea (e.g. smoke hoods and immersion suits).

cleaning agents, contaminants and, of course, old or broken parts. The risk of damage to the hands is always present, of which cuts, trapped fingers, abrasions and skin contamination are merely the most obvious. In some cases gloves' working surfaces will include heavy duty pads for handling hot or sharp objects and include wrist seals to exclude industrial fluids or prevent loss of heat. In all cases, gloves should be provided and always worn.

(4) *Boots and toecaps*
Feet and toes are particularly vulnerable when dismantling used equipment. Heavy metal parts are likely to be slippery and difficult to hold and the risk of dropped or toppling objects to exposed feet, especially in restricted places, is always present. Well-fitting boots with steel toecaps will reduce the risk, protected from bruising and provide good grip on ladder rungs, scaffolds, gratings and other slippery surfaces.

(5) *Eyes and eyesight*
Flying chips, sparks and small objects, chemical sprays, jetted hydraulic fluid, loose wires, internal lights and sharp corners can all damage eyes, which must be protected at all times. Workers even temporarily blinded are in great danger of secondary injury as well as suffering the original damage to sight, and are likely to unwittingly create danger for others. Contact lenses are not recommended for use in the workplace. Many chemicals and infrared and other radiation can cause irritation and possible loss; some workers are extremely short sighted without their contact lenses.

Goggles, visors, screens and safety glasses chosen to suit the application will protect from fluid contamination, burns, lacerations and abrasions. They should be chosen for the application and always worn while work is under way.

Legal requirements for the 'Protection of Eyes' (Regulations 1974) deal with use, availability, marking, availability and duties of the wearer. (Refer also to BS 2092, 'Specification for Eye Protectors for Industrial and Non-Industrial Uses'.)

NB. We should remember that sight may be lost by lack of light, by smoke or failing lighting systems, as well as by damage to the eyes themselves.

It is tempting for maintenance workers in particular to avoid the use of protective clothing and to attempt the work quickly and without restraint. This is one of the reasons why maintenance work takes longer than first impressions suggest, and the pressure for rapid work is always present. This temptation should be avoided and the use of protective clothing required without question by maintenance supervision and the safety officer.

Belts, lifelines and rescues

In the event of an accident it is natural to rush to a colleague's aid without considering the danger to oneself or to rescue personnel. How many times have attempts to rescue children or dogs from drowning resulted in the loss of the 'lifeguard', while the first person in danger has survived?

When a rescue is required, safety procedures must not be overlooked. The safety of the rescuer improves the chances of successful rescue and survival of the initial subject.

Harnesses are used to lift vertically (full harness) or pull horizontally (half harness) confined workers from the workplace, such as from a pressure vessel. Care should be taken to ensure that further injury to unconscious workers is not caused during rescue.

As a protection during normal working, lifelines attached to harnesses worn by workers are used to protect from falling.

11.6 Dangerous cleaning

Many maintenance jobs can only start after cleaning has been completed. The exposure of equipment internals to a contaminating environment can introduce damage through foreign particles not present when the work commenced. The author has witnessed workshop floor concrete being smashed with sledgehammers while reciprocating engines with cylinder heads removed were under repair nearby.

In addition, cleaning is often required as a step in the maintenance or repair process. the most obvious example being the relining of major process pressure vessels. Whenever it occurs, and whatever method is used, it is often associated with maintenance work.

Apparently mundane cleaning introduces unexpected hazards, and this is probably the greatest danger. To the uninitiated, cleaning is domestic activity on an industrial scale. *Such thoughts are dramatically wrong.*

Chemical cleaning, for instance, should only be conducted by trained personnel, and the area to be cleaned clearly signed 'CLOSED TO UNAUTHORIZED PERSONNEL'. Operatives must wear full protective clothing, control the stopping and starting of the process and handle all cleaning materials in marked, undamaged and closed containers before use. Used cleaning fluids should be retained for proper disposal, and drains, leak paths and gratings checked for unwanted ingress of leaking material. Abundant supplies of flowing water, neutralizing compounds and safety supplies, such as eye washes, must be closely available during the cleaning process. Cleaning chemicals can be extremely hazardous in fluid form and release flammable or toxic vapours during use. Gas detectors should be used while fluids are exposed, sources of ignition eliminated and proper ventilation assured.

A more frequently used method is water jetting, and here the dangers are even less obvious. High-pressure jets can create static charges, which can be fatal. Some marine tankers have exploded and sunk when high-pressure jetting of empty tanks created sparks in flammable atmospheres. Also, jets are not always confined to the area being cleaned, and can cause damage and injury on lower decks or other workspaces, erupting from drains and gratings in unexpected places. Jet cleaning of tubes and pipes can be especially dangerous if the target end of an open-ended tube is not protected. Process debris, weld slag, grit, gravel or broken rock can be ejected from the open pipe like grapeshot from an ancient artillery piece. Also, the direct impact of the jet on a person can cause major medical damage. Bearing in mind that water jetting is used to clean steel and to cut or decorate brick, its effect on a human body can be devastating. In any such event, medical assistance must be sought at once.

Sometimes material removed during cleaning may itself be a hazard. Some residues or scale are radioactive, combustible, toxic or corrosive, and can affect other materials or the operator if contact is made.

The build-up of a static charge also occurs when grit blasting is used. This process prepares surfaces for painting or epoxy coating, removes old finish or surface contaminants and results in a 'crystalline' parent metal finish. Warning signs, notices, flow cut-off safeguards and safety clothing should be used, including safety helmets, breathing lines and heavy industrial gloves. A clear unimpeded exit route is needed, and the equipment air intake positioned well away from the work area in an atmosphere free of toxic or flammable gas. Softer materials, such as wood or cloth, should not be grit-blasted and in spite of its early name, sand or silica blast contents should not be used.

11.7 Specific dangers

The following is merely to show typical dangers which have been recognized by previous experience. A list of such dangers can never be complete; there are thousands of equally hazardous possibilities, and new ones are emerging all the time. Even though some dangers disappear with their work or process hosts, many more remain to haunt us. Candidates for the list will continue to increase.

(1) The maintenance manager needs safety support. He requires men to work in dangerous and lonely places.

The work environment is hostile.

(2) Each maintenance task is different. Major maintenance jobs are usually different every time.

Skill and knowledge are more difficult to find.

(3) Obvious dangers are more readily avoided.

But great dangers are often unexpected.

(4) Good safety training expands workers' understanding and decreases the unexpected.

(5) The maintenance man is often the first to confront a new danger.

New circumstances and old features, or vice versa, may fatefully combine.

Hot tapping

This is the 'cutting' into a pipeline or vessel while it remains in operation and at pressure after welding on a branch flange. It should not be considered unless alternatives are unsuitable. It should only be attempted by trained technicians under a safety regimen and full supervision. The API has produced a code for hot tapping.

One of the likely pressures for its use is economic. Careful use of hot tapping may avoid a shutdown, but the dangers associated with this type of work mean detailed attention and great caution.

(1) Residual energy is in place. The release of any high-pressure substance into a low-pressure or normal environment must be prevented by careful procedure and working pressure should be reduced if possible.

(2) Do not tap into lines containing toxic or flammable gases or vapours or any substances which may become so upon decomposition.

(3) Similarly, include lines or vessels containing oxygen, an oxygen-enriched atmosphere or hydrogen.

(4) Avoid compressed air mixed with flammable material.

(5) Any substance which by explosion could rapidly increase the internal pressure should be avoided.

If it is decided to proceed and all precautions have been made, then in addition,

be absolutely sure to choose the right line to tap into and the right position along the line.

Falling objects

Human beings as a species frequently fail to look up, and by not doing so can miss potential dangers overhead. Ironically, when objects are falling looking up can be bad news.

All situations are different, and before the work starts a good maintenance supervisor will check for loose items that can be dropped or dislodged. Workers responsible for their own safety should constantly check. It is a sobering thought that hard hats are the most abundant item of safety clothing, one whose use is endorsed by workers and employers alike. In spite of their proven value, hard hats limit the vision, especially 'up and behind' the wearer. The small peak restricts upward vision forward and when ear defenders are worn head rotation has to be increased to compensate for corner vision limitation.

It is bitter experience that has resulted in existing legislation (SI 1019) requiring new lifting appliances to be tested and certified before they are used followed by regular six-monthly examinations while they are in service. Supplies lifted from the deck of a supply boat by crane, major components such as replacement steel shafts, new motors, waste containers, drill pipes, power transformers plus components lifted out during disassembly, turbine rotors, inspection hatch covers and major subassemblies may all be lifted over the heads of workers, who are often unaware of the danger overhead. We should remember that engineers have used materials in compression from the time of the pyramids and before. Materials in tension, however, are part of recent experience, where practice is still not exact[8].

We also tend to concentrate on the danger of heavy falling objects, but small items dropping great distances with low air resistance reach high velocities, and kinetic

---

8. Conversely, it has been said that at one time railway bridges in the USA were collapsing at a rate of 25 per year.

energy is proportional to the square of the velocity. Small items are more difficult to see and impact can be fatal.

Also, guiding or watching overhead items means that workers are not watching their feet, and trips and falls can follow, especially when the person concerned is walking backwards.

Liquid storage: emptying and filling tanks

Large-volume fluid-containing tanks such as those for fuel oil or water need robust support to cater for the weight they carry. A full tank five feet square by twenty feet long can hold 500 cubic feet or 14 tons of water. Tanks will usually have been checked for leaks by the fabricator prior to despatch to site. Because of the risk of distortion in transit, they should be checked again, full, after installation, when internal pressure and static supports loads are greatest.

These checks should already have been done before maintenance personnel are involved, but several items need to be remembered.

(1) During installation, tanks are often leak tested with water, which is cheap, abundant and will not cause environmental problems on disposal. However, the intended tank contents may not be water and will probably have a higher fluid weight per volume. When overfilling to confirm level controls, previous test weights may be exceeded and tank supports prove inadequate[9].

(2) The nature of the tank's normal contents must be known, even if it is expected to be empty. Fluids can be toxic, corrosive or highly flammable and give off dangerous fumes. Safety precautions should be completed and in use during any work programme. Refer to HSE publication 'The storage of flammable liquids in fixed tanks (up to 10,000 m^3 capacity)', HS(G)50.

(3) Before emptying a tank under normal circumstances, ensure that a ventilation route is open to prevent the formation of an internal partial vacuum and the collapse of the tank because of external atmospheric pressure.

(4) All equipment in use should be waterproof, *especially electrical devices*. Accidentally submerged electrical equipment must not transmit power to the water; for a worker standing in the same liquid results can be fatal.

(5) The metal fabric of the tank may thin because of the effects of corrosion, especially near the fluid surface inside the tank, or because of erosion near fluid entry or exit piping.

(6) Tanks are often mounted high up in buildings or structures to create a pressure head for the stored fluid. This means that the weight and the fluid are above people's heads. Consider the consequences of a tank rupture, the damage likely to follow sudden downrush of fluid, the risk of local flooding, and the effect of the fluid itself on stores, office equipment or other pieces of plant.

---

9. The pressure testing of vessels using pneumatic pressure contains much greater energy at a specific pressure and is much more dangerous. It is not recommended. Refer to HSE Guidance Note (General Series) 4, 'Safety in pressure testing'.

The sudden pressure caused by inrushing fluid can hold doors tight shut, making escape impossible. Although liquid levels will probably fall quickly, depths in a confined volume can be over head height.

The pressure head of a stored liquid is related to its height above a point of discharge. Fluid falling through an open drain system can suddenly erupt upwards at lower floor or deck levels, throwing grills or drain covers many feet into the air. Fluid levels can rise locally when streams are constricted, and the flow can change from large volume slow-moving to reduced volume at much higher velocity as the mass flow is maintained. Closed gangways or stairways can become the focus of great danger, especially when they are in use as routes of escape.

Water, especially when treated sensibly, is regarded as safe. Storage tanks have no moving parts and are not the subjects of sophisticated technical failures. For these reasons, *danger can be overlooked. All maintenance of fluid handling systems must be treated fastidiously*, and such low-risk but high-damage failures treated as serious elements of operational hazard evaluation.

Hot work

A welding torch or flame cutter should not be applied to any vessel, tank or pipeline without first checking the effects of raising the temperature. Even with internal steam cleaning and prolonged venting, it is extremely difficult to remove all traces of the previous contents, and toxic or flammable vapours will often occur. The risks of explosion or combustion in confined spaces are well known and have been caused by the heating of heavy residues which do not produce flammable fumes at normal ambient temperatures.

Work areas should be well ventilated to prevent oxygen deficiency and both breathing and burning difficulties. Gas, oxygen and other detectors should be used to check air quality for breathing and the presence of flammable or toxic fumes. The instruments used must be subject to a maintenance and calibration programme and the renewal date on each calibration label checked before use. The internal and external faces of containers may be treated by painting or plated deposits, and cadmium, zinc, lead and copper finishes or compounds can give off dangerous fumes, while stainless steel or low-hydrogen alloys can produce fluorides when used in welding electrodes.

Gas bottles used for welding or cutting and whose contents are under pressure should be removed from the fire risk area when they are not in use. Apart from the mechanical dangers of ruptured and flying metal, a broken bottle can release large volumes of highly combustible gas rapidly expanding under ambient pressure. Such compressed gas cylinders should not be placed inside a confined volume.

Oxygen can be liberated as a by-product of the flame cutting process and oxygen enrichment can result, particularly if workplace ventilation is inadequate. Excess oxygen causes behavioural changes in human operators, and normally slow-burning substances can burst into flame and some metals will burn steadily when they otherwise would not.

When heat is applied to an object, it will change its shape. Friction-fitted components may become loose, interfaced seals may break down and tightly fitted bolts under high tension may fracture as bolted parts expand. In addition, heat is conducted from one place to another, particularly through metal components such as piping. Cutting or welding safely conducted at one location may cause an explosion at another.

Confined volume

Working inside a container or machine is particularly dangerous, and there are several immediate aspects to remember:

(1) *The risk of fire or explosion*
    The combination of combustible gas or vapour, oxygen enrichment, confined space and flame can easily be lethal.

(2) *Toxic effects*
    Process residues, surface treatment components, some metals and welding rods used can release vapours which harm the operator. Fully effective ventilation is essential.

(3) *Hidden dangers*
    Tanks and vessels often contain internal piping, weirs or vessel dividers which restrict already limited vision and hide fluid or process residue. Never assume that the vessel is clean and empty.

(4) *Poor communication*
    A worker wearing a faceplate or breathing apparatus inside a tank or pressure vessel is almost completely isolated. In addition to the difficulty of working in a confined volume, workers on the inside cannot be readily seen by those on the outside. The use of radios or telephones which are electrically safe will reduce claustrophobia and improve communication when usable, and some people beset by feelings of loneliness and claustrophobia find work in such circumstances impossible.

(5) *Risk of asphyxiation*
    Oxygen levels, ventilation systems and breathing apparatus when in use must be monitored all the time. Breathing difficulty can become a disaster very quickly, as air volumes in confined spaces are limited.
    Ensure that exhaust gases are extracted and breathing apparatus is positively pressurized, i.e. that there is a greater pressure of the breathing air so that external gases are constantly excluded.

(6) *Entry and exit*
    Safety-clad workers will move through narrow hatches and confined spaces slowly. Ensure that trained rescue personnel are on hand whenever a worker is in a confined space[10].

---

10. The permit-to-work procedures are intended to ensure that all hazards are taken into account before a permit is issued and before work begins. They are designed to ensure that dangers like those described above are recognized and prepared for, with all precautions in place at the outset (see Section 11.9).

11.8 Worker resistance

Item 4 in the list above touches on claustrophobia, but the influences on the maintenance worker are wider than this. There is no doubt that some jobs are highly unpopular. Who, for instance, wants to crawl into a foot-high space beneath the floor of the generator room, underneath a running turbine? Such a job is done mainly in darkness, close to very hot metal surfaces that will burn exposed skin, in fetid air, heat, noise and vibration, where head-first is the only way in and feet-first is the only way out.

Similarly, the two-yearly PPM on the flare tips of an offshore oil platform requires a maintenance technician to climb up the access gangway on a tapering flare tower out to the ultimate tip, 250 feet above the water, often in cold winds and poor visibility, hanging on to a cold steel support 18 inches wide.

We have all heard of 'painting the Forth Bridge', probably the one maintenance job that has become part of the spoken culture. This work is not famous because of its danger, the elements, the height above the Firth, the wind and exposure of its men; all that is either overlooked or accepted as part of the job. No, the job is famous because it classically captures a task that is never-ending.

it is work that goes on all the time.

There are thousands of jobs of similar difficulty, like undersea divers routinely inspecting steel structures for cracking or corrosion, or maintenance workers entering the so-called 'oily water caisson' down the inside of a steel support leg around 20 feet in diameter in darkness, in an unbreathable atmosphere, in breathing apparatus, clinging to a narrow steel ladder and lower than the surface of the surrounding sea.

What can a maintenance manager do to ensure that such work will continue to be done and will receive the recognition it deserves? It is sadly ironic that the latter is the more difficult.

(1) Silence anyone who insists that it is all part of the job and compared to other work is nothing special. An invitation to take part personally is usually sufficient.

(2) Include the team in the task of dividing and planning the work. Use the principle of consent, distributing among team members as evenly as possible but taking account of people's limitations. Some would cheerfully climb a chimney or a drill tower, but could not enter a confined space.

(3) Make sure that the maintenance team know that, without hesitation, you will put their safety first. This must not be lip service. Remember the warning: '*In any conflict between what you say and what you do, it is what you do that is believed*'.

(4) Be prepared to listen. That should be said again: *be prepared to listen*, especially to those who will do the job and those who have done it before. Circumstances will change and both work and personal safety will be affected.

(5) Make sure that supervisors attend the workplace, and do this yourself whenever possible. The sharing of difficulties improves understanding and motivation.

(6) Appreciate effort and completed tasks. A simple 'well done', if it is meant, will be sufficient.

(7) Apply pressure thoughtfully, avoiding excitement or agitation, as they will be transmitted to the work itself. Disappointment is legitimate in support of safety standards and targets, in that order. Do not focus on any unless the preceding ones have been satisfied.

Never use authority to cut corners or jeopardize staff

(8) Do not instigate by personal intervention nor encourage, by approval or silence,

any action outside the permit-to-work system.

See footnote 10 and Section 11.9.

The history of maintenance work is littered with horror stories which often relate to the workers themselves and sometimes to innocent parties. Although the law now enshrines the worker's responsibility for his own actions, *the manager is firmly in the chain of responsibility*, which is even more emphatic if he has taken a direct part in the action concerned.

(9) Insist on tight standards. Sloppy work leads to danger and to repeated work. In all these cases:

Work quality is vital.
Work safety is imperative.

11.9 The permit to work

There are many ingredients required before a job can begin, and arguably the most important is the 'authority to proceed'. We have already referred to positive and negative association (Section 8.3) when constructing the maintenance plan and the decisive coupling or separation in time of different maintenance jobs. Permit-to-work systems refer to *all* plant activities, applying such thinking more widely and giving them more force. Initially developed by oil, chemical and energy industries, permit-to-work systems are being adopted by a growing number of industries and are being reinforced by the legal constraints of the Health and Safety at Work Act plus the requirements of the HSE.

The petroleum industry is known for operational risk, regularly handling highly dangerous fluids or materials, and a permit-to-work system is legally required when the assessment of risks can show this to be needed. A permit to work is a formally prepared document sanctioning defined types of hazardous work and giving clear permission for specified tasks to be carried out, with all such work carefully coordinated and controlled. The permit defines the times when the work will be conducted, authorizes a named competent person in charge of the work and specifies the main precautions to be followed for safety.

Objectives of the permit to work

(1) To provide the proper authority for specified types of work to be carried out. (See list of likely work types below.)

(2) To communicate to people engaged in the work its precise scope and nature, together with a clear appreciation of the hazards involved and the precautions to be taken[11].

(3) To define the isolation procedures to be followed, together with all other precautions to be taken (see Section 9.3)

(4) To specify hazards, such as dangerous substances, toxic or corrosive materials and retained energy.

(5) To ensure that the plant manager is aware of work being carried out.

(6) To provide a key addition to the work management system by providing work co-ordination, control and recording of precautions carried out with checks by a competent person.

(7) To control the issue and use of permits relating to different items of work which may affect each other.

(8) To include in the permit system a procedure for the suspension of some tasks before they are completed.

(9) To provide a formal handover procedure for use when a permit is issued for a period longer than one shift.

(10) To provide a formal handback procedure that the part of the plant affected by the work is in a safe condition and ready for reinstatement.

Additional features

Permit-to-work systems are most effective when based on consultations with employees and coordinated and controlled by a competent person. Systems which are imposed without these steps are less well understood, causing confusion among personnel.

Permits should be *displayed prominently*, particularly at the workplace, at the main control room, all local control rooms and at the work site. The permit may also on occasion be suspended if there is a general alarm; for operational reasons, such as a conflict with other preferred work; or the presence of gas or combustible fluid when a hot work permit has been issued; or if the work time allowed has expired, such as the end of the shift.

When this is the case it is prominently displayed, as the plant remains subject to a suspended permit and the security of isolations already made is important – the plant cannot be regarded as safe for operation. The permit can only be reactivated when the responsible person has confirmed that it is safe to do so.

When a permit is prepared it is necessary to check whether any of the actions included in the work permitted interact with steps introduced by other permits already issued and still active. Such permits may have been prepared by different responsible people, and careful coordination will prove necessary, perhaps assisted by extra notes on the permit itself or on the summary of all permits.

An important point to remember is that isolation of a certain valve or section

---

11   Active permits are displayed in the control room, at the points of isolation and the place of work. Other locations apply according to the dictates of the job concerned or the requirements of the safety officer. *This is of added importance when entry permits are in use.*

of pipeline may be required for *more than one* isolation procedure, with separate permits required for each. The completion of one permit-driven piece of work may not release the common point from its isolation required by the other work permit.

What types of work require permits to work?

Permits to work are required when the health or safety of workers or other personnel involved is endangered or critically affected. One of the values attached to them is their use, not in general but in specific cases, and the benefit of using them is reduced if they attempt to control all work. They are best applied to non-production work, which means that maintenance often benefits from this form of work and safety management. Permits are also sensible when more than one person is active on a job, each being a member of different groups or disciplines, or of company and contractor's staff, where the transfer of work or responsibility regularly takes place. The type of jobs requiring permits will include[12]:

(1) Hot work, such as welding, flame cutting or grinding.
(2) Work which may generate incendiary sparks or other sources of ignition (including cameras) in hazardous areas.
(3) Work which may cause an unintended or uncontrolled hydrocarbon release, including any disconnection or opening of any closed pipeline, vessel or equipment containing (or which has contained) flammable or hazardous materials.
(4) Electrical work which may cause danger.
(5) Entry into confined spaces and work inside them.
(6) Work at any place on an offshore installation from which any person will be liable to fall into the sea.
(7) Work involving the use of hazardous/dangerous substances, including radioactive materials and explosives.
(8) Well service operations.
(9) Excavations.
(10) Offshore diving activities.
(11) Pressure testing.
(12) Maintenance operations which compromise critical safety systems or remove them from service, e.g. fire and gas detection systems, public address systems, life-saving equipment and fire-fighting equipment.

The preceding notes have been prepared with reference to the HSE publication 'Guidance on permit-to-work systems in the petroleum industry', which is an Oil Industry Advisory Committee publication. Some of the notes above are abstracts from this document, and the following additional elements are included.

---

12. It is important to note that some of the activities in (1) – (12) will require precautions in addition to a permit-to-work system.

Further points of discussion:

Consideration of the handback procedure.
Regular review and assessment.
Checking whether permit conditions are being met.

Responsibilities under permit-to-work systems of:

Employers/dutyholders.
Offshore installation manager.
Onshore manager.
Contractor/subcontractor.
Responsible person.
Supervisor.
Individual workers.

Outline review of training requirements.

Appendix of legal requirements pertaining to permit to work systems. This appendix lists the statutory acts and regulations and defines the specific clauses or regulations applicable.
Appendix listing types of permit and work that may require them. (This appendix is reproduced below.)
Appendix of system and equipment isolation requirements for safe working (see Sections 9.3 and 11.10).

Check-list for assessment of permit-to-work systems

Contents

The contents of any permit to work will contain certain essential components.

(1) Name of the company and the site, factory or premises where the work is to be conducted.
(2) A unique work location.
(3) The permit reference number.
(4) The applicable date plus the start and finish times[13].
(5) Description of job defining the content and nature of the work and the methods to be used.
(6) Number of persons authorised by the permit[14].
(7) **Authority to proceed**
     Signature of the named person qualified to initiate work, who has checked the readiness of safety precautions and other permits issued or to follow.

---

13. The permit defines a work time-band, which must not be violated. Negatively associated work allowed nearby may have been deliberately restricted to carefully chosen times.

14. This information is also vital, especially when work has been completed inside a confinement which is about to be closed. There are recorded happenings of workers remaining trapped inside the machines or spaces they have been working in. One of the most recent refers to the discovery of the skeletons of a boilermaker and his apprentice trapped unheard behind the riveted steel plates of a steam-driven merchantman returned for reconditioning after a lifetime's work at sea.

(8) **Safety requirements**
    The careful definition of safety systems, precautions and equipment plus the identity of the Safety Officer or Supervisor or both.

(9) **Cautionary reminder**
    For the person accepting the work a reminder of legal force of the procedure and the requirement for safe and responsible action.

(10) **Acceptance of the permit**
    The person responsible for the work and all constraints or conditions attached to it.

(11) **Statement of hazards**
    Reminder of the hazards likely to be encountered in the work situation during the progress of the job.

(12) **Isolations completed**
    Careful definition of all isolations required and carried out cross referred to maintenance work instructions or procedures

In addition to the general permit-to-work documents several supplements often of different colours are used.

Types of permit to work

Appendix 2 of the HSE document referred to, 'Guidance on permit-to-work systems in the petroleum industry', tabulates some of the various types of permit in use. It is reproduced here as an illustration of different work permits which may encountered.

(1) This appendix describes some of the different types of permit-to-work which may be found within the petroleum industry, and some of the activities which may require them. It contains examples only: it is not exhaustive, and it does not imply that every kind of permit mentioned here will be needed at any single workplace or that they are recommended types. They may be used independently or to support a general permit-to-work.

(2) It is important to realise that similar terminology may be used at different sites for types of permits which are fundamentally different. Some of the permits listed here are called certificates. This is not recommended as it can confuse permits-to-work with other kinds of document. It is better for permits-to-work always to be referred to as permits.

Hot work

(3) Hot work is usually taken to apply to operations involving the application of heat to tanks, vessels, pipelines etc which may contain flammable vapour. Hot work permits are often more generally applied to any type of work which involves actual or potential sources of ignition and which is done in an area where there may be risk of fire or explosion, or which involves the emission of toxic fumes from the application of heat. They are normally used for any welding or flame cutting, the use of any tools which may produce sparks and the use of any electrical equipment which is not intrinsically safe or of a suitably protected type. Some sites or installations distinguish between high energy

sources of ignition like naked flames, welding and spark-producing grinding wheels, which are almost certain to ignite flammable atmospheres, and low energy sources like hand tools and non-sparking portable electrical equipment which are likely to cause ignition only if there is a fault. In some cases, to differentiate between these tasks, *fire or naked flame permits* and *spark potential permits* have been introduced.

Cold work

(4)   Cold work permits are frequently used to cover a variety of activities which are not of a type covered by the more specific permits. The activities for which a cold work permit may be appropriate will vary from workplace to workplace but should be clearly defined.

Clearance certificate

(5)   This is the name given to a type of permit, very similar to a cold work permit, used in one company's procedure. The same terminology may be employed by other companies in handback procedures designed to ensure that a plant has been correctly reinstated and all tools etc have been correctly withdrawn before the plant is recommissioned.

Safety certificate

(6)   The term safety certificate is sometimes used in a very loose way to indicate any form of permit-to-work but can also refer to a specific type of certificate covering certain designated types of work.

Preparation/reinstatement certificate

(7)   Permits with this name are sometimes used to cover preliminary work or handback procedures after a job has been done. If used in this latter way, these certificates will normally be employed in conjunction with some form of permit to cover the main job itself.

Equipment disjointing permit

(8)   This type of permit may be used for any operation which involves disconnecting equipment or pipework which has contained any liquid or gas. Certain exemptions may be made at certain sites for pipework for non-hazardous materials such as domestic hot water. This type of permit will normally be used also for the insertion of spades into pipework and for the removal of such spades.

Confined spaces

(9)   A permit is used to specify the precautions to be taken to eliminate dangerous fumes or prevent a lack of oxygen before a person is permitted to enter a confined space. The permit should confirm that the space is free from dangerous fumes or asphyxiating gases. It should also recognise the possibility of

fumes desorbing from residues, oxygen depletion of the atmosphere as a result of oxidation reactions, and the ingress of airborne contaminants from adjacent sources. Precautions should be specified to protect the enclosed atmosphere against these, for example by forced ventilation, or if this is not practicable, by the provision of personal protective equipment including breathing apparatus.

Machinery permit

(10) This type of permit is usually used for work on large complex items of machinery to ensure correct isolation before the work is carried out.

Isolation certificate

(11) This type of permit may be very similar to a machinery permit or an electrical permit. It is usually used as a means of ensuring that the particular equipment is mechanically and electrically isolated before it is worked on. It is possible that a similarly named permit may be used for chemical isolation of plant before work is done on it or entry is made. (See Appendix 3 [not included here] for further advice on system and equipment isolation and Section 9.3).

Electrical work

(12) An electrical permit will normally be used to minimise the risk of electric shock to people carrying out any work on electrical equipment. (A hot work permit might also be required if the work could give rise to a potential source of ignition within a hazardous area.)

Excavation permit/heavy equipment movement permit

(13) This may also be called a ground disturbance permit or something similar. It will typically be required whenever any digging, excavation or boring has to be done to ensure that no underground services or pipework will be dangerously damaged. At certain sites the permit may be required only for excavations below a certain depth. Damage may be done to underground services or pipework by movement or placing of heavy equipment. These activities may be controlled by broadening the scope of excavation permits.

Radioactive material

(14) The use of radioactive materials is covered both onshore and in the offshore petroleum industry by the Ionising Radiations Regulations 1985.

Overhead crane permit

(15) A specific type of permit may be used where overhead travelling cranes are installed and work has to be done on or near the crane tracks. It will be used to ensure compliance with the specific requirements of Section 27(8) of the Factories Act 1961 or to ensure the operator's safety where the Factories Act does not apply.

Roof access

(16) A specific type of permit may be used by some companies where access to roofs of any sort is required, particularly roofs which are of fragile construction or where it is necessary to control the risk of falling from roof edges. This sort of work may also be controlled by a cold work permit or some other general permit procedure.

Permit for removal of equipment from work site

(17) It may be necessary to remove equipment from the work site to enable specialised work to be done or to enable work to be done in a special environment. In certain cases it may not be possible to completely free the equipment from hazardous substances before removal. An equipment removal permit may be used to communicate the precautions required for transporting the equipment safety and for making it safe to work on at its new location. The permit may also be used to certify that equipment may be disposed of for scrap.

Special area work permits

(18) Some companies have developed special permits for work in areas where a serious inherent hazard exists, for example in alkylation units in which hydrofluoric acid is used. They list the special precautions required for work in the defined areas, thereby reducing the risk of errors or omissions which could result from the use of general permits.

Offshore diving permit

(19) An offshore diving permit can be used to control the diving activity itself and to ensure that there are no other activities taking place nearby which create extra hazards.

11.10 Electrical safety

Unlike mechanical work, problems with electrical items or systems are more difficult to spot, and electrical work has particular problems, especially when work is required on degraded equipment.

As in the preceding safety sections, the following notes do not attempt to cater for the full scope and complexity of such a dangerous and specialized subject. The intention is to underline the vital nature of electrical safety and help ensure that a formal and comprehensive review of statutory requirements and expert commentary has been made before electrical work is undertaken especially for the first time. In many work situations electrical engineers and safety specialists maintain an awareness and enforce a range of work practices designed to adhere to both the regulations and the special safety needs of the company position[15].

Other smaller organizations do not have this advantage and, conducting electrical work occasionally, they concentrate on project and contractor management. However, even when the maintenance or repair work is carried out by a third party a general knowledge of this and any other specialist area should be regarded as essential.

Electrical injury can be brutal, unexpected and often fatal. Most readers of this manual will have experienced an electric shock during their lifetime, indicating how pervasive is the electrical presence and how easy it is to forget. In the Western world electrical power is everywhere: it is installed in residences, in places of work or transit, and is provided on site by often temporary installations or mobile devices. Even specialists, trained and experienced in the ways of electricity, injure or kill themselves with sobering regularity by a variety of gruesome means.

Electrical power is endemic: sometimes invisible, always dangerous

There is a wide range of injury causes, of which these are a few:

Electric shock

Shock often kills; it can be swift and fatal. In addition, shock causes muscle contraction, which may increase the conduction of current through the victim's body, making the shock worse. Muscular reaction caused by the shock can 'throw' people from ladders, buildings or pylons, creating indirect injuries which can be even more painful or final.

Electric burns

Intense heat caused by arcing and high or localized current can attack limbs or deeper parts of the body causing long-lasting and painful burns.

Electric arcs

Arcs are sometimes overlooked as a cause of injury, but the results can break and damage the skin, penetrate the body and create severe burns.

Ultraviolet radiation

Intense ultraviolet radiation causing damage to unprotected eyes can be an electrically caused injury, which comes as a nasty surprise.

These are merely some of the consequences of electrical effects. Resulting from the disruption other hazards may occur, either as part of the same overall failure or caused by the electrical breakdown, such as fire and explosion, or the loss of

---

15. Many industries have specific guidelines to be followed, with additional and special elements included, while other sectors do not, but in all cases the relevant standards and safety publications should be regularly checked.

electrically powered safety devices, communications equipment, computers and lights[16].

Regulations

Discussed below are some of the key headings which appear in the regulations and which will feature among others in an electrical safety review.

Almost all work situations are included under the electricity at Work Regulations 1989[17] and are clarified through guidance notes produced by the HSE[18].

There are also standards which apply to electrical work in almost all places (see electrical safety publications listed in Chapter 14 and electrical safety standards shown in Chapter 16).

Equipment

Although the design and construction of most electrical equipment is deliberately safe there is some equipment which is intended to be beyond the contact of untrained personnel. Safety in these is maintained by allowing access only under specific authority. In addition, equipment may be in a poor state, so that the original safe state no longer applies and restraints of use and access are necessary.

Voltages

The international definition of high voltage is 1000 volts (In the UK extra precautions are required above 650 volts). Refer to Electricity at Work: Safe Working Practices, HSE publication HS(G)85 ISBN 0 7176 0442 X.

Injuries of the types described above are more serious at high voltage, although danger is still present and often disregarded at low voltages. Electrical energy can lead to overheated conductors or can be stored at very low figures of current and voltage which can steadily accumulate to cause damage or injury when suddenly and unexpectedly discharged.

Environmental

Electrical applications are affected by the environment in which they are located or through which they are distributed. Adverse environmental features should be included in any assessment and be considered when inspection routines or instructions are produced and any work is to be undertaken, such as corrosion on terminals and conductors causing arcing or overheating.

Installation

Safe electrical installations feature designs that can be made safe by specific circuit isolation, which can be taken out of service for inspection or maintenance while

---

16. Many critical systems have separate or duplicate power supplies. Refer to 'Electricity at Work, Safe Working Practices', HS(G)85 ISBN 0 7176 0442 X.

17. Electricity at Work Regulations SI 1989/035 HMSO ISBN 0 11 096635 X.

18. HS(R)25 Memorandum of guidance on the Electricity at Work Regulations 1989 HMSO ISBN 0 11 883963 2.

other circuits remain in operation. Equipment is usually insulated and shrouded so that live parts are not accidentally accessible, with powerful and controlling elements separated where possible.

Proposed work assessment

Because electrical hazards are more difficult to identify than many others, safety procedures recommend a carefully conducted assessment of the steps required and the likely hazards present before the work is attempted, even when, or perhaps, especially when, unplanned sudden intervention is necessary. Such checks include **careful judgement about whether dead or live working** is to be adopted and the strict sequence of planned steps to be followed. An assessment of the risks and a careful judgement as to whether 'dead' circuits are essential or whether 'live' circuits are permitted must be made.

Management and supervision

An inherent requirement at the outset is the necessary involvement of managers and supervisors who are requiring and controlling the work being done. Of particular importance is the early inclusion of the safety regulations and the arrangements applicable to contract personnel when outside workers are to be used. It is a vital work management requirement that all participants have an accurate and comprehensive understanding of what is required, the steps to be followed and most importantly the safety precautions to be followed.

To ensure that safety regulations are being properly carried out, regular check visits to the workplace by supervisors are essential.

Prepare a plan

The management framework requires work to be carried out to a clear plan, identifying all work activities, the sequence of attention and the formal supervisory aspects of delegation, coordination and control. The plan should also cater for clear fault-finding tasks to be followed after a breakdown. This provision should be followed carefully to avoid frequent accidents which occur during fault-finding but before corrective work is started.

In addition, the plan should identify the people assigned to the different tasks, the hazards that they face and the precautions to be followed. The plan, forming the foundation of the work programme, should specify the correct rules, written instructions and procedures, together with a simple policy statement supplemented by the steps required to cater for the following:

Clear identification

Work conducted on live circuits or equipment thought to be safely switched off or contact with items which should have been isolated is a regular cause of accidents. Correct identification will need reference to drawings and other documents, all of which must be checked for **current validity,** as old or inaccurate drawings can themselves cause accidents. Also, electrical information is frequently presented in schematic or diagrammatic form and careful translation to the physical situation is needed when circuits or networks are accessed.

Prevention of injury

The chances of injury can be reduced even when live working has been permitted:

(1) By erecting temporary insulating and protective screens. Even when disconnection, isolation and earthing have been completed, avoid contact with other live equipment nearby.

(2) By ensuring that required safety clearances are present where work is to be done. Figures required are 915 mm (for 415 volts) or 1375 mm if parts are live on both sides of a work position. In addition there must be sufficient headroom and adequate lighting.

(3) By ensuring that all personnel engaged on the work are properly trained, can be regarded as competent persons, and are capable of interpreting signs of damage, deterioration and danger applicable to equipment under their attention.

(4) By ensuring that all electrical safety requirements are meticulously followed.

(5) By ensuring that properly insulated tools are used where needed which are proof against damage and regularly inspected[19]. Any other tools or instruments must be provided in good condition and calibrated where necessary.

(6) By ensuring that protective clothing and safety equipment are provided and used.

(7) By ensuring that the tops of instruments, cable trays, power units and other surfaces are not used for temporary storage of tools and replacement or reject parts.

(8) By ensuring that prompt action occurs in the case of shock and that workers are accompanied whenever safety precautions recommend.

(9) By ensuring that all information is clearly provided in support of the rules and procedures to be followed and workers are fully instructed in their use.

Dead system procedures

The procedures applied to dead circuits and equipment vary from one situation to another but certain elements are needed in most cases.

As referred to above, correct circuit identification and labelling are essential. The conduct of work on live systems incorrectly assumed to be dead has been mentioned as a frequent cause of accidents.

Disconnection: Regulation 12 in the statutory requirements[17] specifies that equipment is disconnected from every source of electrical energy before work commences on parts which have been live or are likely to be live.

Isolation

Any isolation device should be secure and undamaged and carefully locked in the OFF position, preferably using a lock with a unique key. If a fuse or plug has been removed ensure that such safety steps cannot be bypassed by replacing the removed

---

19. Electrotechnical Commission IEC 900: 1987 Handtools for live working up to 1000 volts AC and 1500 volts DC.

item or inserting a similar item in its place. Where several devices and lock OFF keys are used, a whole enclosure or cabinet can be locked closed or removed items placed in a lockable container which is itself held secure with a single unique key. Also, gaps between disconnected sections should be sufficient for the voltages being disconnected, and once the work has been satisfactorily completed make sure that all locks have been removed before systems are powered back up. (Refer to Section 9.3.)

Display warning signs

Display a caution notice at points of disconnection stating clearly that the system or apparatus is de-energized and men at work will be injured if power is restored. Also show items which still carry energy and remain dangerous to those at work and others in the vicinity. It is most important that notices which are easily understood are in place as soon as the work begins and are removed as soon as it is complete[20].

Dead proving

Once disconnection or isolation are complete check that the circuits, equipment and apparatus to be worked on are certainly dead. The device used to check this should be regularly tested as functioning correctly before these safety checks are done. The instruments themselves should protect the user against electric shock, prevent short circuits, be constructed with fusing or energy limiting elements, provide indication such as lamps or meters and be constructed with proper insulation.

(Refer to HSE note GS 38 (revised), Electrical Test Equipment for use by Electricians ISBN 011 883533 5.)

Earthing

The risk to the technician can be further minimized by earthing conductors using proper leads or earthing devices applied at those points where the equipment is isolated from the supply and are sufficient for the current or energy which could arise with failure. Further earthing may be applied after dead proving at the work point has been done if it is remote from the point of isolation. Such steps are essential for high voltage circuits or stored energy equipment, such as capacitors, and where practical, for low voltage equipment where there is a risk of re-energization.

Work with high voltages

Electric shock, burns and other injuries can be caused without making direct contact with live apparatus when high voltages arc through an air gap. Isolation devices must include safe gaps between different live elements and between live elements and those made dead. The earthing of conductors and the use of locking off and unique keys, plus the safety methods discussed above become even more important and should be backed by a formal 'Permit To Work System' (Section 11.9).

---

20. When warning signs or notices are left in place after work has been completed and the need for them has gone, they can become degraded in the eyes of observers, who may fall into the habit of disregarding current warnings.

Chapter 12

Safety-critical systems

We have already met the concept of equipment criticality in its general sense, referring to that equipment which in the event of its failure causes production downturn.
The question of safety criticality is similar, but it is a more recent concept and
definitions vary. We could argue, for instance, that any equipment which by its
failure introduces a safety problem is clearly 'safety-critical'. This overstates the
case: the failure of a breathing set, for instance, would merely be followed by its
replacement, or if a lifeboat was suddenly taken out of service then the number
of permitted platform workers would be reduced. In both cases the response to
the equipment failure would be less severe than the safety-critical title would suggest. We might alternatively consider equipment which is only required when human beings are present. Such a description would include life support and
accommodation facilities, which also cannot be correct.

More realistically:

> A 'safety system' required in support of work on site, which in the event of failure
> could not be rapidly rectified or replaced, putting site personnel in immediate danger,
> would be classified as a 'safety-critical system'.

This is the author's description rather than a definition, and it assumes the option
of rapid response, which is not always available. A more specific definition is to be
found in SI 1996, No. 913, The Offshore Installation and Wells (Design and Construction, etc.) Regulations 1996, Regulation 26, Schedule 2 (see Section 12.2). See also
'critical defect' and 'critical state' in the glossary pages 360 and 361.

Definition of safety-critical elements

'Safety-critical elements' means such parts of an installation and such of its plant
(including computer programmes), or any part thereof.

(1)   the failure of which could cause or contribute substantially to;

or

(2)   the purpose of which is to prevent or limit the effect of, a major accident;

'Major Accident' means –

(1) a fire, explosion or the release of a dangerous substance involving death or serious personal injury to persons on the installation or engaged in an activity on or in connection with it;

(2) any event involving major damage to the structure of the installation or plant affixed thereto or any loss in the stability of the installation;

(3) the collison of a helicopter with the installation;

(4) the failure of life support systems for diving operations in connection with the installation, the detachment of a diving bell used for such operation or the trapping of a diver in a diving bell or other subsea chamber used for such operations;

or

(5) any other event arising from a work activity involving death or serious personal injury to five or more persons on the installation or engaged in an activity in connection with it

This definition of 'major accident' is referred to in SI 1995, No. 743, The Offshore Installations (Prevention of Fire and Explosion, and Emergency Response) Regulations 1995, but is recorded in full detail as above in SI 1992, No. 2885, The Offshore Installations (Safety Case) Regulations 1992.

The definition is also repeated in the HSE booklet L65, 'Prevention of fire and explosion response on offshore installations, approved code of practice and guidance'[1,2].

12.1 The basic dilemma

Major incidents, like Piper Alpha in the UK North Sea, the Bhopal Chemical complex accident in India and the oil terminal and processing site explosion in Bantry Bay,

---

1. This definition of safety criticality deals with offshire oil or gas production platforms; references to helicopters, diving operations and damage to the structure have less relevance to most onshore plants.

2. For the maintenance man there is an interesting conundrum: if critical system maintenance requires a shutdown, how is the worker protected while the maintenance work is done? There is clearly some latitude with regard to timing, and a distinction between 'failed' and 'shutdown' equipment. The following would be applied for work of this sort.

(1) Whenever possible, major maintenance work on safety-critical systems would be done when the site was vacated. An annual shutdown would be ideal.

(2) When (1) was not possible, system shutdown would only occur for a known duration agreed by the safety department and subject to permit-to-work requirements (see Section 11.9).

(3) Access to the affected area would be restricted while normal safety arrangements were affected. Severity of the steps applied by the safety group would reflect how rapidly the relevant safety system could be restored to full operational order.

(4) Work would require full safety monitoring throughout, plus the application of safety procedures during preparation and completion.

(5) The restoration procedure would be formalized and agreed in advance with copies posted with safety, the control room and at the workplace.

(6) Particular attention would be paid to the bypassing of alarms, trips and safety devices during shutdown and isolation, work in progress, reconnection and start-up.

We should never forget that many of the items described feature in existing regulations because of parts they have played in accidents that have occurred before.

Ireland, plus an apparently endless parade of many others in a variety of industries, point to the never-ending necessity for professional safety vigilance and relevant legislation[3]. This is important to remember, and before we become gripped by the fateful consequences of accidents we should recall what has already been achieved.

This can be done by looking back to industrial operations in factories, mines or shipping and other areas like farming and transport, or even military activity, for merely a hundred years or so, to demonstrate quickly the positive influence of safety thinking over the way that work is done and human beings are employed. Unfortunately, progress does not eliminate accident, and if working men are both perpetrators and sufferers, a truly accident-free environment may only be possible when processing is fully automatic and human beings are not present. We may take further comfort from the fact that many areas of safety attention apply to technologies and plant that have changed little and that 'safety considerations', included in the design process, are delivering improvements in product quality, operational safety and technical performance.

Unfortunately it is also argued that had operations been safe, the accidents referred to above, and others, would never have occurred, and that the sources of such occurrences lie in a mixture of faulty implementation and safety regulations that are out of date. It is the essence of safety work that, while there is some truth in such allegations, safety officers are constantly faced with the problems of human nature and technical change, and may implement regulations which subsequent events can show to be inadequate or incorrect.

Such difficulties strike at the heart of safety implementation. Human operators and technicians are sometimes reluctant to follow safety rules, while their managers underline safety requirements at large and emphasize different targets at home. There is also the question of group or social understanding and the team's perception of the regulations they follow. How else can we explain different safety performance by different countries or cultures, where identical machines or technical processes are at work? There is also the matter of reporting. There is a current legal requirement that dangerous events or accidents are reported, and although the responsibility is limited to selected 'Reportable Incidents' it is both a chore and an admission of fault[4]. To counter the suspicion that incidents are not always reported, penalties for such omissions can be severe, *but even more importantly it is the principal means by which we learn.*

---

3. The recommendations of Lord Cullen's report of the explosion and events on Piper Alpha on 6 July 1985 were influenced by HSC/E's experience of regulating major hazards onshore under the Control of Major Accident Hazards Regulations 1984 (CIMAH) (SI 1984/1902). These regulations were themselves a response to major incidents that took place in the 1970s, including Flixborough in the UK (1974) and Seveso in Italy (1976).
  Refer to the HSE booklet L30, 'A guide to the Offshore Installations (Safety Case) Regulations 1992'.
4. Recent requirements are defined under 'The reporting of Injuries, Diseases and Dangerous Occurrences Regulations 1995' (RIDDOR 1995), which came into force on 1 April 1996. The regulations are made under the Health and Safety at Work etc. Act 1974 (HSW Act). Guidance is provided in the HSE booklet L73, 'A guide to the Reporting of Injuries, Diseases and Dangerous Occurrences Regulations 1995'.

There is also the ever-present question of change. In the present Western climate, rapid technical changes allow higher quality and lower cost to be immediately implemented by fundamental changes in the operating organization. A major safety preoccupation is the growing reliance on new machines and processes with dramatically fewer personnel. We do not have to look far to see this at work. Major heavy industries like shipbuilding and steel operate as shadows of their earlier selves, sometimes using old equipment whose safety and maintenance is inadequate. The railways, the previous employers of hundreds of thousands, retain their general targets and much of their ageing equipment but with far fewer staff. In contrast there are complete industries erupting to prominence in rapid time, like microcomputing and biotechnology whose processes are only fully understood by a few who already work for company practitioners, not for regulatory authorities.

We seem to be faced with three major problem areas:

(1) Old industries or installations cutting back on maintenance procedures and extending the working life of aged equipment in an attempt to push economic decline to breakeven further into the future, or to snatch a quick profit from circumstances that more responsible operators regard as too dangerous.

(2) Industries so new that proper procedures and correct working practice have yet to be established, which, coupled with the real risks of commercial failure, limit knowledge to a few overstretched individuals and deny the chance of regularizing workplace operations, particularly maintenance.

(3) Those companies or whole industries suffering severe hardship during periods of economic depression making survival economies in both jobs done and the methods used and inevitably in the safety of their maintenance operations[5].

At least group (3) features company management aware of the correct actions required, who will limit dangers as far as possible and return to full safety processes as soon as practicable. The main problem is of course that it is a nose-dive difficult to pull out of; standards painfully established over long years of patient work are very difficult to restore once they have been abandoned, and group (3) can easily slide into group (1).

Group (2) is another matter. The correct procedures may simply be unknown, and it may take the misfortune of accident to establish those practices which have to be avoided. By professional care and good fortune, major accidents are infrequent, arising from an unusual combination of circumstances, which means regrettably that

opportunities to learn are few

and, when thinking in advance of the large numbers of mathematically unlikely events which combine to cause an accident,

are difficult to find or to predict.

---

5. While such steps are known to be economically inefficient, they are seen as inevitable defence of cash flow.

The general point is fundamental: the creation of effective standards and regulations *follows* the development of the technology and the gathering of operating experience. Whether such regulations are written by a separate body or by the industry itself makes the little difference to the question of timing. If the pace of change applicable to the workplace is faster or as fast as the regulatory output, *then the regulations in place will never be up to date* and it is the clear duty of a manager to think and act in advance of events, not merely to apply existing regulations, but, on a foundation of current professional thinking, to look for unexpected events often beyond our previous experience.

The dilemma then is twofold: (a) how can the difficulties of accident prediction and correct procedure be met while providing the certainty that such procedures will be faithfully applied, and (b) how can we best anticipate the unexpected when it is often unknown in our previous experience? It is now certain that both the working and the legal environment are changing in response to these and similar questions arising with such severity from documents like the Cullen Report into the accident on Piper Alpha. We refer to some resultant regulations in Section 12.2.

12.2 Changing regulations

Major industrial alarm and the detailed recommendations of the Cullen Report following Piper Alpha, plus questions similar to those in Section 12.1, resulted in the reassessment of regulations and of the way that they were applied.

In the first days of North Sea operations, directly applicable regulations were few, although those from other industrial areas and marine experience which were regarded as closely relevant were adopted and applied. The result, like a patchwork quilt, saw some operations better covered than others and a corresponding number of applicable authorities, for example:

Department of Energy	For oil and gas processing and pipeline transfer
Lloyd's Marine	For supply boats and shipment of floating structures
Department of Transport	Lifeboats
Department of the Environment	Sea Pollution
Certifying Authorities (CAs)	For example, Lloyd's Register of Shipping; Det Norske Vertas (DNV); American Bureau of Shipping (ABS)
Civil Aviation Authority	Helicopter operations

This list is by no means exhaustive, and if the relevant platform construction project is added the numbers rise to nearly 50 overall, including Fishing Associations, the Coastguard, Coastal Command and many others. These organizations are professional and conscientious, but the overlap of authority and responsibility required careful attention, making some jobs more complex than necessary and allowing both conflicts of interest and subjects of inadequate cover.

Recent statutory changes addressing such questions are being introduced by the Secretary of State exercising powers included in The Health and Safety at Work Act 1974, responding to proposals submitted to him by the Health and Safety Commission. The parallel organization charged with interpreting, implementing and verifying industrial adherence and the focus of much recent activity is the Health and Safety Executive.

From the late 1970s to the early 1990s, the general maintenance philosophy was captured by the initials ALARP (as low as reasonably practicable) and the concept of 'criticality' was based on PPM (pre-planned maintenance) and arose from backlog, one thought being that the longer planned maintenance for an item remained outstanding, the more critical it became. During this period, operations and maintenance followed a prescriptive framework of regulation which defined the nature and frequency of maintenance jobs, safety checks and plant or equipment testing, also injecting into maintenance procedures detailed actions required which were backed by statutory obligation.

For example, under the requirements of The Offshore Installations (Operational Safety, Health and Welfare) SI 1976/1019, lifting cables and appliances were tested by an approved test house prior to supply and subjected to a thorough examination every six months of subsequent service. Such examinations were conducted by an approved third party organization with ratified expertise and required a visit by the inspecting engineer to the workplace, where the equipment was examined[6]. In this particular case regulations followed previous unhappy experience of lifting equipment, where its accident potential was well known. Although most of the data is still required under the updated legislation the emphasis[7] has moved from 'reporting' to 'recording'.

Prescriptive methods like these are being amended or replaced, becoming proactive or target-setting in approach[8,9] as the operating companies themselves are required to identify and quantify risk, demonstrating that such assessments have been made. These assessments form the foundation of the 'safety case' required by the HSE for all installations and were extended with the addition of regulation 11 in November 1995. The new regulations do not stop at the question of risk; they require the operator to show how he proposes to cater for each one, to implement actions which arise and to constantly verify that such actions are being carried out. It does not even

---

6. It is worth remembering that, in spite of our growing technical confidence, experience of material in tension is a comparatively recent phenomenon, measured in hundreds of years, whereas experience of materials in compression extends to thousands of years and is as old as human construction itself.

7. Regulations 2(1), 4(6) and 18(4) of the SI 1976/1019 regulations have been revoked.

8. Refer to The Lifting Plant and Equipment (Records of test and Examation) Regulations 1992 and HSE publication L20 guidance notes on this subject. Data-recording requirements are also specified; however, the competent person will be able to use a method of recording that he finds suitable and the inspector has the necessary power to examine records held electronically.

9. The HSE booklet L82, 'A guide to the Pipelines Safety Regulations 1996', guidance on regulations comments in its introduction that 'The Pipeline Safety Regulations replace earlier prescriptive legislation on the management of pipeline safety with a more integrated, goal setting, risk-based approach . . .'.

end here, because the authorities require that as well as implementing and verifying the actions required, they have to be demonstrated also; like blind justice, it must be seen to be done[10].

Expanding the approach used in the safety case regulations, the HSE has deliberately targetted worker involvement, increasing the prospects of both acceptance and operation of the new regulations. The following outlines the principal vehicles involved, together with HSE supporting publications.

1 Offshore Installations (Safety Case) Regulations 1992

These came into force, in general, on 31 May 1993 (SI 1992/2885) (see Note 3).

During the investigation following Piper Alpha it was recognized that offshore operations included certain features which made them particularly hazardous, such as the routine presence of toxic or combustible substances, the physical confinement of the workplace, the fact that working crews lived on the installation during their tour of duty and the increased risk of accident being heightened by greater difficulty of escape in an emergency. The CIMAH regulations (Control of Industrial Major Accident Hazards Regulations 1984 (SI 1984/1902) had resulted from onshore incidents and did not wholly cater for offshore work. As a result, recent regulations require the demonstration of safe operation, and for offshore installations the submission of a safety case to the HSE. The following are the consequential features of the safety case regulations:

(1) that the duty holder's standards for management of health and safety and the control of major hazards should be subject to formal acceptance and that, beyond a transitional period, any installation whose safety case has not been accepted will not be allowed to operate;
(2) that measures to protect the workforce should include arrangements for temporary refuge from fire, explosion and associated hazards during the period for which they may need to remain on an installation following an uncontrolled incident, and for enabling their evacuation, escape and rescue;
(3) that suitable use should be made of quantitative risk assessment (QRA) as part of the demonstration of the adequacy of preventive and protective measures, including temporary refuge, and
(4) formal requirements relating to safety management and audit.

The preceding notes draw heavily from the documents referred to in Note 10.

2 Offshore Installations (Prevention of Fire and Explosion, and Emergency Response) Regulations 1995 (SI 1995 No. 743)

These regulations complement other health and safety regulations, particularly some of those in this chapter, and deal with:

(1) preventing fires and explosions and protecting personnel from their effects;

---

10. Refer to SI 1992, No. 2885, The Offshore Installations (Safety Case) Regulations 1992, and to HSE publication L30, 'A guide to Offshore Installations (Safety Case) Regulations 1992'.

(2) securing effective response to emergencies affecting persons on the installation or engaged in activities in connection with it, and which have the potential to require evacuation, escape and rescue from the installation.

The PFEER regulations support the requirement of The Health and Safety at Work etc. Act 1974 (HSW Act), which places duties on employers to ensure, as far as reasonably practicable, the health and safety of their employees *and others who might be affected by their activities* (author's italics). These general requirements are supported by:

(1) specifying specific goals for preventive and protective measures to manage fire and explosion hazards and secure effective emergency responses;
(2) recognizing that one person, the primary duty holder, is best made responsible for the measures described above.

PFEER assessment can support safety case regulations which require demonstration in the safety case that major accident hazards have been identified, their risks evaluated, and action taken to reduce the risks as low as is reasonably practicable. PFEER regulation 5 specifies requirements for fire and explosion, and evacuation, escape and rescue assessment.

Many of the preceding notes are drawn from the HSE document L65, 'Prevention of fire and explosion, and emergency response on offshore installations', approved code of practice and guidance.

3 Offshore Installations and Pipeline Works (Management and Administration) Regulations (SI 1995/738) June 1995

These regulations also complement other regulations which deal with the safe management of offshore installations, such as the safety case regulations referred to above and the Management of Health and Safety at Work Regulations 1992 (SI 1992/2051). In addition, they replace earlier prescriptive legislation with more broadly based requirements, setting out the objectives to be achieved. *They revoke various requirements which had become outdated and unnecessary* (author's italics).

These notes are largely reproduced from the HSE publication L70, Guidance on the Regulations.

The following subject areas are covered in the regulations.

(1) Notification to the HSE of change of ownership of mobile or fixed installations, and the movement of mobile installations out of UK waters.
(2) Appointment, functions, duties, and powers of offshore installation managers (OIMs).
(3) Permit-to-work systems and the provision of written instructions (see Sections 9.4 and 11.9).
(4) The keeping of records of persons on board.
(5) Various operational matters (including communications, meteorological information, identification by sea and air, and providing information to workers on location of the relevant HSE office).

(6)   The provision of health surveillance.
(7)   The provision of food and water supplies.
(8)   The need for cooperation among duty holders.
(9)   Arrangements for helideck operations.
(10) Amendments to the Offshore Installations (Safety Representatives and Safety Committees) Regulations (SI 1989/971).
(11) Consequential amendments to other regulations (e.g. the Offshore Installations and Pipeline Works (First-Aid) Regulations (SI 1989/1671) to ensure consistency.

The preceding notes draw heavily from the HSE document L70, 'A guide to the Offshore Installations and Pipeline Works (Management and Administration)', as well as the quoted statutory instrument.

4 Pipeline Safety Regulations 1996 (SI 1996 No. 825)

This statutory instrument complements both offshore (SI 1992/2885) and onshore regulations (SI 1996/551, Gas Safety (Management) Regulations). Based on the approach outlined above, earlier prescriptive regulations are replaced with a more targetted risk-based approach while revoking previous requirements, which had become unnecessary.

Applicable to all pipelines in Great Britain, in territorial waters and the continental shelf (with some exceptions), these regulations refer to pipeline definition, duties/purpose, need for operator cooperation, arrangements to prevent damage, consequential amendments to other legislation and, for major accident hazard pipelines:

Description of a dangerous fluid.
The requirement for emergency shutdown valves at onshore installations.
The notifications structure.
The major accident prevention document.
The arrangements for emergency plans.
The transitional arrangements.

Regulation 7 specifically refers to the need for safe access to the pipeline to carry out examination and maintenance (see Section 11.2). In addition, the relation with both offshore and onshore regulations is noted. From a maintenance perspective, pipelines include many undesirable features of maintenance and repair work.

As noted, pipelines relate to both onshore and offshore statutory requirements.
Most work is conducted in remote locations with limited local support.
Workers, materials and tools needed have to be conveyed to site.
Repair/maintenance facilities brought to the location are usually basic.
Pipelines convey dangerous fluids.
Work coordination/communication with affected installations and pipeline stations requires strict attention.
Emergency equipment is usually installed.

Refer to HSE publication L82, 'A guide to the Pipelines Safety Regulations 1996', from which many of these notes have been drawn.

5 The Offshore Installations and Wells (Design and Construction etc.) Regulations 1996 (SI 1996, No. 913)

The standard instrument defines the amendments to safety regulations applicable to design and construction activity for offshore installations and wells.

Under Part 11, 'Integrity of Installations', item 5c, the SI requires that maintenance and repair of the installation may proceed without prejudicing the integrity of the installation, recognizing a maintenance question in the design process.

Schedule 2, regulation 26, also defines 'safety-critical elements' (referred to at the start of this chapter).

Regulation 12 also specifically refers to the significance of doors and gates, traffic routes, outdoor workplaces, danger areas, floors, walls and ceilings of rooms, roofs, windows and skylights, plus many others which influence the nature of maintenance work (see Section 11.7).

12.3 Safety-critical elements

As indicated in Section 12.2, the regulations are applicable to all areas of offshore safety management and refer to a comprehensive range of issues. The HSE publication L30, 'A guide to the Offshore Installations (Safety Case) Regulations 1992', under 'Content of Safety Cases', refers to 'written schemes providing for systematic examination, maintenance and where appropriate testing, by a competent person, all parts of the installation and its equipment; for the intervals at which this should be done; and the arrangements to ensure that defects are suitably acted upon'. The following list looks at the equipment focus only. Safety-critical elements can be identified for all installations and can also emerge from HAZOP (hazards in operations) or similar analysis. The following list shows the areas of likely emphasis. Safety-critical aspects will vary for each installation, but selected elements will typically include the following (the term component is used in some commentaries in relation to the whole platform or the whole plant referring to a part or component of the whole):

(1) Overall layout, design and construction philosophy and specification. Review of safety issues.
(2) Primary structure.
(3) Risers and pipelines (within 500 m of an offshore platform).
(4) Production system: Pressure vessels
                         Fired heat exchanges
                         Pressure piping
                         Pressure system protective devices
                         Process issues
(5) ESD system
    Fire and gas detection
    Active fire-fighting

      Passive fire-fighting
      Life-saving
      Escape routes
      Hazardous areas
      Temporary refuge

(6) Lifting appliances.
(7) Safety-critical aspects of electrical distribution or control systems.
(8) Wells: well control equipment, BOPs, Christmas trees, well kill equipment
(9) Safety-critical aspects of machinery
(10) Safety-critical software

The preceding list is abstracted from a note on the subject of forthcoming offshore regulations ('New and Forthcoming Offshore Regulations', Lloyd's Register, July 1995), not from the regulations themselves, and represents one interpretation; they are intended here as a guide. (Refer also to the regulations outlined in Section 12.2 and the definition of safety-critical elements at the start of this chapter.)

Notice the inclusion of safety-critical software; most items on the list are to be expected, as is software with a moment's thought, but it introduces some difficulty. Any software used on a computer system to control any other item on the list would qualify. For example, the use of automatic deluges or water jets as part of active fire-fighting can injure vulnerable personnel caught by a software-controlled response. Similarly, escape routes can be blocked if fire doors automatically close with escaping personnel on the wrong side of the barrier, and there are many other examples. The testing of software is notoriously difficult, given the large number of logical combinations that influence the system response, and some combinations may lie dormant without ever being triggered. If faults reside here the danger is that it becomes known for the first time when an accident is in progress.

Software is not the only element that supports another safety-critical item. A failure of a pressure relief valve (pressure system protective device) can cause explosion at the weakest, but remote, point of the pressure system. Similarly, lifeboats, key elements of the life-saving system, need full and immediate functionality but depend on lifting appliances when they have to be launched.

12.4 Ladder of criticality

Based on the experience of responsible operators and safety authorities, the key steps in criticality assessment have been identified and the ladder is used here as a simple illustration of the whole criticality concept. There are two particularly appropriate features of a ladder which are reflected in the overall question.

(1) It is possible to navigate on the ladder in two directions, up or down, which allows progress to be considered as 'implementation' from the bottom up and 'strategy' from the top down (see Figure 12.1).
(2) Navigation up or down the ladder can only be accomplished if the preceding step has been completed.

There are, of course, differences of approach, and operators naturally emphasize the opportunities arising from safety-critical assessments, which, combined with

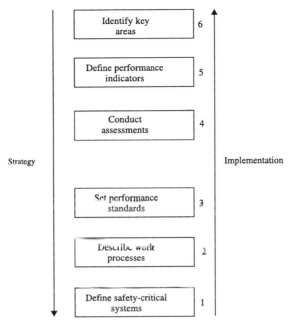

Figure 12.1
The ladder of safety criticality.

reduced accident costs, provide economic justification for the entire exercise. This is not to imply that the regulations can be avoided; there are only two real options: to operate or not, and even this choice is only available before major capital investments are made. Apart from the proactive intent of the new regulations, the main difference is the determined emphasis on verification and steps on the ladder are increasingly subject to scrutiny. This has not always been the case, the maintenance programme particularly suffering from some earlier short-cuts and omissions. There is a story of an inch-thick computer print-out of maintenance history tossed into a waste paper bucket by an examining surveyor with the comment 'a thick and useless document that owes more to fiction than to reality'.

Regulation 15A of SI 1996, No. 913, details the requirement for a verification scheme and lists its purposes (it is repeated as item 5 in Section 12.2.)

Verification is inherent in the process and, referring to Figure 12.1[11],

(1) *Define safety-critical system*
Using the plant or platform safety-case, plus inputs from all operational groups, including maintenance and safety, identify, confirm and agree the critical safety

11. These notes 1–6 and Figure 12.1 refer to a paper on this subject by Lloyd's Register outlining the 'LR WSE Service, Profile Strategy and Capability – May 1995'. We should understand that, attractive as the Lloyd's approach can be seen to be, it is one of a raft of different commercial possibilities, as the statutory foundation of previous activities is steadily being removed by legislation.

systems applicable to the installation. Ensure verification by a competent and independent person.

(2) *Describe work processes*

Maintenance work described in work instructions and standard procedures exists in effective maintenance programmes already. Subsequent steps on the ladder stiffen work requirements, whose progress achievement and backlog are subject to verification.

(3) *Set performance standards*

Safety performance is referred to here. 'Performance standards' identify the assumptions that have been defined within the safety case and related formal safety analysis, and the 'assessment' performance standards identify the functionality, survivability, reliability and availability standards applicable to each plant type.

(4) *Conduct assessment*

Many of the inputs required for assessment will be available from installation-specific safety documentation and supported by maintenance history. Assessments often employ a written scheme of examination[12,13,14] to suit the requirements of the PFEER regulations.

Two types of examination are used, both part of the assessment process and specific to each item of plant.

An *initial* examination carried out before an item is brought into service refers to design review or type approval to confirm suitability for purpose.

A *subsequent* examination conducted periodically confirms fitness for purpose and makes quantitative judgements against performance standards.

(5) *Define performance indicators*

A refinement of business strategy. Key performance indicators define quantitatively required levels of performance of each selected safety-critical system.

---

12. Written schemes of examination are a requirement of Regulation 19(2), 'Offshore Installations (Prevention of Fire and Explosion, and Emergency Response) Regulations 1995: '. . . the duty holder shall ensure that there is prepared and operated a suitable written scheme for the systematic examination, by a competent and independent person, . . .'

13. It would also be feasible to enhance selected sections of a conventional PPM system to provide for the requirements of WSEs, but this is specifically prohibited by the legislation: 'In addition the written scheme of examination required by regulation 19(2) requires the duty holder to establish a system to check that certain plant provided in compliance with the relevant requirements of PFEER and remains fit for purpose. *This scheme is above and beyond any checks carried out as part of routine maintenance and testing programmes.*'

The guidance notes go on to say: '*The scheme replaces examinations of life-saving appliances and fire-fighting equipment required under previous law; the scope of the scheme is similar, but now the duty holder is responsible for preparing the scheme, determining both the frequency of examinations and the competent and independent person required to undertake them.*' Note (author's italics).

14. *Written schemes of examination*. We have already noted that equipment performance will vary according to its duty and location as well as its type or design. WSEs may be tailored to suit such variations and competent persons will review these under fitness for purpose questions in subsequent examinations.

TABLE 12.1 THE VERIFICATION SCHEME

Owner's/operator's obligation	Safety case regulations
To have scheme prepared. To have scheme approved. To implement scheme.	15A(1) & (2)
Definition of a verification scheme Applies to safety-critical items. To ensure by means described below: – have been correctly identified. – are or will be suitable. – continue to be safe.	Regulation 7A
The means (referred to in 7A) Examination, including testing of safety critical elements Examination of designs, specs, certificates, CE marking, document or standard relating to it. Taking appropriate actions. Other such steps etc.	Regulation 7B
Definition of who can perform work referred to in (15A) and (7B) Independent and competent persons.	Regulation 7C

(6) *Identify key result areas*
The opening step in the top-down strategy, which relates acceptable overall business risk to achievable performance figures.

12.5 Verification

Regulation 15A of SI 1996, No. 913, requires the inclusion of a verification scheme for safety-critical elements to ensure that:

(1) a record is made of the safety critical elements;
(2) comment on the record by an independent and competent person is invited;
(3) a verification scheme is drawn up by or in consultation with such person;
(4) a note is made of any reservation expressed by such person as to the contents
    of: (i) the record; or
            (ii) the scheme; and

(5) such scheme is put into effect.

(See Section 12.2.)

The requirement for a verification scheme and its approval is also part of the Safety Case and is shown in Table 12.1. Regulation numbers shown refer to 'Offshore Installations (Safety Case) Regulations', SI 1992, No. 2885.

Chapter 13

Performance evaluation

Because maintenance is affected by so many influences beyond the control of the maintenance team, performance monitoring can become a contentious issue and in difficult situations quickly make problems more serious. A repeated allegation is that performance monitoring highlights work problems while ignoring genuine difficulties faced by maintenance teams. There are many of the latter, and good monitoring will underline such problems so that they can be corrected. First, however, we should look at the purposes of performance monitoring.

(1) To specifically monitor equipment failures which give rise to safety reports or failures of safety-critical equipment

(2) To act as a vehicle for the recognition of effort, achievement and motivation of the maintenance team.

(3) The indication of general operational health of the production process and the equipment population.

(4) To quantifiably report the two key values of plant availability and process downtime caused by the production-critical equipment failure.

(5) The assessment of maintenance work management and the system employed, including the cogence of the plans used and the performance indicators themselves.

(6) The identification of opportunities highlighting future work required and the effectiveness of previous changes.

At a more detailed level:

(7) The identification of equipment type problems, including performance limitations, repeated failures and design aberrations.

(8) Location of work quality shortfalls, amount of rework and equipment and performance failures.

(9) Pinpointing of work output limitations and under-performance against schedules.

(10) Magnifying of maintenance work constraints, including access denial, resource limitations, late delivery of materials and many others.

All maintenance computer systems include a range of computer-generated reports, many of which address questions of performance. Additionally, customer reports can often be designed and prepared for a specific need. Such reports are too varied and too numerous to consider in detail, and it is assumed here that the performance indicators described already exist or can be readily provided. The reports referred to are mainly prepared by collection and arrangement of data from the equipment database (see Section 5.2) and from the maintenance work history, which is described in Section 5.1 as the collection of closed-out work orders which are retained on the database for future interrogation. The data fields included in work order creation in progress and closeout, are listed in Section 5.4 and become referent data in the work order database. As both plant and maintenance system get older the wealth of data available increases. In spite of this, not all performance indicators derive from maintenance computer records, as the following will indicate.

Structure of the indicator

Each indicator should be expressed numerically by a single number and the units of that number defined. Also, the mean value of the previous 13 work periods should be shown. Other values, such as the direction of the trend, can be included where necessary. Some indicators will refer to incidents which are few in number, and month by month changes are so volatile as to have little meaning.

A report containing the performance indicators should be prepared monthly and circulated up the command chain and to the managers of departments that interface with maintenance affairs and to maintenance department personnel. When indicators are derived from non-maintenance systems the provenance and reliability of the numbers should be regularly verified.

The indicator summary

Summary number (1–13)
Issue date
Data sources
Circulation

Indicator Number	Purpose	Description	Source
1	Safety-critical	The number of defect work orders (Type 2) raised in respect of safety-critical equipment	Plant safety report from failure/problem causes or work order history
2	Maintenance of safety equipment	The number of man-hours assigned to maintenance inspection and repair of equipment	Work orders closed out complete in two previous work tours

Indicator Number	Purpose	Description	Source
3	Operational health	The number of production hours lost in downtime to unscheduled equipment stoppages	Production downtime report and closed work order history
4	Operational health	Critical production plant availability range of values	Daily status reports and closed work order history
5	Operational health	Planned to unplanned ratio (performance value)	Calculation from total man-hours recorded on work orders closed out in two preceding work tours
6	Equipment	All equipment availability range of values	Status reports raised during the period
7	Maintenance system check	Ratio of labour man-hours scheduled to labour man-hours available	Maintenance schedules for the period plus personal time records
8	Maintenance work check	Number of man-hours rescheduled into backlog	Information output from computer rescheduling run
9	Maintenance work check	Total number of man-hours of period work orders closed out 'not done' in the period	Work order history
10	Maintenance work check	Number of defect work orders or equipment failures	Closed out Type 2 WOs in period
11	Maintenance quality check	Number of repeated failures	Number of items with more than one Type 2 work order in preceding 12 months from work order history
12	Maintenance performance	Overall progress percentage Progress to date percentage	Progress reports
13	Maintenance performance	Maintenance cost performance	From cost/progress report pounds ± from correct progress cost figure
14	Equipment performance	Identity of major plant or tag number items of major unplanned failures	From work order history; closed out Type 2 during period

13.1 The progress position

Figure 7.1 shows *expected* and *actual* costs plotted against the time to which they refer. Actual costs are taken from the results data recorded on each work order and the time of closeout, and expected costs are calculated from the costs of jobs in the maintenance plan plus the unplanned allowance. Calculation of the progress position is a little different; its main currency is man-hours and it compares what has been done with what was forecast to be done, the latter using expected man-hours only.

Calculation of the expected figure

At any period in the maintenance year, say weeks 30–32, the expected figure is the sum of planned man-hours for all maintenance jobs and the man-hour allowance for unplanned work, both forecast for completion by the end of the period. In this way the expected figure can be calculated for any work tour and plotted for each two-week period during the year.

The type of curve used can of course vary. A chosen plot could show, for example, a total timespan of six months, or could be used to display a single discipline only. When similar graphs are prepared for single projects, the time base used will often be chosen as a little longer than the expected project duration, so a seven month project could be shown on an eight month time base to allow for overruns to be presented on the same diagram structure. In the case of a rolling plan, like that described here for maintenance, the 12 month time base is convenient and is chosen to match the company's working year. In addition, it is usual to plot the figures as cumulative, showing the values of the relevant figures rising over the span of the graph. (see Table 13.1, columns 5 and 6). Also, the total expected man-hours in column 5 can be simply expressed as a percentage of the overall cumulative figure, shown here as 961.

At the end of the year's work, if 100% progress has been achieved, the number of expected man-hours achieved is the number of expected man-hours originally forecast, in this case 961 (i.e. the period 13 cumulative figure at the bottom of column 6).

The calculation of hours recovered

When a planned job has been completed the man-hours originally planned for its completion are attributed to achieved progress. Each completed job is 'worth' the number of man-hours estimated for it in the plan. The term used is 'recovered'; if a plan estimates 1000 man-hours for a job, when the job is finished 1000 planned man-hours have been recovered. Similarly, for the same job when it is half done, i.e. it has a progress position of 50%, 500 planned man-hours have been recovered, and estimates of this kind are used when progress is estimated for jobs or projects that are incomplete. These calculations disregard the actual number of man-hours used to achieve this position, concentrating on progress against the plan. In the model described here, progress is compared with the expected man-hours cumulative figure to that point on the time-base where the progress is drawn, i.e.

TABLE 13.1 ESTIMATED PLANNED AND UNPLANNED MAN-HOURS

Main period	Mechanical expected man-hours (× 100)	Electrical expected man-hours (× 100)	Instrumentation expected man-hours (× 100)	Total expected man-hours (× 100)	Cumulative expected man-hours (× 100)	Cumulative (%)
(1)	(2)	(3)	(4)	(5)	(6)	(7)
1	28	21	7	56	56	5.8
2	38	28	10	76	132	13.7
3	42	32	11	85	217	22.6
4	36	27	9	72	289	30.1
5	38	28	10	76	365	38.0
6	46	35	11	92	457	47.6
7	24	18	6	48	505	52.5
8	40	30	10	80	585	60.9
9	38	28	10	76	661	68.8
10	38	28	10	76	737	76.7
11	44	33	11	88	826	85.8
12	40	30	10	80	905	94.2
13	28	21	7	56	961	100
	480	359	122	961		

- *For planned work*

  The number of planned hours recovered is simply the sum of the estimated man-hours for Type 1 work orders closed out in the period.

- *For unplanned work*

  The number of unplanned hours recovered is:

  Actual man-hours spent when the sum of Type 2 work order actual man-hours is less than the total allowance for the period.
  or
  The man-hours allowance when the sum of Type 2 work order actual man-hours is greater than the unplanned allowance.

This means that the maximum which can be recovered is the total allowance for the period[1]

---

1. The calculation of hours recovered for unplanned jobs is a little tricky, and various methods are used. In some progress calculations, planned hours only are considered. For maintenance work the unplanned portion is a fact of life, so it makes sense to make provision for it, but it is difficult to provide estimates for the unexpected. In the author's view, for maintenance work the method described here is the best compromise, whereby an allowance for unplanned work is included in the expected figure.

This means that the hours recovered for unplanned work completed in the period must be calculated with care.
  (a) Hours recovered should not exceed the hours actually used.
  (b) Total unplanned hours recovered for the period must not exceed the hours in the unplanned allowance.
  (c) Provided (a) or (b) above are not violated in this situation only, actual hours and estimated hours are regarded as the same.

There is a subjective feel to some of these numbers. In the first place, forecast or planned time is an acknowledged estimate, the progress position is rarely as clear cut as the example above indicates and if the actual data is inaccurate it would initially appear that the results could be misleading. Surprisingly perhaps, this is not the case, and with careful use such methods and presentations can be extremely useful.

Firstly, all planning and estimating improves with practice, and even when judging disparate work useful comparisons exit. Also, errors tend to be self-cancelling. An overestimate in one period is followed by an underestimate in the next. In addition, because maintenance plans consist of many jobs throughout the year, the statistical influence of each one is small, so for most maintenance jobs monitored by this method a progress estimate of 0 or 100% is sufficient[2]

The estimated man-hour figures for these disciplines have been used to calculate the costs shown in Table 7 3 (page 152).

The progress statement

Most major projects and work programmes regularly report their progress against a plan, and the progress position is usually described as a percentage. Because input work order data can be compared with the forecast tabulations, calculation of the progress position is simple. By summing all expected discipline man-hours recovered from all work orders closed out the period in question, the figure can be produced. For example, to the end of period 8:

Man-hours recovered $= 49800$
Man-hours expected $= 58500$
Progress percentage to date $= \frac{498}{585} \times 100$ $= 85\%$

This is the progress position to date, i.e. 85% of expectation has been achieved; in other words, 85% of the expected man-hours so far have been recovered (see column 6, Table 13.1). The more usual figure is the progress overall, calculated as

Man-hours recovered $= 49800$
Overall man-hours expected $= 96100$
Progress position overall $= 52\%$

and shown in column 7 of table 13.2. The figures are plotted in figure 13.1.

The progress position may be lower than the target because of technical difficulties, work inefficiency or simple lack of manpower. For this reason, efficiencies are sometimes calculated. For example, work efficiency if to achieve 498 hours of recovery 520 actual man-hours were recorded:

Work efficiency $= \frac{498}{520} \times 100$ $= 96\%$

Figures of work efficiency derived in this manner should be treated with caution because there are many influences at work over which the maintenance technician

2. Estimates of the progress position for incomplete jobs would be required at the end of the working year.

TABLE 13.2 THE PROGRESS TABLE

	Expected (Columns 2, 3 and 4)			Recovered (Column 5)		
Main period	Total expected man-hours (× 100)	Cumulative expected man-hours (× 100)	Cumulative expected progress (%)	Total man-hours recovered (× 100)	Progress to date (%)	Progress overall (%)
(1)	(2)	(3)	(4)	(5)	(6)	(7)
1	56	56	5.8	55	92	5.7
2	76	132	13.7	125	95	13.0
3	85	217	22.6	180	83	18.7
4	72	289	30.1	281	97	29.2
5	76	365	38.0	354	97	36.8
6	92	457	47.6	404	88	42.0
7	48	505	52.5	461	91	47.9
8	80	585	60.9	498	85	51.8
9	76	661	68.8			
10	76	737	76.7			
11	88	826	85.8			
12	80	905	94.2			
13	56	961	100.0			
	961					

has no control, and they should certainly not be used as a means of personal perform-ance monitoring. If this is ever practised, manipulation of recorded data input will result. Their best use is as a device for trend monitoring, to see if the results of effort

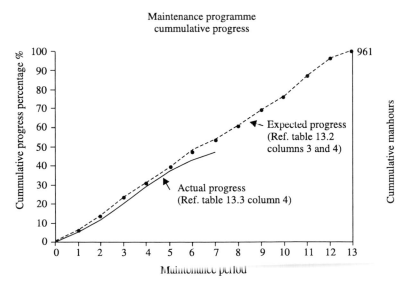

Figure 13.1

are getting better or worse and whether changes such as training or new tools are beneficial to work efficiency. It may also be helpful to compare machine groups by type or location with the overall figure to look for problem areas or star performers.

13.2 Cost of progress

Sections 7.2 and 13.1 describe the preparation of maintenance costs and progress, both related to the time-scale or time-base applicable to the maintenance plan, and shown in both as a full year of 13 maintenance periods of four weeks each. The presentation of expected and actual costs, together with expected and actual progress, gives a very powerful assessment of the way in which the work programme or project is performing. Both curves, or versions of them, are used by managers to help keep a general understanding of what is happening, though of the two it is the progress curve which is used most often. This curve can be quickly calculated from a maintenance plan, which is often being prepared or updated anyway, and cost information is already included in the financial management system. When time is limited this is an effective approach, but there are difficulties with maintenance costs. The financial system is more detailed and ponderous, and short small projects may be complete before first cost reports are available. These may be enclosed in much wider aggregations that are not directly helpful to the manager concerned with an individual programme. The maintenance manager is normally better off than some other managers, as maintenance is often defined, monitored and reported as a separate area of the financial system.

There is a further dilemma, that of the cost of the progress achieved. It is quite possible for high progress to be correctly reported which has been achieved through the penalty of heavy cost or for the costs to remain comfortably within target while progress is falling short. What is more difficult to spot is when both cost and progress show a modest shortfall, when the examination of each curve would not produce undue alarm. The problem is better illustrated when both cost and progress are displayed together, so that the case of higher costs for less progress stands out. It is also difficult for the manager to decide whether the costs charged to the work programme are giving good value for money, and both of these questions are answered by using a different presentation: see Table 13.3 and Figure 13.2.

13.3 The planned to unplanned ratio

There is one useful indicator which justifies early and separate attention. This is the ratio of planned to unplanned work. There are several reasons for its use, but probably the most important is that in a single number it captures the fundamental condition of the entire maintenance work management operation. The ratio is calculated from anticipated work that is carried out according to a plan and work that is carried out immediately and without prior planning. It is usually presented with planned work expressed as a percentage of the whole:

$$\text{Planned work percentage} = \frac{\text{Planned work total}}{\text{Planned + Unplanned total}} \times 100$$

TABLE 13.3

Maintenance period	Expected total cumulative cost	Expected cumulative progress (%)	Actual cumulative costs	Actual overall progress (%)
(1)	(2[a])	(3[a])	(4[b])	(5[b])
1	728	5.8	715	5.7
2	1716	13.7	1625	13.0
3	2822	22.6	2655	18.7
4	3758	30.1	3934	29.2
5	4746	38.0	4727	36.8
6	5942	47.6	6015	42.0
7	6566	52.5	6591	47.9
8	7606	60.9	7591	5.18
9	8594	68.8		
10	9582	76.7		
11	10726	85.8		
12	11766	94.2		
13	12494	100.0		

[a]Column 2 refers to Table 7.3 and column 3 refers to Table 13.2.
[b]Columns 4 and 5 show fictitious actual values for which periods 9–13 inclusive have yet to be reported.

Here are the main reasons for its widespread use:

(1) Considered as a simple percentage, the health and status of many maintenance organizations can be recognized in a single number.

(2) The number is calculated as a quotient, producing a non-dimensional result which makes comparisons between machines, groups or whole and otherwise separate organizations easier and more realistic.

(3) The number is easy to calculate. When a computer-based system is in use the additional processing required would pass unnoticed.

(4) Although organizational, operational and maintenance changes inevitably effect the number, it is still valid and direct before-and-after comparisons can be made.

The simplest approach would be to compare Type 1 (PPM) and Type 2 (defect) work orders, and this does give a useful result. However, this is imprecise. Some jobs which start life as an equipment failure or reported defect can be attended to as planned work, work can be prepared for a shutdown, or the original defect WO may be closed undone and a corresponding PPM advanced instead. In some cases, equipment may be changed out following a switch to 'standby': a short-term but planned response to an equipment failure. Selection of the jobs to be included in each category should follow careful rules which will vary slightly with each different system. The ratio is based on the work order numbers, and the method of choice for each group will affect the result.

It should also be clear that the figures used can be based on the number of jobs or the number of hours required to do them, and it is the latter (the number of man-hours) that mainly concerns us here. Consider, for instance, the effect on the work

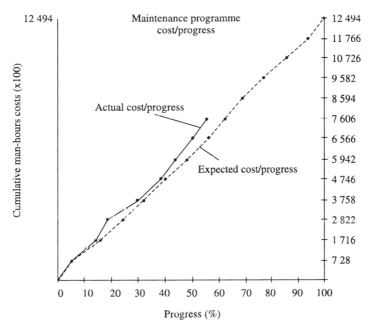

Figure 13.2
Maintenance cost progress: expected and achieved.

schedule. How many hours were diverted to deal with the unexpected, with prepared tasks cancelled or deferred to meet the unexpected need? Additionally, were these unplanned tasks prepared but deferred previously because of other unplanned work, only to become unplanned defects in their turn? Some of these deferrals could be major jobs which have attracted considerable investment in job preparation already, or be regular inspections, modest in their work content, but which when delayed weaken the detection of danger and impending failure.

The whole maintenance work management process is influenced by this simple figure. If the unplanned work element is too high, then the maintenance team are devoted to fire-fighting, to dealing with the unexpected, working in unprepared and inefficient ways. If the planned work element is too high 'over-maintenance' is almost certain and the chance to learn about equipment failure is absent. Opinions about the ideal ratio vary according to the site and to circumstances, but a figure of 80% has much support as a target. The unplanned total includes all unexpected work. Some, like jobs in response to statutory or safety instructions, may be just as urgent and unplanned but not relate to actual or anticipated failure. The result is a wide variety of unexpected jobs all contributing to the unplanned total.

There is a touch of irony here. For the ratio to be calculated, some failures must occur, which is of course contrary to one purpose of having a maintenance system. Maintenance personnel will sometimes soberly comment that 'failure is the only certainty and some failures are essential'. If, when the PPM work is being steadily carried out, a high incidence of failures (a low planned to unplanned ratio) is evident then too little maintenance is being done. Conversely, if the ratio is high then some equipment is being over-maintained. Notice the reference to 'some equipment'. Measurement is being based on a large population, and there will be a spread of results applicable to different groups of equipment. The way to identify the culprits is to compare different disciplines, different groups of equipment, and even different locations, and remember that small variations applicable to a large number will have a similar effect on mean numbers as would a large variation over small numbers.

There are two other influences to bear in mind:

(1) The value of the ratio varies according to the age of the plant, tending to rise following installation and initial operational problems and falling much later as the machine approaches the end of its working life and repair work increases. These changes should be gradual where equipment populations are large. (Refer to 'ageing failure' in the Glossary.)

(2) As observed above, calculation of the ratio requires data which only arises with equipment use. The result is that the ratio is a more useful and confident number after a period of plant operations and careful data acquisition.

There is a further consideration: how else do we finally learn about failure, which is a dominant element of unplanned work? In spite of the increasing abundance of analytical techniques which effectively model the maintenance scene, it is the failing machine itself which exactly describes when failure will occur. Unfortunately, the tolerance of failure is acceptable only within strict limits, and it is the management of these limits which makes the ratio useful. Even when failure is permitted we can still prepare, and we can still be selective. In RCM terms selection is based on the consequences of failure. Such failure does not have to be catastrophic: with care and attention we can power down at one minute to twelve, close to exact knowledge but far from major damage[3].

The figures used in calculating the ratio are based on location and time, such as the 'figures for Site X during Maintenance Weeks 16–20'. Useful as this is, it does not have to be this limited; the analysis can be far more selective, for example, examining a particular group of machines, such as all lifting equipment or all rotary pumps more than two years old. We can make some valuable comparisons and use a simple figure to do it. We could, for example, compare the ratio for a group of machines with that of the whole site or one machine with other machines of the same type. The periods of analysis may also be changed. Apart from a regular (perhaps monthly) assessment, we may wish to consider the ratio value at different times of the year, or compare one month with another. The variations are almost unlimited, depending mainly on the requirements of the situation and the imagination of the

---

3. In some situations equipment is deliberately allowed to deteriorate (see Sections 3.1 and 3.2), not to examine failure but as the chosen management method for some equipment. Subsequently, if the failed item is of low importance or a member of a large equipment population, it may be left unattended until it can be included in the next convenient work schedule as a planned job (refer to the work box figure in Section 2.3).

maintenance engineer. The purpose behind such effort is firstly to monitor the performance of the maintenance system in general and secondly, by using a different method of assessment, to identify areas needing attention which might otherwise escape detection.

Almost any work management system can produce this figure, and its preparation depends on the following main ingredients:

(1) The clear distinction between the two types of work, either by different work order types or by use of a simple code.
(2) The careful recording of the labour man-hours used for all disciplines taking part.
(3) Summation of the different figures and calculation of the final value.

Having underlined the usefulness and method of this simple technique we must also underline its limitations and there is one major point to emphasize:

the ratio is primarily a tool of planned maintenance.

When used with PPM the incidence of unplanned work can be reduced by increasing the number or the scope of PPM jobs or by merely increasing the maintenance frequency. Conversely, when the period between planned jobs is extended, usually intended to achieve a maintenance cost saving, the ratio is one of the indicators watched by maintenance managers to ensure that period extensions are not followed by unplanned penalties.

Not all philosophies are susceptible to this simple analysis, which is not a criticism of such methods (see Chapter 3) but points to the need for other forms of performance analysis. Condition-based maintenance, for example, extends the maintenance period of monitored machines until signs indicate that maintenance is required. This is a very effective technique, but it usually means that the maintenance event is deliberately taken closer to the point of failure, and sudden or unplanned action is more likely. This is normal for condition-based maintenance, which to be fully effective needs rapid maintenance reaction. This in turn means that the planned to unplanned ratio will be different for machines managed under a CBM regime, and comparisons with PPM managed equipment can be misleading.

13.4 Management reporting

One duty of maintenance management is to manage the interface between the company maintenance function and the board of senior management. The key method of accomplishing this is through management reporting. There are two things to observe at once:

(1) *The maintenance manager usually holds the initiative.*
    It is unrealistic to expect other directors or senior managers to anticipate the specialist requirements of the maintenance function, but if the maintenance manager does not, then opinions and speculation will arise from other sources and maintenance will shoulder unjustified blame for many technical failures.

(2) *Reporting refers to many actions*

Maintenance is a key company consideration, a subject which must remain foremost in the commercial eye, which in parallel with production management reporting will employ a battery of different vehicles, including:

Daily action outlines
Circulated planning abstracts
Monthly management reports
Performance analysis
Asset management commentary
Special interpretations
Training for senior personnel

These reports will vary according to the company situation and be supplemented by other communications, such as telephone calls, personal visits, presentations, films, videos and site inspections. The emphasis here is on:

maintenance department-instigated reporting rather than responses to other initiatives.

It is sadly true that maintenance in general circles is not the most attractive subject, and when preparing reports of the kind noted above a deliberate effort is needed to clarify the work required and the effects of delay.

It is, of course, tempting to follow the easy path, to relax when things are going well and making sure that company interests are protected and the maintenance function is well managed. Such an approach will evidence a well-run maintenance operation, but eventually an emergency will occur and protracted delay will follow any lack of readiness.

Efficient maintenance is built on initiative. It is essential to be prepared.

And probably more importantly,

maintenance should never become a predominantly reactive activity.

Equipment failures and non-revenue-earning work are parts of the maintenance scene. Although company senior managers do not like this, they do understand it and some of the poor maintenance image in so many industries arises from the

failure of maintenance management to promote its own message.

Company senior management, like all human beings, respond for the better when they understand, and it is part of the maintenance manager's role to keep them prepared and informed.

The list below assumes that the non-reactive or proactive approach is more effective and gives some aims and methods of effective reporting.

- *Report facts and opinions under separate headlines*
  This will reduce that part of a report which may be subject to argument. If opinions are also kept short and sharp they will discourage disagreement.

- *Keep to a regular reporting format*
  This will help to make the maintenance subject more familiar and less of a communication problem. But choose the format carefully: make sure that it is clear, precise and short.

- *Make reports pertinent and punctual*
  Reports which refer to known incidents in familiar terms are more likely to be accepted, provided they are current.

- *Make sure that reports are complete*
  Reports returned or labelled as incomplete have little value, and the valuable content of the submission could be lost.

- *Be first with the news*
  With good news this is important; with bad news it is essential Whichever reporting method is used, make sure that the current value is retained.

- *Balance bad news with good news*
  The readership is human: give them some encouragement.

- *Put things in perspective*
  A wider picture will help balance opinion and understanding.

- *Remember the trend*
  Use of the trend smooths the short-term effects and retains the link to the wider picture.

- *Anticipate breakdowns*
  When they occur, report factually and refer back to previous predictions. Do not understate or exaggerate.

- *Forecast the wider future*
  Consider the plant overall, the age of the equipment population, its current value, the value compared to revenue generated, resources employed and overall costs.

- *Underline successes*
  Wise spending can be a source of satisfaction and good preventive work can often be proved. Do not let maintenance spending be an *extra* cost; report job costs, highlight those under budget and include the 'year to date' figure.

- *Avoid the unexpected*
  Even equipment failures can be anticipated by referring to plant or equipment age, current levels of production, frequency of failures in general, timing of shutdowns, major works etc. Failure frequently follows change, so reference to duty cycles, load patterns and high output can be useful.

- *Tell the truth*
  Do not be tempted to minimize or ignore the unpalatable; while understandable, it will diminish the value of all maintenance reports.

- *Ask for the moon*
  You never know, you might get it! If you need replacement equipment or extra resources ask for them; it is most unusual for others to make the offer. You may have to time the request, but it should be well researched and argued. Do not anticipate negative decisions. Even refusal or delay requires a company response and can still be useful.

- *Do not attribute blame or make excuses*
  If there are mitigating circumstances, state them, but if others are at fault use such information only to defend the maintenance team, not to gain an advantage. There will be times when you will be grateful for others' restraint.

- *Give credit elsewhere*
  Maintenance work requires the support and understanding of other groups. If it is clear that their actions are appreciated, they are far more likely to be given, and maintenance activity will be more likely to be recognized by others.

The purpose of these recommendations is not to disguise or to deceive. It is merely to stimulate logical as opposed to emotional thinking and to demonstrate the significance of current events in the wider context. It is very easy to see only the negative picture and to be depressed by thinking magnified out of scale.

Chapter 14

Health and safety publications

The following publications are listed by the HSE as currently available and can be obtained from the address shown in the Bibliography. ISBN should be quoted so that documents can be clearly defined. The list of inclusions is not exhaustive and those selected have been used in the preparation of this manual or have a bearing on maintenance or the working environment in which maintenance is carried out.

HS(G) series

Number	Title	Date	ISBN
HS(G)48	Human Factors in Industrial Safety	1989	0 7176 0472 1
HS(G)50	The Storage of Flammable Liquids in Fixed Tanks	1990	0 11 885532 8
HS(G)58	Evaluation and Inspection of Building and Structures	1990	0 11 885441 0
HS(G)85	Electricity at Work: Safe Working Practices	1993	0 7176 0442 X
HS(G)97	A Step by Step Guide to COSHH Assessment	1993	0 11 886379 7
HS(G)131	Energetic and spontaneously Combustible Substances: Identification and Safe Handling	1995	0 7176 0893 X
HS(G)155	Slips and Trips: Guidance For Employers on Identifying Hazards and Controlling Risks	1996	0 7176 1145 0

Legal L series

Number	Title	Date	ISBN
L1	A Guide to the Health and Safety at Work etc. Act, 5th edition	1992	0 7176 0441 1
L20	A guide to Lifting Plant and Equipment (Records of Test and Examination)	1992	0 7176 0488 8
L24	Workplace Health, Safety and Welfare Approved Code of Practice	1992	0 7176 0413 6

L25	Personal Protective Equipment at Work Regulations. Guidance on the Regulations	1992	0 7176 0415 2
L30	A guide to the Offshore Installations (Safety Case) Regulations 1992. Guidance on the Regulations	1992	0 11 882055 9
L65	Prevention of Fire and Explosion, and Emergency Response on Offshore Installations Offshore Installations Regulations 1995 and Approved Code of Practice and Guidance	1995	0 7176 0874 3
L70	A Guide to the Offshore Installations and Pipeline Works (Management and Administration) Regulations 1995	1995	0 7176 0938 3
L73	A Guide to RIDDOR 95	1996	0 7176 1012 8
L82	A Guide to the Pipeline Safety Regulations 1996	1996	0 7176 1182 5

Guidance notes in the general (GS) series

Number	*Title*	*Date*	*ISBN*
GS4	Safety in Pressure Testing (Rev)	1992	0 7176 0811 5
GS5	Entry into Confined Spaces	1995	0 7176 0787 9
GS31	Safe Use of Ladders, Step Ladders and Trestles	1984	0 7176 1023 3

Guidance notes in the plant and machinery (PM) series

Number	*Title*	*Date*	*ISBN*
PM(7)	Lifts: Thorough Examination and Testing	1982	0 11 883546 7
PM(26)	Safety at Lift Landings	1981	0 11 883383 9

Subject-specific guidance (1997 price list)

Page	*Title*	*Date*	*ISBN*
38	Quantified Risk Assessment: its Input to Decision Making	1989	0 7176 0520 5
41	Oil Industry Guidance on Permits to Work systems in the petroleum industry	1997	0 7176 1281 3

Appendix 2 From 'Electricity at Work' Booklet.

Electricity at Work Regulations 1989 SL 1989/635, HMSO, ISBN 0 11 096635 X

HS(R)25, Memorandum of Guidance on the Electricity at Work Regulations 1989, HMSO, ISBN 0 11 883963 2.

GS 24, Electricity on Construction Sites, HMSO, ISBN 0 11 883570 X.

GS 27, Protection against Electric Shock, HMSO, ISBN 0 11 883583 1.

GS 37, Flexible Leads, Plugs, Sockets etc., HMSO, ISBN 0 11 883519 X.

GS 39 (rev), Electrical test equipment for use by Electricians, HMSO, ISBN 0 11 883533 5.

GS 47, Safety of Electrical Distribution Systems on Factory Premises, HMSO, ISBN 0 11 885596 4.

PM 29 (rev), Electrical Hazards from Steam/Water Pressure Cleaners etc., HMSO, ISBN 0 11 883538 6.

PM 32 (rev), Safe Use of Portable Electrical Apparatus, HMSO, ISBN 0 11 885590 5.

PM 65 Electrical Safety in Arc Welding, HMSO, ISBN 0 11 883938 1.

PD 6519, Parts 1–2, Guide to Effects of Current Passing Through the Human Body (IEC 479, Parts 1–2).

Chapter 15

Acknowledgements

The author gratefully acknowledges assistance both moral and material given for manual preparation and during many years of work and experience as part of the (then) Chevron Ninian Maintenance team and to specific contributors noted below.

Ms A Barnes
Marketing & Communications
Manager
PSDI (UK) Ltd.
Woking Eight
Forsyth Rd.
Woking
Surrey
GU21 5SB

Mr Paul Chandler
Communications Coordinator
Chevron UK Ltd.
No 2 Portman St.
London
W1H 0AN

Mr G Edwards
Managing Director
Matrix Resource Ltd.
Matrix House, Bradford Rd.
Wrenthorpe, Wakefield
West Yorkshire
WF2 0QH

Mr D Gillard
Sales Manager
Soft Solutions Ltd.
Matrix House, Bradford Rd.
Wrenthorpe, Wakefield
West Yorkshire
WF2 0QH

Mr S Gilligan
Senior Surveyor
Lloyds Register
Offshore Division
11 Golden Square
Aberdeen
AB10 1RB

Mr Dennis R Krahn
Director,
European Offshore Affairs,
International Association of Drilling
Contractors
PO Box 202
Aberdeen
AB12 3AG

Mr R Leighton
Systems Consultant
Environmental Equipments Ltd.
Fleming Rd.
Newbury
Berkshire
RG14 2 DE

Mr J J O' Connor
Terminal Manager
Bantry Terminals Limited
Reenrour
County Cork
Ireland

Mr P Shrieve
Managing Director
ATL Consulting Ltd.
36 – 38 The Avenue
Southampton
Hants.
SO17 1XN

Mr D Southerland
Department of Trade & Industry,
Standards & Technical Regulations
Directorate
151 Buckingham Palace Rd.
London
SW1 9SS

Mr R C Storey
Managing Director
Environmental Equipments Ltd.
Fleming Rd.
Newbury
Berkshire
RG14 2 DE

Mr J Woodhouse
Managing Director
Asset Performance Tools Ltd.
Headly Common Rd.
Newbury
Berkshire
RG19 8LT

Chapter 16

Standards and related organizations

The following list is a selection of British, American and International standards often referred to in connection with maintenance activities.

CP 1011	Maintenance of electric motor control gear
BS 1129	Specification for portable timber ladders, steps, trestles and light staging
BS 2037	Specification for aluminium ladders
BS 2091	Specification for respirators for protection against harmful dusts, gases and scheduled agricultural chemicals
BS 2092	Specification for eye protectors for industrial and non-industrial uses
BS 2830	Construction industry lifting regulations
BS 3811	Terms used in terotechnology
BS 4001	Parts 1 & 2 Care and maintenance of underwater breathing apparatus. Standard diving equipment
BS 4211:1994	Specification for ladders for permanent access to chimneys, other high structures,silos and bins.
BS 4275	Recommendations for the selection, use and maintenance of respiratory protective equipment
BS 4430	Recommendations for the safety of powered industrial trucks, operation and maintenance
BS 4778	Quality vocabulary Part 1 International terms Part 2 Quality concepts and related definitions Part 3 Availability, reliability and maintainability terms

BS 5228	Part 2 Noise control on construction and open sites. Guide to noise control,legislation for construction and demolition including road construction and maintenance.
BS 5233	Glossary of terms used in metrology (incorporating BS 2643)
BS 5306	Fire extinguishing installations and equipment on premises: Part 3 Code of practice for selection, installation and maintenance of portable fire extinguishers.
BS 5345	Code of practice for selection, installation and maintenance of electrical apparatus for use in potentially explosive atmospheres (other than mining applications of explosives processing and manufacture) – General recommendations Parts 1 and 3–8 inc.
BS 5430	Periodic inspection, testing and maintenance of transportable gas containers (excluding dissolved acetylene containers). Parts 1,2,3 & 6
BS 5502	Part 80 1990 Buildings and structures for agriculture. Code of practice for design and construction of workshops, maintenance and inspection facilities.
BS 5671 1979, IEC 545 1976	Guide for commissioning, operation and maintenance of hydraulic turbines.
BS 5730 1979	Code of practice for maintenance of insulating oil
BS 5760	Reliability of systems, Equipment and components
BS 5980	Code of practice for site investigations
BS 6031	Code of practice for earthworks
BS 6071	Specification for periodic inspection and maintenance of transportable gas containers for dissolved acetylene
BS 6180:1995	Code of practice for protective barriers in and about buildings.
BS 6351	Part 3 1983 Electric surface heating. Code of practice for the installation, testing and maintenance of electric surface heating systems.
BS 6423 1983	Code of practice for maintenance of electrical switchgear and controlgear for voltages up to and including 1kV
BS 6467	Part 2 1988 Electrical apparatus with protection by enclosure for use in the presence of combustible dusts. Guide to selection, installation and maintenance

BS 6521	ISO 7592–1983 Guide for proper use and maintenance of calibrated round steel link lifting chains
BS 6548	Part 1: Maintainability of equipment: Guide to specifying and contracting for maintainability.
	Part 2: Maintainability of equipment: Guide to maintainability studies during the design stage.
	Part 3: Maintainability of equipment: Guide to maintainability, verification, and the collection, analysis and presentation of maintainability data.
	Part 4: 1993 IEC 706–4: 1992 Maintainability of equipment. Guide to the planning of maintenance and maintenance support
	Part 5: Maintainability of equipment:Guide to diagnostic testing.
BS 6570 1986	Code of practice for the selection, care and maintenance of steel wire ropes
BS 6263	Care and maintenance of floor surfaces. Code of practice for resilient sheet and tile flooring
BS 6626 1985	Code of practice for maintenance of electrical switchgear and control gear for voltages above 1 kV and up to and including 36 kV
BS 6661 1986	Guide for design, construction and maintenance of single-skin air supported structures
BS 6704 1996	Code of practice for selection, installation and maintenance of intrinsically safe electrical equipment in coal mines
BS 6755 Part 1	Testing of valves.Specification of Production Pressure Testing Requirement.
BS 6867 1987	Code of practice for maintenance of electrical switchgear for voltages above 36 kV
BS 6880	Part 3 1988 Code of practice for low temperature hot water heating systems of output greater than 45 kW. Installation, commissioning and maintenance
BS 6959 1989	Code of practice for selection, installation, use and maintenance of apparatus for the detection and measurement of combustible gases (other than for mining applications or explosives processing and manufacture)
BS 6968 1988	ISO 3056–1986 Guide for use and maintenance of non-calibrated round steel lifting chain and chain slings
BS 7000	Design management systems
BS 7020 1988	Guide for selection, use and maintenance of eye-protection for industrial and other uses

BS 7117	Part 3 1991 Metering pumps and dispensers to be installed at filling stations and used to dispense liquid fuel. Guide to maintenance after installation
BS 7184 1989	Recommendations for selection, use and maintenance of chemical protective clothing
BS 7258	Part 3 1994 Laboratory fume cupboards. Recommendations for selection, use and maintenance
BS 7270	Part 2 1994 Grounds maintenance. Recommendations for the maintenance of hard areas (excluding sports surfaces)
BS 7373	Guide to preparation of specifications.
BS 7375 1991	Code of practice for the distribution of electricity on construction and building sites
BS 7812 1995	ISO 4266.1994 Guide for selection, specification, installation, operation and maintenance of automatic liquid level and temperature measuring instruments on petroleum storage tanks
BS 8210 1986	Guide to building maintenance management
BS ISO 2382–14 1978	Information technology, Vocabulary, Reliability, maintenance and availability
BS EN 458 1994	Hearing protectors. Recommendations for selection, use, care and maintenance. Guidance document
BS EN ISO 9000 – 1 1994	Quality management and quality assurance standards: guidelines for selection and use.
BS EN ISO 9001 : 1994	Quality systems: Model for quality assurance in design, development, production, installation and servicing.
BS EN ISO 9002: 1994	Quality systems: Model for quality assurance in production, installation and servicing.
BS EN ISO 9003 · 1994	Quality systems: Model for quality assurance in final inspection and test.
BS EN ISO 9004 – 1 : 1994	Quality management and quality system elements : Guidelines.
PD 6519 Parts 1–2 (IEC 479 parts 1–2)	Guide to effects of current passing through the human body International Electrotechnical Commission IEC
900:1987	Hand tools for live working up to 1000 v AC and 1500 v DC IEE Regulations for Electrical installations 16th edition ISBN O 852965 109
ANSI B 28.10 1986–00–00	Rubber machinery-Endless Belt Building Machines-Safety Requirements for Construction Installation, Operation and Maintenance

BSI DIN 1982–00–00	Safety Requirements for Design, Use and Maintenance of Metal Scrap Processing Equipment
ASME A 112.21 3M* ANSI A 112.21.3M 1985.00.00	Hydrants for utility and maintenance use
ASME A 120.1* ANSI A 120.1 1996–00–00	Safety requirement for powered platforms for building maintenance
ASME B 133.12* ANSI B 133.12 1981–00–00	Gas Turbine maintenance and safety
ASME OM* ANSI OM 1995–00–00	Code for operation and maintenance of nuclear power plants
ASME OM-S/G* ANSI OM-S/G 1994–00–00	Standards and guides for operation and maintenance of nuclear power plants
ASME OMa Addenda* ANSI OMa Addenda 1996–00–00	Code for operation and maintenance of nucklear power plants; Addenda
ASTM E 1008 1984	Installation, inspection and maintenance of valve body pressure relief methods for geothermal and other high-temperature liquid applications
ASTM F 1449 1992	Care and Maintenance of Flame Resistance and Thermal Protective Clothing
ASTM F 1716 1996	Standard Guide for Transition and Performance of Marine Software Systems Maintenance
IEEE 56*ANSI 56 1977	Guide for insulation and maintenance of large ac rotating machinery (10,000 kVA and larger)
IEE 67*ANSI 67 1990	Guide for operation and maintenance of turbine generators
IEE 432 1992	Guide for insulation maintenance for rotating electric machinery (5 hp to less than 10,000 hp)
IEEE 450 1995	Recommended practice for maintenance, testing and replacement of vented lead-acid batteries for stationary applications
IEEE 492*ANSI 492 1974	Guide for operation and maintenance of hydro-generators
IEEE 515*ANSI 515 1989	Recommended practice for the testing, design, installation, and maintenance of electrical resistance heat tracing for industrial applications
IEEE 516 1995	Guide for maintenance methods on energized power lines
IEEE 625*ANSI 625 1990	Recommended practice for improved electrical maintenance and safety in the cement industry

IEEE 978*ANSI 978 1984	Guide for in-service maintenance and electrical testing of live-line tools
IEEE 1067 1996	Guide for in-service use, care, maintenance and testing of conductive clothing for use on voltages of up to 765 kV ac and <+ -> 750 kV dc
IEEE1106 1995	Recommended practice for installation, maintenance, testing and replacement of vented nickel-cadmium batteries for stationary applications
IEEE 1145 1990	Recommended practice for installation and maintenance of nickel-cadmium batteries for photovoltaic (PV) systems
IEEE 1149.5 1995	Module test and maintenance bus (MTM-Bus) protocol
IEEE 1188 1996	Recommended practice for maintenance, testing and replacement of valve-regulated lead-acid (VRLA) batteries for stationary applications
IEEE C 37 35* ANSI C 37.35 1976	Guide for the application, installation, operation and maintenance of high-voltage air disconnecting and load interrupter switches
IEEE C 37.48* ANSI C 37.48 1987	Guide for application, operation and maintenance of high-voltage fuses, distribution enclosed single-pole air switches fuse disconnecting switches, and accessories
IEEE C 37.61* ANSI C 37.61 1973	Guide for the application, operation and maintenance of automatic circuits reclosers
IEEE C 57.94* ANSI C 57.94 1982	Dry-type general purpose distribution and power transformers
IEEE C 57.106* ANSI C 57.106 1991	Guide for acceptance and maintenance of insulating oil in equipment
IEEE C 57.111 1989	Guide for acceptance of silicone insulating fluid and its maintenance in transformers
IEEE C 57.121* ANSI C 57.121 1988	Guide for acceptance and maintenance of less flammable hydrocarbon fluid in transformers
BS 5671:1979; IEC 545:1976 1979	Guide for commissioning, operation and maintenance of hydraulic turbines
BS 5760 Pt II:1994 IEC 300–3–2:1993 1994	Reliability of systems, equipment and components. Collection of reliability, availability, maintainability and maintenance support data from the field
BS 6228:1985; ISO 6750–1984 1985	Guide to format and content of manuals on operation and maintenance for earth-moving machinery
BS 6521:1984; ISO 7592–1983 1984–08–31	Guide for proper use and maintenance of calibrated round steel link lifting chains

BS 6968:1988; ISO 3056–1986 1988–06–30	Guide for use and maintenance of non-calibrated round steel lifting chain and chain slings
BS 7713:1993; IEC 944: 1988 1993–11–15	Guide for the maintenance of silicone transformer liquids
BS 7812:1995; ISO 4266: 1994 1995–06–15	Guide for selection, specification, installation, operation and maintenance of automatic liquid level and temperature measuring instruments on petroleum storage tanks.
BS EN 61203:1995 IEC 1203:1992	Synthetic oprganic testers for electrical purposes. Guide for maintenance of transformer testers in equipment.
IEC 31L/45/CDV*IEC-PN 31L/1779–6*-CEI 31L/45/ CDV*IEC-PN 31L?1779–6	Guide for the selection, installation, use and maintenance of apparatus for the detection and measurement of flammable gases
IEC 81/86/CDV*IEC-PN 81/1024–1–2*CEI 81/86/ CDV*IEC-PN 81/1024–1–2	Protection of structures against lightning – Part 1: General principles-Section2: Guide B: Design, construction, maintenance and inspection of lightning protection systems.
IEC 79–17*CEI 79–17 1996–12–00	Electrical apparatus for explosive gas atmospheres-Part 17: Inspection and maintenance of electrical installations in hazardous areas (other than mines)
IEC 422*CEI 422 1989–04–00	Supervision and maintenance guide for mineral insulating oils in electrical equipment
IEC 545*CEI 545 1976–00–00	Guide for commissioning, operation and maintenance of hydraulic turbines
IEC 706–4*CEI 706–4 1992–09–00	Guide on maintainability of equipment: Part 4, section 8: maintenance and maintenance support planning
IEC 805*CEI 805 1985–00–00	Guide for commissioning, operation and maintenance of storage pumps and of pump-turbines operating as pumps
IEC 944*CEI 944 1988–00–00	Guide for the maintenance of silicone transformer liquids
IEC 962*CEI 962 1988–00–00	Maintenance and use guide for petroleum lubricating oils for steam turbines
IEC 978*CEI 978 1989–06–00	Maintenance and use guide for triaryl phosphate tester turbine control fluids
IEC 1203*CEI 1203 1992–12–00	Synthetic organic testers for electrical purposes; guide for maintenance of transformer testers in equipment
IEC 1241–1–2*CEI 1241–1–2 1993–08–00	Electrical apparatus for use in the presence of combustible dust; part 1; electrical apparatus protected

	by enclosures; section 2: selection, installation and maintenance
IEC/TR 1360–3*CEI/TR 1360–3 1995–10–00	Standard data element types with associated classification scheme for electric components-Part 3: Maintenance and validation procedures
ISO 2382–14 1978–04–00	Data processing; vocabulary; Section 14: Reliability, maintenance and aviailability bilingual edition
ISO/DIS 2710–2 1996–05–00	Reciprocating internal combustion engines – Vocabulary – Part 2: Terms for engine maintenance
ISO 3056 1996–05–00	Non-calibrated round steel link lifting chain and chain slings: Use and maintenance
ISO 3929 1995–02–00	Road vehicles – Measurement methods for exhaust gas emmissions during inspection or maintenance
ISO 8152 1984–09–00	Earth-moving machinery; Operation and maintenance; Training of mechanics
ISO 8331 1991–09–00	Rubber and plastics hoses and hose assemblies; guide to the selection, storage, use and maintenance
ISO 10419 1993–08–00	Petroleum and natural gas industries; drilling and production equipment; installation, maintenance and repair of surface safety valves and underwater safety valves offshore
ISO/DIS 10452 1996–05–00	Hearing protectors – recommendations for selection, use, care and maintenance – Guidance document
ISO 10462 1994–12–00	Cylinders for dissolved acetylene – periodic inspection and maintenance
ISO/FDIS 12478–1 1997–03–00	Cranes – Maintenance manual Part 1: General
ISO/DIS 12944–8 1996–03–00	Paints and varnishes-Corrosion protection of steel structures by protective paint systems – Part 8: Development of Specifications for new work and maintenance
ISO/DIS 13534 1997–06–00	Petroleum and natural gas industries – drilling and production equipment – Inspection, maintenance, repair and remanufacture of hoisting equipment
ISO/DIS 14224 1996–12–00	Petroleum and natural gas industries – Collection and exchange of reliability and maintenance data for equipment

The following names and adresses are major international offices responsible for standards in the countries concerned.

ARGENTINA (IRAM)

Instituto Argentino de
Racionalization de Materiales
CHILE 1192
1098 BUENOS AIRES
Tel: +54 1 383 37 51
Fax: +54 1 383 84 63

AUSTRALIA (SAA)

Standards Australia
1 The Crescent
HOMEBUSH
NEW SOUTH WALES 2140
Postal address

PO Box 1055
STRATHFIELD
NEW SOUTH WALES 2135
Tel: +61 2 746 47 00
Fax: +61 2 746 84 50

AUSTRIA (ONI)

Osterreichisches Normungsinstitut
Heinestrasse 38
Postfach 130
A-1021 WIEN 2
Tel: +43 222 267535
Fax: +43 222 267552

BELGIUM (IBN)

Institut belge de normalisation
Av de la Brabanconne 29
B-1040 BRUXELLES
Tel. 1 32 2 734 92 05
Fax: + 32 2 733 42 64

BOSNIA AND HERZEGOVINA (BASMP)

Institute for Standardization
Metrology and Patents (BASMP)
c/o Permanent Mission of Bosnia
and Harzegovina
22 bis rue Lamartine
CG-1203 GENEVE
Tel: +387 71 67 06 55
Fax: +387 71 67 06 56

BRAZIL (ABNT)

Associacao Brasileira de Normas
Tecnicas
Av 13 de Maio no 13.27o andar
Caixa Postal 1680
20003–900 RIO DE JANEIRO -RJ
Tel: +55 21 210 31 22
Fax: + 55 21 532 21 43

BULGARIA (BDS)

Committee for Standardization
and Metrology at the Council of
Ministers
21 6th September Street
1000 SOFIA
Tel: +359 2 85 91
Fax: +359 2 80 14 02

CANADA (SCC)

Standards Council of Canada
45 O'Connor Street
Suite 1200
OTTAWA
Ontario K1P 6N7
Tel: +1 613 238 32 22
Fax: +1 613 995 45 64

CZECH REPUBLIC (COSMT)

Czech Office for Standards
Metrology and Testing
Biskupsky dvur 5
113 47 PRAHA 1
Tel: +42 2 232 44 30
Fax: +42 2 232 43 76

DENMARK (DS)

Dansk Standard
Baunegaardsvej 73
DK-2900 HELLERUP
Tel: +45 39 77 01 01
Fax: +45 39 77 02 02

EGYPT (EOS)

Egyptian Organisation for
Standardisation & Quality Control
2 Latin America Street
Garden City CAIRO
Tel: +20 2 354 97 20
Fax: + 20 2 355 78 41

FINLAND (SFS)

Finnish Standards Association SFS
PO Box 116
FIN-00241 HELSINKI
Tel: +358 0 149 93 31
Fax: +358 0 146 49 25

FRANCE (AFNOR)

Association francaise de
normalisation
Tour Europe
F-92049 PARIS LA DEFENSE
CEDEX
Tel: +33 1 42 91 55 55
Fax: +33 1 42 91 56 56

GERMANY (DIN)

DIN Deutsches Institut for Normung
Burggrafenstrasse 6
Postfach 1107
D-10787 BERLIN
Tel: +49 30 26 01-0
Fax: +49 30 26 01 12 31

GREECE (ELOT)

Hellenic Organisation for
Standardisation
313 Acharnon Street
GR-111 45 ATHENS
Tel: +30 1 201 50 25
Fax: +30 1 202 07 76

HUNGARY (MSZT)

Magyar Szabvanyugvi Testulet
Ulloi ut 25
H-1450 BUDAPEST 9
Pf 24
Tel. +36 1 218 30 11
Fax: +36 1 218 51 25

ICELAND (STRI)

Icelandic Council for
Standardisation
Keldnaholt
S-112 REYKJAVIK
Tel +354 587 70 00
Fax +354 587 7409

INDIA (BIS)

Bureau of Indian Standards
Manak Bhaven
9 Bahadur Shah Zafar Marg
NEW DELHI 110002
Tel: +91 11 323 79 91
Fax: +91 11 323 40 62

IRELAND (NSAI)

National Standards Authority
of Ireland
Glasnevin
DUBLIN-9
Tel: +353 1 837 01 01
Fax: +353 1 836 98 21

ISRAEL (SH)

Standards Institution of Israel
42 Chaim Levanon Street
TEL AVIV 69977
Tel: +972 3 646 51 54
Fax: +972 3 641 96 83

ITALY (UNI)

Ente Nazionale Italiano di
Unificazione
Via Battistotti Sassi 11/b
1–20133 MILANO
Tel: +39 2 70 02 41
Fax: +39 2 70 10 61 06

JAPAN (JISC)

Japanese Industrial Standards
Committee
c/o Standards Department
Agency of Industrial Science
and Technology
Ministry of Internaitonal Trade and
Industry
1–3–1 Kasumigaseki, Chiyoda-ku
TOKYO 100
Tel. +81 3 35 01 92 95
Fax: +81 3 35 80 14 18

MALAYSIA (SIRIM)

Standards & Industrial Research
Institute of Malaysia
Persiaran Data'Menten Section 2
PO Box 7035 40911 Shah Alam
SELANGOR DARUL EHSAN
Tel: +60 3 559 26 01
Fax: +50 3 550 80 95

MEXICO (DGN)

Direccion General de Normas
Calle Puente de Tecamachalco
No 6
Lomas de Tecamachalco
Seccion Fuentes
Naucalpan de Juarez
53 950 MEXICO

NETHERLANDS (NNI)

Nederlands Normalisatie-instituut
Kalfjeslaan 2
PO Box 5059
NL-2600 GB DELFT
Tel: +31 15 2 69 03 90
Fax: +31 15 2 69 01 90

NEW ZEALAND (SNZ)

Standards New Zealand
Standards House
155 The Terrace
WELLINGTON 6001

Postal address

Private Bag 2439
WELLINGTON 6020
Tel: +64 4 498 59 90
Fax: +64 4 498 59 94

NIGERIA (SON)

Standards Organisation of Nigeria
Federal Secretariat
Phase 1, 9th Floor
Ikoyi
LAGOS
Tel: +234 1 68 26 15
Fax: +234 1 68 18 20

NORWAY (NSF)

Norges Standardlseringsforbund
?? mmensveien 145
Postboks 353 Skoyen
N0212 OSLO
Tel: +47 22 04 92 00
Fax: +47 22 04 92 11

PAKISTAN (PSI)

Pakistan Standards Institution
39 Garden Road, Saddar
KARACHI-74400
Tel: +92 21 772 85 27
Fax: +92 21 772 81 24

POLAND (PKN)

Polish Committee for
Standardisation
ul Elektoraina 2
PO Box 411
00–950 WARSZAWA
Tel: +48 22 620 54 34
Fax: +48 22 620 07 41

ROMAINA (IRS)

Institutul Roman de Standardizare
Str Jean-Louis Calderon Nr 13
Cod 70201
BUCURESTI 2
Tel: +40 1 211 32 9
Fax: +40 1 210 08 33

PORTUGAL (IPQ)

Instituto Portugues da Qualidade
Rua C a Avenida dos Tres Vales
P2825 MONTE DE CAPARICA
Tel: +351 1 294 81 00
Fax: +351 1 294 81 01

RUSSIAN FEDERATION (GOST R)

Committee of the Russian Federation
for Standardisation Metrology and
Certification
Leninsky Prospekt 9
MOSKVA 117049
Tel: +7 095 236 40 44
Fax: +7 095 237 60 32

SAUDI ARABIA (SASO)

Saudi Arabian Standards Organisation
Imam Saud Bin Abdul Aziz Bin
Mohammed Road West End
PO Box 3437
RIYADH 11471
Tel: +966 1 452 00 00
Fax: +966 1 452 00 86

SINGAPORE (SISIR)

Singapore Institute of Standards and
Industrial Research (SISIR)
1 Science Park Drive
SINGAPORE 118 221
Tel: +65 778 77 77
Fax: +65 778 00 86

SLOVAKIA (UNMS)

Slovak Office of Standards
Metrology and Testing
Stefanovicova 3
814 39 BRATISLAVA
Tel: +42 7 49 10 85
Fax: +42 7 49 10 50

SLOVENIA (SMIS)

Standards and Metrology Institute
Ministry of Science & Technology
Kotnikova 6
SI-61000 LJUBLJANA
Tel: +386 61 131 23 22
Fax: +386 61 31 48 82

SOUTH AFRICA (SABS)

South African Bureau of Standards
1 Dr Lategan Rd, Groenkloof
Private Bag X191
PRETORIA 0001
Tel: +27 12 428 78 11
Fax: +27 12 344 15 68

SPAIN (AENORI)

Asociacion Espanola de
Normalisacion y Certificacion
Fernandez de la Hoz.52
E-28010 MADRID
Tel: +34 1 432 60 00
Fax: +34 1 1310 49 75

SWEDEN (SIS)

SIS-Standardiseringen-Sverige
St Eriksgatan 115
Box 6455
S-113 82 STOCKHOLM
Tel: +46 8 610 30 00
Fax: +46 8 30 77 57

SWITZERLAND (SNV)

Swiss Association for Standardisation
Muhlebachstrasse 54
CH-8008 ZURICH
Tel: +41 1 254 54 54
Fax: +41 1 254 54 74

SYRIAN ARAB REPUBLIC (SASMO)

Syrian Arab Organisation for
Standardisation and Metrology
PO Box 11836
DAMASCUS
Tel: +963 11 445 05 38
Fax: +963 11 441 39 13

TURKEY (TSE)

Turk Standardiari Enstitusu
Necatibey Cad 112
Bakanliklar
06 100 ANKARA
Tel: +90 312 417 83 30
Fax: +90 312 425 43 99

UKRAINE (DSTU)

State Committee of Ukraine for
Standardisation Metrology and
Certification
174 Gorky Street
GSP KIEV-6 252650
Tel: +380 44 226 29 71
FAX: +380 44 226 29 70

UNITED KINGDOM (BSI)

British Standards Institution
389 Chiswick High Road
GB-LONDON W4 4AL
Tel: +44 181 996 90 00
Fax: +44 181 996 74 00

URUGUAY (UNIT)

Instituto Uruguayo de Normas
Tecnicas
San Jose 1031 P.7
Galeria Elysee
MONTEVIDEO
Tel: +598 2 91 20 48
Fax: +598 2 92 16 81

USA (ANSI)

American National Standards
Institute
11 West 42nd Street
13th Floor
NEW YORK N.Y 10036
Tel: +1 212 642 49 00
Fax: +1 212 398 00 23

YUGOSLAVIA (SZS)

Savezni zavod za standardizaciju
Kneza Milosa 20
Post Pregr 933
YU-11000 BEOGRAD
Tel: +381 11 64 35 57
Fax: +381 11 68 23 82

Chapter 17

BIBLIOGRAPHY

The following titles are those referred to during the preparation of this Handbook. Where standards are noted they will also be found in the overall statement of Chapter 16.

Health and Safety Council (HSE)
1994

Oil Industry Advisory Committee
HSE Offshore Safety Division (OSD)
'Play your part!'
ISBN 0–7176–0786–0

Peter Willmott – 1994
Butterworth-Heinemann
Linacre House
Jordan Hill OXFORD OX2 8DP

Total Productive Maintenance (TPM)
The Western Way
ISBN 0–7506–1925–2

Joseph D Patton Jr
1995

Preventive Maintenance
The International Society for Measurement and Control
ISBN 1–55617–533–7

B.K.N Rao
Elsevier Advanced Technology

Handbook of Condition Monitoring
ISBN 1 85617 234 1

Arthur Townsend
1980 and 1992
(2nd Edition)

Maintenance of Process Plant
I.Chem.E (Institution of Chemical Engineers
ISBN 0–85295 2929

Health & Safety Executive
1992 (HSE)

COSHH and Peripatetic Workers
H S(G) 77 Series
ISBN 0–11–885733–9

Health & Safety Executive
(HSE) 1993

The Assessment of Pressure
Vessels Operating at Low Temperature
HS(G) Series
ISBN 0–11–882092–3

Lloyds Register Executive Lloyds Register (Under the UK PFEER Regs)	Pressure Vessel Inspection Periodicity Profile, Strategy and Capability LR Project Ref OD 8073/2000/089
HSE Guidance Note GS4	Safety in Pressure Testing
HS (G) 155	Slips and trips ISBN 07176–1145–0
Health & Safety Commission (HSC) Oil Industry Advisory Committee (OIAC)	Guidance on permit-to-work systems in the Petroleum Industry ISBN 0–7176–1281–3
HS (L) 70 Guidance on Regulations	A guide to the Offshore Installations and Pipeline Works (Management and Administration Regulations 1995 ISBN 0–7176–0938–3
HS (G)85 Safe Working Practices	Electricity at Work ISBN 0–7176–0442 X
HS (G) Series	Storage of Flammable Liquids in Fixed Tanks (up to 10,000 m³ total capacity)
HSC (Health & Safety Commission) Approved Code of Practice & Guidance L24	Workplace health, safety and welfare regulations 1992
HSE (G) 48	Human factors in industrial safety ISBN 0–7176–0472–1
HSC Guidance on the Act	A guide to the Health and Safety at Work etc Act 1974
HSE (L) 30 Guidance on Regulations	A guide to the Offshore Installations (Safety Case) Regulations 1992 ISBN 0–11–882055–9
HSE (L) 20 Guidance on Regulations	A guide to the Lifting Plant and Equip- ment (Records of Test and Examina- tion etc) Regulations 1992 ISBN 0–7176–0145–2
HSE (L) 82 Guidance on Regulations	A guide to the Pipelines Safety Regula- tions 1996 ISBN 0–7176–1182–5
HSC (L) 65 Approved Code of Practice and Guidance	Prevention of fire and explosion and emer- gency response on offshore installations ISBN 0–7176–1386–0
HSE (L) 73 RIDDOR 95	A Guide of the Reporting of Injuries, Diseases and Dangerous Occurrences Regulations 1995 ISBN 0–7176–1012–8

BSI Standards Formerly Handbook 22 Part 1	Quality Management Handbook Quality Assurance
BSI Standards Formerly Handbook 22 Part 2	Quality Management Handbook Vol I Reliability
BSI Standards Formerly Handbook 22 Part 2	Quality Management Handbook Vol II Maintainability
Statutory Instrument 1996 No 913 Wells	The Offshore Installations and (Design and Construction etc.) Regulations 1996.
Statutory Instrument 1995 No 738	The Offshore Installations and Pipeline Works (Management and Administration) Regulations 1995
Statutory Instrument 1995 No 743 1995	The Offshore Installations (Prevention of Fire and Explosion, and Emergency Response) Regulations
Statutory Instrument 1992 No 2885	The Offshore Installations (Safety Case) Regulations 1992.
Statutory Instrument 1996 No 825 1996	The Pipelines Safety Regulations

Chapter 18

Glossary

The terms in this glossary are either used in the manual or are relevant to the topics discussed. A few, included in the interests of completeness, have no direct cross-reference. The definitions below are to be found in several source documents, many of which are included in the entries themselves.

References are made to:

BS3811   Terms Used in Terotechnology
           The majority of definitions given here are restatements of those given in this British Standard; others are noted where they apply.
BS4778   Quality Vocabulary
BS5233   Terms Used in Metrology
BS5760   Reliability of Systems Equipment and Components
BS7000   Design Management Systems

Author's definitions are notated *AN*.

Acceptable condition	The condition of an item agreed for each particular usage. *Note*: attention is drawn to the fact that statutory requirements may exist governing minimum acceptable conditions. Acceptable condition would be implied when an item is able to deliver its different functions when called upon to do so. *AN*.
Accessibility	A qualitative or quantitative measure of the ease of gaining access to a component for the purposes of *maintenance*.
Active corrective maintenance time	That part of the *active maintenance time* during which *actions* of *corrective maintenance* are performed on an *item*.

Active maintenance time	That part of the *maintenance time* during which a *maintenance action* is performed on an *item*, either automatically or manually, excluding *logistic delays*. *Note*: a *maintenance action* may be carried out while the *item* is performing a required function.
Active preventive maintenance time	That part of the *active maintenance time* during which *actions* of *preventive maintenance* are performed on an *item*.
Administrative delay	For *corrective* maintenance. The accumulated time during which an *action of corrective maintenance* on a faulty *item* is not performed due to administrative reasons.
Ageing failure	A *failure* whose probability of occurrence increases with the passage of time, as a result of processes inherent in the *item*. A very illuminating expansion of the question of age-related failure is given in BS 5760, Part 2, 1994, of which the first part is reproduced below. *AN*.

Variation of failure rate with time: the bath-tub curve
Failure rates can vary with time. Observation of many systems shows that the occurrence of failures with time may follow various patterns. The figure below shows three basic patterns of failure, which when combined can generate an overall failure rate pattern with time known as the bath-tub curve. The three components of the bath-tub curve are as follows:
(a) a period of decreasing failure rate in which quality related and learning effects predominate;

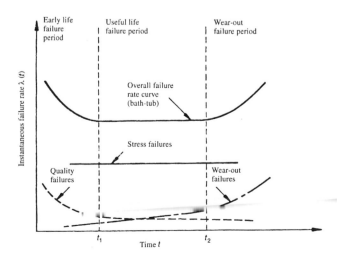

(b) a period of effective constant failure rate in which failures are due primarily to externally induced high stresses such as shock loads, electrical overstress, etc. or marginal design, which appear at a constant average rate throughout the life of the equipment;

(c) a wear-out period, in which failures due to wear-out phenomena such as fatigue, corrosion and wear predominate.

The bath-tub curve is frequently invoked as a conceptual rather than a mathematical model for the instantaneous failure rate of a system over its whole life. It has been shown that the concept does not apply to a number of electronic components and it is doubtful whether it can apply to component parts generally. There is also some doubt about its applicability to unmaintained systems.

Assessment	The determination of values to be ascribed to *items*, activities or *elements of cost*.
Asset	The buildings, plant, machinery and other permanent *items* required by the user to produce and supply the *product*. *Note*: Terotechnology clearly distinguishes this specialized meaning from its more general meeting. *Author's note*: the terms *asset* and *asset register* are relevant to financial management, where the value of all assets is monitored and reported in company financial affairs. Major items of equipment and machinery are included in such financial calculations, but smaller items, which still concern the maintenance engineer are frequently omitted (see *equipment database* Section 5.2).
Asset register	A record of items, including information such as constructional and technical details about each. *Note 1*: the source of this definition is BS 3811 No. 6120 labelled as 'Physical Asset Register'. *Note 2*: refer to *asset* above. Some systems include the closely similar 'equipment database' (see Section 5.2). *AN*.
Audit	A systematic *examination* of, for example, documents, reports, accounts, *stock* holdings or quality attributes. *Note*: audits are usually performed to confirm a stated performance, quantify a particular situation or conform to a statutory requirement under company law.

Authorized person

A *competent person* who is given authority to perform a given *task* or *tasks*.

Availability (performance)

The ability of an *item* to be in a state to perform a required function under given conditions at a given instant of time or over a given time interval, assuming that the required external resources are provided.

Note 1: this ability depends on the combined aspects of the *reliability performance*, the *maintainability* performance and the *maintenance support performance*.

Note 2: required external resources, other than *maintenance resources*, do not affect the availability performance of the *item*.

The preceding definition can be found in BS 3811. A useful definition of 'system availability' sourced from BS 5760 is reproduced below. *AN*.

Available state

The state of an *item* when it is capable of performing its required functions in the defined condition of use.

Bar chart

A chart on which activities and their durations are represented by lines drawn to a common time-scale showing a sequence of *operations*.

Note 1: this definition is similar to 10021 in BS 4335, 1987.

Note 2: the term 'bar diagram' is used to signify a means of depicting statistical information. See BS 5532, Part 1, for the relevant definition.

Bath-tub curve

See *Ageing failure*.

Breakdown

Failure resulting in the non-*availability* of an *item*.

Refer to *availability (performance)*.

Budget centre

A section of an organization for which control may be exercised through budgets.

Budgeting control

The establishment of financial accountability, relating the responsibilities of executives to the requirements of a policy, and the continuous comparison of actual financial expenditure with that planned, either to secure by individual *action* the objective of that policy or provide a basis for its revision.

Calibration

All the *operations* for the purpose of determining the values of the errors of a measuring instrument (and, if necessary, to determine other metrological properties).

	Note 1: the metrological usage of the term calibration is often extended to include *operations* such as adjustment, gauging and scale graduation.
Capability	The ability of an *item* to meet a *service* demand of given quantitative characteristics under given conditions. *Note 1*: this may be any combination of faulty and correctly functioning sub-items. *Note 2*: for telecommunications *services* this ability is called trafficability performance.
Certification	A procedure which by a third party gives written assurance that a *product*, process or *service*, conforms to specified requirements. *Note 1*: the term certification should not be used in the context of second or first parties; see suppliers declaration *Note 2*: certification is an area where *quality assurance* impinges on regulations, approvals and requirements for manufacturers to satisfy legal obligations. It is a means by which a producer can demonstrate *compliance* with these constraints.
Certifying authority	In this manual the certifying authority refers to specific organizations responsible for interpreting and authorizing operations and ensuring that operators meet certain statutory obligations. *AN*. Organizations such as Lloyd's Register of Shipping and Det Norske Veritas are two examples. Recent changes to British legal requirements means that the role of the certifying authorities in British affairs is steadily changing. See Chapter 12. *AN*.
Check-out time	That part of *active corrective maintenance time* during which *function check-out* is performed.
Clean	To reduce contamination to an acceptable level.
Commissioning	Advancement of an installation from the stage of static completion to full working order and achievement of the specified operational requirements.
Competent person	A person who, by virtue of training and experience, can perform specified *tasks* satisfactorily and safely.
Condition appraisal	A formal and systematic appraisal of the condition of an *item* in respect of its ability to perform its required function.
Condition-based maintenance	The *maintenance* carried out according to the need indicated by *condition monitoring*.

Condition monitoring	The continuous or periodic measurement and interpretation of *data* to indicate the condition of an *item* to determine the need for *maintenance*. *Note*: condition monitoring is normally carried out with the *item* in operation, in an operable state or removed but not subject to major dismantling.
Conformity	An affirmative indication or judgement that a *product* or *service* has met the requirements of the relevant *specifications*, contract or regulation; also the state of meeting the requirements. *Note*: in *quality control* usage, conformity customarily refers to an *assessment* not dependent on the passage of time in *product* use, as contrasted to *reliability* that has a time connotation.
Corrective maintenance	The *maintenance* carried out after *fault recognition* and intended to put an *item* into a state in which it can perform a required function.
Cost-benefit analysis	An *assessment* of the desirability of *projects* where the indirect effects on third parties outside those affecting the decision-making process are taken into account.
Cost centre	A location, person or *item* of equipment (or group of these) in respect of which *costs* may be ascertained and related to *cost* units, e.g. process cost centre, production cost centre, service cost centre.
Cost control	The regulation by management *action* of the *costs* of operating an undertaking, particularly where such *action* is guided by *cost* accounting.
Cost function	An expression of the way *cost* varies with a given parameter.
Costs	The expenditure (actual or notional) incurred on, or attributable to, a given thing. *Note*: the basis of cost can be so variable that whenever possible it should be qualified as to its nature or limitations; for example, historical or variable. Costs should also be related to a particular thing or 'object of thought'; for example, a given quantity or unit of goods made or *services* performed.
Critical defect	A defect that analysis, judgement and experience indicate is likely to result in hazardous or unsafe conditions. See BS 4778, Part 2.
Criticality analysis	A quantitative analysis of events or *faults* and the ranking of these in order of the seriousness of their consequences.

Critical path analysis (CPA)	A technique that is used to determine the minimum time to complete a project by mapping the shortest continuous path through the set of sequential activities within the project. *Note*: In the author's view this is an incomplete definition. CPA is also concerned with the identification of all main activities contained in a project and their logical dependencies and relationships to each other. It also specifies the effects on time available of late working and overruns and defines those activities which can be safely conducted in parallel. *AN*.
Critical state	A state of an *item* assessed as likely to result in injury to persons, significant material damage or other unacceptable consequences. *Note*: a critical state may be the result of a critical *fault*, but not necessarily.
Data	A representation of facts, concepts or instructions in a formalized manner suitable for communication, interpretation or processing by human beings or by automatic means.
Defect	The non-fulfilment of an intended requirement or an expectation for an *entity* including one concerned with *safety*. *Note 1*. the requirement or expectation should be reasonable under the existing circumstances. *Note 2*: this term has been defined in recent legislation, where its application is restricted to those *faults* that may cause the *risk* of harm or death to persons. In order to avoid the chance of the term being misunderstood or misinterpreted it is strongly recommended that the term 'defect' should not be used except where this legal interpretation is unarguably applicable. See also 3.1 in BS 4778, Section 3.1, 1991 and the foreword of this standard. Alternative terms that accurately represent the state or event should be used, e.g. *blemish, imperfection, nonconformity* or *failure*. Notes 3 and 4 are taken from the source document on which the above definition is based. *Note 3*: the definition covers the departure or absence of one or more quality characteristics from intended usage requirements. *Note 4*: the basic difference between *nonconformity* and defect is that specified requirements may differ from the requirements for the intended use.

Author's note: the word 'defect' is widely used by maintenance personnel, and although subject to recent and more precise definition it will be met abundantly with various meanings, as described above.

Defect work order	A work order detailing the job required to correct an equipment defect (defect work orders and PPM work orders are described in Section 5.4). Refer also to *work order* definition below.
Deferred maintenance	*Corrective maintenance* which is not immediately initiated after a *fault recognition* but is delayed in accordance with given *maintenance* rules.
Design audit	A formal and systematic analysis of a proposed design for, or on behalf of, the *user* or user organization to ascertain that the user's requirements are, or will be, met.
Design failure	A *failure* due to inadequate design of an *item*.
Design review	A documented, comprehensive and systematic critique of a design, for or on behalf of the *supplier*, to: (a) evaluate the design, its requirements and implementation; (b) identify problems, if any, and to propose solutions. *Note*: the design requirements include all those that concern an *entity* at all stages of the quality loop and at all phases of its *life cycle*.
Diagnosis	The art or act of deciding from symptoms the nature of a *fault*.
Diagnosis time	The time during which *fault diagnosis* is performed.
Disabled state outage	A state of an *item* characterized by its inability to perform a required function for any reason.
Discounted cash flow	The *concept* of relating cash inflows and outflows over the life of a *project* or *operation* to a common base value (for example present value), using discount rates based on compound interest.
Disposal costs	The net total cost of dispoising of an *item* of material when it has failed or is no longer required for any reason. *Note*: these *costs* may be either positive or negative.
Disposal instructions	The document that describes in detail the method and precautions to be observed in discarding or otherwise disposing of an asset when it has failed or is no longer required for any reason. *Note*: this document forms part of the *technical manual*.

Down state	A state of an *item* characterized either by a *fault*, or by a possible inability to perform a required function during *preventive maintenance.*
Downtime	The time during which an *item* is in a *down state.*
Economy quality level	The economic level of quality at which the *cost* of securing higher quality would exceed the benefits of the improved quality.
Efficiency	(1) The ratio of useful work performed to the total energy expended.
	(2) A term expressed qualitatively to reflect the relationship between the *output* from, and the input to, an *item* or an activity.
Emergency maintenance	The *maintenance* that it is necessary to put in hand immediately to avoid serious consequences.
Equipment database	A comprehensive set of data describing all equipment which receives maintenance attention (including non-operational equipment and items awaiting removal), usually held on the maintenance computer system. *AN* (see Section 5.2 and *asset register*).
Ergonomics	The study of the relationship between workers and their occupation, equipment and environment, and particularly the application of anatomical, physiological and psychological knowledge to the problems arising therefrom.
Evaluation	Ascertainment, as closely as possible, of the numerical value of a given parameter or quantity.
Examination	A comprehensive *inspection* supplemented by measurement and physical testing in order to determine the condition of an *item*.
External disabled state	That subset of the *disabled state* when the *item* is in an *up state*, but lacks required external resources or is disabled due to other planned *actions* than *maintenance*.
External failure cost	The *cost* arising outside an organization due to the failure to achieve the quality specified. *Note*: the term can include the *costs* of claims against warranty, replacement and consequential losses within the organization.
Fail-safe	A design property of an *item* which prevents its *failures* from resulting in critical *faults*.
Failure classification	The allocation of a *failure* to one of the *failure* types Ref 04 of BS 4778, Section 3.2, 1991.

Failure mechanism	The physical, chemical or other process which has led to a *failure*.
Fault	The state of an *item* characterized by the inability to perform a required function, excluding the inability during *preventive maintenance* or other planned actions, or due to lack of external resources. *Note*: a fault is often the result of a *failure* of the item itself, but may exist without prior failure.
Fault correction	Actions taken after fault localization to restore the ability of the faulty item to perform a required function.
Fault correction time	That part of active corrective maintenance time during which fault correction is performed.
Fault diagnosis	Actions taken for fault recognition, fault *localization* and cause identification.
Fault modes and effects analysis	A qualitative method of *reliability* analysis which involves the study of the *fault* modes which can exist in every sub-item of the *item* and the determination of the effects of each *fault* mode on other sub-items of the item and on the required functions of the item.
Fault report	A document reporting the detection of a *fault* or the occurrence of an incident, giving details of its nature, symptoms and consequences as far as is known.
First line maintenance	Basic maintenance actions such as oil changes, equipment cleaning and air filter replacement, usually conducted by the operator rather than the maintenance technician. *AN*.
Forced outage duration	Within a specified period of time, the period of time during which an item is incapable of performing its function because of a fault.
Function	RCM refers to the primary and secondary functions of an item of equipment, the former detailing the main reason or purpose for its use and the latter to other non-primary purposes or uses which depend on its inclusion. *AN*. It is always necessary to define the function or functions of the product in terms of the performance required. For complex systems this can be difficult, particularly where the performance gradually degrades (refer to BS 5760, Section 4.2).
Function-affecting maintenance	A maintenance action during which one or more required functions of the item under maintenance are interrupted or degraded.

Function-degrading maintenance	Function affecting maintenance that degrades one or more of the required functions of a maintained item, but not to such an extent as to cause complete loss of all the functions.
Gantt chart	A bar chart used as a means of control on which work planned and work done are represented, showing their relation to each other and to time.
Hardware	Physical equipment, as opposed to programs, procedures, rules and associated documentation.
Hazard	A situation that could occur during the lifetime of a product, system or plant that has the potential for human injury, damage to property, damage to the environment or economic loss.
Hazard analysis	The identification of hazards and the consequences of credible accident sequences of each hazard.
Idle state	A non-operating up state during non required time.
Information technology	The technology of recording, storing, transmitting, processing and displaying data
Inspection	Examination of a product design, product service or plant and determination of their conformity with specific requirements, or (on the basis of professional judgment) general requirements. *Note 1*: Inspection of a process includes personnel, facilities, technology and methodology. *Note 2*: the results of inspection may be used to support certification.
Inspection	Activities such as measuring, testing, gauging one or more characteristics of a product or service and comparing these with specified requirements to determine conformity. BS 4778; Part 1.
Installation instructions	The document that describes in detail the procedure for installing the product, including, if necessary, the procedure for unpacking and preparation prior to installation.
Interactive	Having the ability to question and analyse information according to current needs. Communicating with the computer on the basis of action and response.
International system of units (SI)	The coherent system of units adopted and recommended by the General Conference on Weights and Measures (CGPM).

metre, length
kilogram, mass
second, time
ampere, electric current
kelvin, thermodynamic temperature
mole, amount of substance
candela, luminous intensity

Intrinsic Qualifies the value, determined when *maintenance* and operational conditions are assumed to be ideal.

Invisible failure A failure which, having occurred, is not in evidence because the failed part or equipment has not been called into use. *AN*.

Job (1) All the *tasks* carried out by one or more workers and/or units of equipment in the completion of their prescribed duties and grouped to either under one title or definition (e.g. as on a *work order*).
(2) A defined area of accountability within an organization (e.g. as in 'my job').
Author's note: in this handbook *tasks* are treated as *components* of the *job*. The definition above is to be found in BS 3811, No. 4111, where the job refers to *all the tasks* and the source is quoted as BS 3138.

Job card Deprecated; see work order.

Life cycle The time interval that commences with the initiation of the *concept* and terminates with the *disposal* of the *asset*.

Life cycle cost The total cost of ownership of an *item* taking into account all the *costs* of acquisition, personnel training, *operation, maintenance, modification* and *disposal*.

Localization Actions taken to identify the faulty subitem at the appropriate indenture level.

Loss prevention A systematic approach to preventing hazardous events or minimizing their effects (BS 4778, Section 3.1, 1991).

Maintainability The probability that a given active *maintenance action* for an *item* under given conditions of use can be carried out within a stated time interval, when the *maintenance* is performed under stated conditions and using stated procedures and resources.
Note: the term 'maintainability' is also used to denote

the maintainability performance quantified by this probability.

Maintainability requirements

A statement of the principal means and frequency of preventing an *item* from failing or of restoring its function when it has failed.
Note: compare with 4239.

Maintenance

The combination of all technical and administrative actions, including *supervision actions*, intended to retain an *item* in, or restore it to, a state in which it can perform a required function.

Maintenance action

A sequence of *elementary maintenance activities* carried out for a given purpose
Note: examples are fault diagnosis, fault localization or function check-out, or a combination thereof.

Maintenance cost

The total cost of retaining an *item* in, or restoring it to, a state in which it can perform its required function.

Maintenance history

A *record* of past *maintenance tasks* that is used for the purpose of *maintenance planning*.
Maintenance history is used for analytical purposes, such as assessment of failure causes (leading to preventive action), man-hours and disciplines used, forecasts to failure, time taken and job costs. *AN*.

Maintenance-induced failure

A failure directly caused by preceding maintenance actions. *AN*.

Maintenance instructions

The document that describes in detail the procedure and circumstances for carrying out *maintenance*.

Maintenance management

The organization of *maintenance* within an agreed policy.

Maintenance man-hours

The accumulated durations of the individual *maintenance times*, expressed in hours, used by all *maintenance* personnel for a given type of *maintenance action* or over a given time interval.

Maintenance philosophy

A system of principles for the organization and execution of the *maintenance* (see Chapter 3).

Maintenance planning

Deciding in advance the *jobs*, methods, materials, tools, machines, labour, time required and timing of *maintenance actions*.
Note: the maintenance plan and maintenance planning are described in Section 5.3.

Maintenance programme (schedule)

A time-based plan allocating specific *maintenance tasks* to specific periods.

Note: the source of this definition is BS 3811, No. 4238, and it is included because it can be misleading. The manual makes a clear distinction between tasks and jobs, the latter being itemized in both plan and schedule (refer to the definition of *job* above). In addition, 'plan' and 'schedule' should not be confused. The latter, being partly derived from the plan, is a tabulation of jobs to be tackled in a short-term time period, such as a work tour. *AN*.

Mean	(1) The value obtained as the expectation of a random variable.
	(2) An integral whose magnitude depends on a time during a given interval divided by the time interval itself. Formal statistical usage refers to several different types of mean.
	The items below are based on the 'arithmetic mean' most simply calculated from the sum of a number of values divided by the number of values taken. *AN*.
Mean operating time between failures	The expectation of the operating time between *failures*. Usually abbreviated to MTBF. *AN*.
	We must beware of the assumption that periods of non-operation also mean periods of no decay. Environmental effects, such as corrosion or temperature changes, can also cause failures. *AN*.
Mean time to failure	The expectation of the time to *failure*.
	Note: the term is normally used in connection with non-repairable *items*.
	Usually abbreviated to MTTF. *AN*.
Mean time to first failure	The expectation of the time to first *failure*.
Mean time to recovery	The expectation of the time to *restoration*. Sometimes referred to as the mean time to repair and abbreviated to MTTR. *AN*.
Measurement	The set of operations having the object of determining the value of a quantity.
Metrology	The field of knowledge concerned with measurement (BS 5233, 1986).
	Note: Metrology includes all aspects, both theoretical and practical, with reference to *measurements*, whatever their level of accuracy, and in whatever fields of science or technology they occur.
Minimum facility plant	A plant designed to include minimum facilities only, without standby equipment or built-in surplus capacity. *AN*.

Misuse failure	A *failure* due to the application of stresses during use which exceed the stated capabilities of the *item*.
Module	A subassembly within an *item* of equipment or an item of equipment within a larger *system* which may be removed, replaced or inter-changed in one piece, usually without the need to adjust or dismantle other adjacent items.
	The term 'module' is used in a larger sense on offshore installations to describe the modules of functional speciality (e.g. well-head module, water injection module) and refers back to the main design and the ability to lift modules during construction. They are of concern to the maintenance team and often define the location of an item. *AN*.
Monitoring	Activity performed either manually or automatically, intended to observe the state of an *item*.
Network techniques	A group of techniques, for the description, analysis, planning and control of activities, that consider the logical interrelationships of all activities.
	Note 1: the group includes techniques concerned with time, resources, *costs* and other influencing factors, e.g. uncertainty.
	Note 2: the terms 'programme evaluation and review technique' (PERT), 'critical path analysis' (CPA), 'critical path method' (CPM) and 'precedence method' refer to particular techniques and should not be used as synonyms for network analysis.
Non-operating state	The state when an *item* is not performing a required function.
Observed data	Values related to an *item* or a process obtained by direct observation.
	Note 1: values referred to could be events, time instants, time intervals etc.
	Note 2: when observed data are recorded, all relevant conditions and criteria should be stated.
Obsolescent	Becoming, or about to become, obsolete, out of date or unobtainable.
Obsolete item	An *item* is no longer being manufactured or supplied.
Off-site maintenance	*Maintenance* performed at a location different from that where the *item* is used.

Note: an example of off-site maintenance is the *repair* of a subitem at a *maintenance* centre.

On-site maintenance — *Maintenance* performed at the location where the *item* is used.

On-stream — The state of the *asset* in which it is performing its intended function.
Note: the term originates from, and is particularly applied in, processes.

Operating instructions — The document that describes in detail the methods of starting up, shutting down, controlling and *monitoring* the *asset* under all foreseeable conditions.
Note: the documents forms part of the *technical manual*.

Operating state — The state when an *item* is performing a required function.

Operation — The combination of all technical and administrative *actions* intended to enable an *item* to perform a required function, recognizing necessary adaptation to changes in external conditions.
Note: by external conditions are understood, for example, *service* demand and environmental.

Opportunistic — *Maintenance* of an *item* that is deferred or advanced in time when an unplanned opportunity becomes available.

Optimized — Balanced between maximum benefit and minimum *costs* by selection of the best combination of characteristics (see Section 7.4).

Optimized life cycle costs — The least *costs* that provide the maximum benefit from an *asset* over the *life cycle* period.

Original equipment manufacturer (OEM) — That company or organization that originally manufactured the item of equipment and is often the primary source of operating and maintenance information and replacement parts. *AN*.

Outage rate — For a particular class of *outage* and a specified period of time. The quotient of the number of outages to the *up time* for an *item*, for example *scheduled outage* rate, *forced outage* rate.
Note: this term is used in the electricity supply industry.

Output — A quantitative and qualitative measure of the *product* produced.

Overhaul — A comprehensive *examination* and *restoration* of an item, or a major part thereof, to an *acceptable condition*.

Parts list	A definitive list of all *items* that form the *asset*. *Note*: this document forms part of the *technical manual*.
Permit to work	A signed document, authorizing access to an *item*, that defines conditions, including *safety* precautions, under which work may be carried out. This may include a document, signed on completion of *maintenance*, stating that an item is safe and ready for use.
Physical asset register	A *record of items* including information such as constructional and technical details about each. *Note*: the physical asset register may be combined with an inventory (see note 2 to 4265 *item*). Refer to *equipment database*.
Planned maintenance	The *maintenance* organized and carried out with forethought, control and the use of *records* to a predetermined plan. *Note*: *preventive maintenance* is always part of planned maintenance; *corrective maintenance* may or may not be. *Author's note*: many of the jobs detailed in maintenance plans specify inspection activities, often called fault finding tasks, wich will result in corrective maintenance work.
Potability	The quality of a liquid (usually water) of being capable of being drunk without adverse effects on health.
Predicted	Qualifies a value, assigned to a quantity, before the quantity is actually observable, computed on the basis of earlier observed or *estimated values* of the same quantity or of other quantities using a mathematical model.
Preventive maintenance	The *maintenance* carried out at predetermined intervals or according to prescribed criteria and intended to reduce the probability of *failure* or the degradation of the functioning of an *item*.
Primary failure	A failure of an *item* not caused either directly or indirectly by a *failure* or a *fault* of another item.
Procurement	The process of obtaining goods, materials and *services* from an internal or external *supplier*. This process includes the managerial, technical, contractual and physical *actions* required to control the *availability* and ordering of such requirements.
Product	The specific material *item*(s) or *service*(s) to be supplied.
Production-critical	A reference in this manual only, to an equipment or

	service whose failure results in production downtime. *AN*.
Profit centre	A part of a business responsible for costs and revenues.
Project	The planning and implementation of an *approved concept* from initial acceptance, through the *life cycle* until the *asset* is disposed of. *Note*: in other contexts, a project may last only until the delivery or *commissioning* of the *asset*.
Quality	The totality of features and characteristics of a product or service that bear on its ability to satisfy stated or implied needs (BS 4778, Part 2, 1991, subsection 4.1.1).
Quality control	The operational techniques and activities that are used to fulfil requirements for quality (BS 4778, Part 2, 1991).
Reaction time	The time that elapses between the recognition of a need for *repair* and its execution.
Record	(1) A set of related data or words treated as a unit. *Note*: as an example, in *stock* control each invoice could constitute one record. (2) To preserve information in a file.
Redundancy	In an *item*. The existence of more than one means for performing a required function.
Rehabilitation refurbishment	Extensive work intended to bring plant or buildings up to current acceptable functional conditions, often involving *modifications* and improvements.
Reliability	The ability of an *item* to perform a required function under given conditions for a given time interval. *Note 1*: it is generally assumed that the *item* is in a state to perform this required function at the beginning of the time interval. *Note 2*: the term 'reliability' is also used as a measure of reliability performance.
Reliability	The probability that an item can perform a required function under given conditions for a given time interval (BS 4778, 191–12–01).
Reliability assessment	Reliability is an important but elusive characteristic in the sense that it is often more difficult to specify than many major performance characteristics and is certainly more difficult than most to measure (BS 5760, Part 2, 1994, Section 2). (Refer also to *failure classification* above. *AN*).

Remote maintenance	*Maintenance* of an *item* performed without physical access of the personnel to the item.
Repair	That part of *corrective maintenance* in which manual *actions* are performed on the *item*.
Repaired item	A repairable *item* which is in fact repaired after a *failure*.
Repair time	That part of *active corrective maintenance time* during which *repair actions* are performed on an *item*. That event when the *item* regains the ability to perform a required function after a *fault*. Refer to *mean time to repair* above. *AN*.
Replacement theory	A study comparing various options for replacing, renewing or retaining an *asset* or component reaching the end of its life.
Required time	The time interval during which the *user* requires the *item* to be in a condition to perform a required function.
Risk	A combination of the probability, or frequency of occurrence, of a defined *hazard* and the magnitude of the consequences of the occurrence.
Risk analysis	The integrated analysis of the *risks* inherent in a *product*, system or plant and their significance in an appropriate context.
Running costs	The total *costs* to the owner or *user* of the *operation*, *maintenance* and revenue-funded *modification* of an *item*, system or *asset*. *Note*: the term 'costs in use' is normally applied to buildings.
Running maintenance	*Maintenance* that can be carried out whilst the *item* is in *service*.
Safety	The freedom from unacceptable *risks* of personal harm. *Note*: safety is defined in the context of *risk* of personal harm. It is traceable quantitatively in decision-making on acceptable *risks*.
Safety-critical item	Component or system that by its failure can endanger human life or property.
Safety-critical system	System in which failure could result in severe injury or death to a person dependent on it (BS 4778, 3.18, Section 3.2, 1991, and BS 7000, Part 10).
Safety rule	A regulation designed to enforce a safety code (Refer to Chapter 11).
Schedule	A tabulation of jobs to be tackled in a short term time

period such as a work tour. It is derived from the maintenance plan and rescheduled work from backlog. It will also often restate work already in progress.

Note: 'plan' and 'schedule' should not be confused (see Sections 2.6 and 5.3 and definitions in this glossary).

Scheduled maintenance

The *preventive maintenance* carried out in accordance with an established time schedule.

Scheduled outage

Outage due to the programmed taking out of *service* of an *item*.

Scheduled outage duration

Within a specified period of time, the period of time during which an *item* is not available to perform its function because it is withdrawn from *service* according to programme.

Note: the term is used in the electricity supply industry.

Sensitivity analysis

An *examination* of the effects of selected values of varying parameters within a system (see Section 7.4).

Service

(1) The supply to a *customer* of a *product* or utility that is not of a physical nature, this could include public, social and commercial assistance.

(2) The results generated by activities at the interface between the *supplier* and the *customer*, and by supplier internal activities to meet the customer's needs.

(3) In the context of connections to plant or equipment or any other *asset*, the supplies of energy (e.g. electricity) or fluids (e.g. water, gases, compressed air) needed for the functioning of the plant etc.

(4) To carry out routine activities necessary to keep an *item* in operating condition.

Shutdown maintenance

Maintenance that can be carried out only when the *item* is out of *service*.

Spares policy

A declared basis by which the holding of a *stock* of spares is determined.

Spares stock

Items that are held available for *maintenance* purposes or for the replacement of defective parts.

Note: if spares stock is associated with a saleable *product* it is regarded as direct *stock*, whereas if associated with the fixed assets (e.g. plant, vehicles) it is regarded as indirect *stock*.

Specification

The document that prescribes the requirements to which the *product* or *service* has to conform.

Note: a specification should refer to, or include, drawings, patterns or other relevant documents and should

	also indicate the means and the criteria whereby conformity can be checked.
Standby state	A non-operating *up state* during the *required* time, BS 4778, Section 3.1, 1991, 12.2.4, Refers to standby: 'The state of an item when it is available but not required to be operating'.
Stock	All the tangible material *assets* of a company other than the fixed assets; comprising all the finished or saleable *products*, all the *items* to be incorporated into the finished products and all the items to be consumed in the process of manufacturing the product or in the carrying out of the business.

Note 1: inventory, when used as a generic term, is synonymous with stock. This use is common in the USA and extensive in the UK.

Note 2: an inventory, when used specifically, is defined as a list of tangible material assets. For production control purposes this can be limited to being a list of stock.

Storage life	The specified length of time prior to use for which *items* that are inherently subject to deterioration are deemed to remain fit for use under prescribed conditions.
Strategic spares	Spares held against circumstances that are not expected to arise routinely or frequently during the life of the *asset* but which would have serious consequences if they did occur.
Supplier	The provider of the *asset* to the *user*.

Note 1: *terotechnology* clearly distinguishes this specialized meaning from its more general meaning. The supplier, as used in the wider sense, may be defined as 'the organization' that provides a *product* to the *customer*.

Note 2: in a contractual situation the supplier may be called the contractor.

Note 3: the supplier may be, for example, the producer, distributor, importer, assembler, or *service* organization.

Note 4: the supplier can be either external or internal.

System availability	The concept of availability combines *reliability* and maintainability. For example, steady state or long-term average availability may be defined as:

$$MTBF + MTTR + MTPM$$

MTBF = mean time between failures,

MTTR = mean time to repair
MTPM = mean time for preventive maintenance
(Refer to *System maintainability* below.)

System maintainability

Mean time to repair (MTTR) and repair rate are used in much the same way as mean time between failures (MTBF) and failure rate. The implication is that repair times vary even for narrowly defined work and that they have a statistical distribution. To take account of preventive maintenance MTPM is also required (refer to BS 5760, Part 0, section 8).

Task

(1) The smallest indivisible part of an activity when it is broken down to a level best understood and performed by a specific *user*.

Technical manual

A document that communicates appropriately and effectively specific direction, data and information about the *product* to cover the subjects of:
(a) purpose and planning (what the product is for);
(b) purpose and planning (what the product is for); *operating instructions* (how to use the product);
(c) purpose and planning (what the product is for); technical description (how the product works);
(d) purpose and planning (what the product is for); handling, installation, storage, transit (how to prepare the product for use);
(e) purpose and planning (what the product is for); *maintenance instructions* (how to keep the product working);
(f) purpose and planning (what the product is for); *maintenance schedules* (what is done and when);
(g) purpose and planning (what the product is for); *parts list* (what the product consists of);
(h) purpose and planning (what the product is for); *modification* instructions (how to change the product);
(i) purpose and planning (what the product is for); *disposal instructions* (how to dispose of the product)
Note 1: technical manuals are provided for those involved with managing, operating, maintaining and provisioning for the complete asset.
Note 2: see also BS 4884.

Tender

An offer, normally in writing, to supply a commodity

	or to execute works, giving price and *compliance* with a *specification*.
Terotechnology	A combination of management, financial, engineering, building and other practices applied to physical *assets* in pursuit of economic *life cycle costs*.
	Note 1: terotechnology is concerned with the *specification* and design for reliability and *maintainability* of physical assets such as plant, machinery, equipment, buildings and structures. The application of terotechnology also takes into account the processes of installation, *commissioning, operation, maintenance, modification* and replacement. Decisions are influenced by feedback of information on design, performance and *costs*, throughout the *life cycle* of a *project*.
	Note 2: terotechnology applies equally to both assets and *products* because the product of one organization often becomes the asset of another. Even if the product is a simple consumer *item* its design and *customer* appeal will benefit from terotechnology, and this will reflect in improved market security for the product.
Test	Technical *operation* that consists of the determination of one or more characteristics of a given *product*, process or *service* according to a specified procedure.
Threshold limit	A limiting concentration of a substance or intensity of a radiation in an environment beyond which adverse effects may be expected.
	Note: this term specially applied to limits set by statutory bodies to ensure that the adverse effects are prevented.
Total quality management (TQM)	Management philosophy and company practices that aim to harness the human and material resources of an organization in the most effective way to achieve the objectives of the organization.
Traceability	The property of a result of a measurement whereby it can be related to appropriate standards, generally international or national standards, through an unbroken chain of comparisons (BS 5233, 'Glossary of terms used in metrology', Section 6, Measurement Standards).
Treatment	The *action* of reprocessing a material or *effluent* to render it less harmful or more suitable for its intended purpose.
Tribology	The science and technology of interacting surfaces in relative motion and of related subjects and practices.

Unplanned maintenance	The *maintenance* carried out to no predetermined plan.
User	The individual who, or organization that, operates or uses the *product*. *Note:* *terotechnology* clearly distinguishes this specialized meaning from its more general meaning.
Utilisation	The actual usage of a resource compared with the maximum possible while it is available for use during a given period.
Vendor	See *supplier*.
Vendor assessment	The *assessment* by or for a *customer* of potential *suppliers* with regard to their technical ability and financial ability/stability, and, in other respects, to determine their suitability as suppliers.
Verification	Confirmation by examination and provision of objective evidence that specified objectives have fulfilled (BS 7000, Part 10).
Warning	A statement advising of the need to take care lest there be serious consequences resulting in death of personnel or in a *hazard* to health (see *failure warning*).
Waste storage	The storage of material no longer required prior to disposal.
Wearout failure	A *failure* whose probability of occurrence increases with the passage of time, as a result of processes inherent in the *item* (see *ageing failure*).
Work history	Usually based on closed work orders held on the computer system for work that has been completed, such work orders include details of time spent, resources used and the job

Editorial Index

Index to Advertisers

LK Global Construction   Facing page 12
Market House, Lengen Street, St. Alton,
Hants, G34 1HG
United Kingdom
Tel: +44 (0)1420 89898

Maintenance Journal   Facing page 84
Fax: +61 359 75 5735

Maintenance & Technical Management (MTM) Ltd   Facing page v (Contents)
6, Churchbridge, Oldbury,
W. Midlands, B69 2AP
United Kingdom
Tel: +44 (0)121 525 3525
Fax: +44 (0)121 500 5387

PSDI (UK) Ltd (MAXIMO)   Facing page 85
Unit 5, Woking Eight, Forsythe Road, Woking,
Surrey, GU21 5SB, United Kingdom
Tel: +44 (0)1483 727000
Fax: +44 (0)1483 727979

Status Green Engineering   Facing page iv
Rose Court, Bramley Green, Nr Basingstoke
Hants, RG26 5AJ
Tel/Fax: +44 (0) 1256 880726